EXS 84

Orientation
and Communication
in Arthropods

Edited by M. Lehrer

Birkhäuser Verlag
Basel · Boston · Berlin

Editor:

Dr. Miriam Lehrer
Institute of Zoology
University of Zurich
Winterthurerstrasse 190
CH-8057 Zurich
Switzerland

Library of Congress Cataloging-in-Publication Data
A CIP catalogue record for this book is available from the library of Congress,
Washington D.C., USA

Deutsche Bibliothek Cataloging-in-Publication Data
Orientation and communication in arthropods / ed. by
M. Lehrer. - Basel ; Boston ; Berlin : Birkhäuser, 1997
  (EXS: 84)
  ISBN 3-7643-5693-6 (Basel ...)
  ISBN 0-8176-5693-6 (Boston)
84. Orientation and communication in arthropods. – 1997
EXS. - Basel ; Boston ; Berlin : Birkhäuser
  Früher Schriftenreihe
  Fortlaufende Beil. zu: Experientia

© 1997 Birkhäuser Verlag, PO Box 133, CH-4010 Basel, Switzerland
Printed on acid-free paper produced from chlorine-free pulp
Printed in Germany
ISBN 3-7643-5693-6
ISBN 0-8176-5693-6
9 8 7 6 5 4 3 2 1

Contents

List of Contributors

F.G. Barth, Biozentrum, Institut für Zoologie, Universität Wien, Althanstr. 14, A-1090 Wien, Austria.

R. Campan, Laboratoire d'Ethologie et Psychologie Animale, UMR CNRS
5550, Université Paul Sabatier, 118 route de Narbonne, F-31062 Toulouse
Cedex, France.

T.S. Collett, Sussex Centre for Neuroscience, School of Biological Sciences, University of Sussex, Brighton BN1 9QG, UK.

R. Gadagkar, Centre for Ecological Sciences, Indian Institute of Science,
Bangalore 560012, India, and Animal Behaviour Unit, Jawaharlal Nehru
Centre for Advanced Scientific Research, Jakkur, Bangalore 560064,
India.

K.-E. Kaissling, Max-Planck-Institut für Verhaltensphysiologie, D-82319
Seewiesen/Starnberg, Germany.

W.H. Kirchner, Universität Konstanz, Fakultät für Biologie, Postfach 5560
M657, D-78457 Konstanz, Germany.

K. Kirschfeld, Max-Planck-Institut für biologische Kybernetik, Spemann-
strasse 38, D-72076 Tübingen, Germany.

M. Lehrer, Institute of Zoology, University of Zurich, Winterthurerstrasse 190, CH-8057 Zurich, Switzerland.

M.V. Srinivasan, Centre for Visual Sciences, Research School of Biological Sciences, Australian National University, P.O. Box 475, Canberra,
ACT 2601, Australia.

D. von Helversen, Institut für Zoologie II, University of Erlangen, Staudt-
strasse 5, D-91058 Erlangen, Germany.

M.M. Walker, Experimental Biology Research Group, School of Biological Sciences, University of Auckland, Private Bag 92019, Auckland,
New Zealand.

R. Wehner, Institute of Zoology, University of Zurich, Winterthurerstrasse
190, CH-8057 Zurich, Switzerland.

M.J. Weissburg, School of Biology, Cherry-Emerson Building, Georgia Institute of Technology, 310 First Ave., Atlanta, GA 30332-0230, USA.

J. Zeil, Centre for Visual Sciences, Research School of Biological Sciences, Australian National University, P.O. Box 475, Canberra, ACT 2601, Australia.

S.W. Zhang, Centre for Visual Sciences, Research School of Biological Sciences, Australian National University, P.O. Box 475, Canberra, ACT 2601, Australia.

Orientation and Communication in Arthropods
ed. by M. Lehrer
© 1997 Birkhäuser Verlag Basel/Switzerland

Introductory remarks

Animals move about for many reasons: to search for food, prey, or a potential mate; to find a shelter, or a suitable nesting site; and to escape predators, explore territories, and interact with competitors. With some probability, an animal may encounter food, a shelter, or a mate even if it searches in random directions, and even if it moves passively, being carried by air or water currents. However, the probability of arriving at a particular site or of finding a relevant target will be very much higher if the animal uses information that may reliably guide it to its goal. Oriented locomotion is particularly important to animals that keep returning, from every excursion, to a particular place, be it a profitable food source, or the place where they dwell or have their nests, or a site that offers favourable microclimate conditions at particular times of the day or the year.

The mode of active locomotion clearly depends on the properties of the medium in which the animal has developed its motor organs in the course of evolution. Some animals can only swim or only walk, whereas others have developed the capacity to swim as well as walk, or to walk as well as fly, or to swim as well as walk and fly. In the context of the present topic, however, the role of the environment in shaping the mode of locomotion is less important than is its role in determining the cues that are best suitable to guide locomotion. It is the properties of the environment, as well as those of the goal itself, which determine the sensory modality used in a particular oriented behaviour.

Arthropods have built up their niches in every conceivable type of habitat, and are therefore most suitable for studying various types of oriented behaviour. It is a most striking phenomenon that arthropods, no less than vertebrates, have developed sensory systems for exploiting information of every possible sensory modality. The perception and use of visual, chemical, acoustical, vibratory, tactile, and magnetic cues for orientation are universal in the animal kingdom.

The present volume is concerned with the use of each of these types of signal by insects, crustaceans and spiders. This volume has been conceived so as to include the particular role that intraspecific communication plays in oriented behaviour. The individual chapters, or individual sections within a chapter, will be presented according to the particular sensory modality that is used in the performance under consideration. The purpose of choosing this criterion is to highlight the evolutionary and ecological aspects of the use of particular signals.

Apart from the sensory modality involved, several further aspects of orientation and communication could be used for classifying oriented behaviours. One type of classification could be based on the function of the observed behaviour. In any one particular behavioural context, e.g., foraging, hunting, homing, escaping, courting and mating, cues of different sensory modalities may be used. Another criterion for classifying oriented locomotion would be the distance over which locomotion takes place. Depending on the distance to be travelled, the cues that guide the animal to its goal may be provided by the goal itself, or else by the environment on the route to that goal. Furthermore, over relatively long distances, compass information is needed for selecting the appropriate direction of locomotion, and this information may rely on celestial or terrestrial cues and thus on different types of sensory performance. On short journeys, on the other hand, compass-independent cues would usually be sufficient, and these may be of a different sensory modality than the cues used during a long-distance journey. Still, even for short-distance orientation, directional information based on compass cues is often used, e.g., for acquisition of compass-independent information. Thus, the use of a compass is clearly not restricted to long-distance navigation, and, at the same time, compass orientation is assisted, in many cases, by the use of compass-independent cues. Another aspect of successful orientation is that, regardless of whether locomotion is over long or short distances, and whether or not it requires the use of compass mechanisms, it requires, in all cases, that the animal be capable of identifying the target once it has arrived there. Even in the identification task, various sensory modalities may be used, and these need not be the same as those used during the journey. And, finally, during oriented locomotion, as well as for recognizing and pinpointing the goal, some cues are only useful based on a learning process, whereas the use of others may be independent of previous experience. All of these aspects will be highlighted in due context, despite the fact that the organization of this volume is based mainly on the sensory modality involved in the observed behaviour.

Although we focus here on results of work conducted on arthropods, we will not refrain from an occasional side look at what is known on orientation performances in vertebrates. When it comes to oriented behaviour, the parallels between the two groups of animals are too conspicuous to be overlooked. We believe that workers on orientation in vertebrates may learn from the work on arthropods as much as may workers on arthropods profit from work on oriented behaviours in vertebrates.

This volume presents 13 chapters by 15 authors. All of the chapters are concerned with the animal's behaviour, but some deal, in addition, with further aspects, such as the physical parameters of the adequate signal, the signal-propagation properties of the medium, the anatomy and physiology of the sensory systems involved, and the possible neural mechanisms underlying the observed behaviour. Accordingly, a variety of experimental methods, behavioural, electrophysiological, anatomical, and physical, will be described and, in some cases, methods involving modelling, computer simulations, and robotics will be introduced in addition.

I have chosen the chapter by Raymond Campan ("Tactic components in orientation") to be the first chapter in this volume, not only because it addresses all of the sensory modalities that will be considered in further chapters, but, mainly, because it deals with the most fundamental question concerning the development of oriented behaviours. Based on many examples taken from work on numerous arthropod species, the author shows how various types of complex oriented behaviour develop on the basis of inherited (tactic) orientation capacities that are modified during ontogeny due to individual experience. The role of tactic tendencies may have been crucial even in the *phylogenetical* evolution of complex oriented behaviours.

The five chapters that follow deal with *visually* guided locomotion. Due to the large variety of visual cues that can be used for guidance, there are several aspects to visual orientation. The chapter by Thomas Collett and Jochen Zeil ("The selection and use of landmarks in insects") is concerned with the use of landmarks for route finding and target localization. This chapter shows that, at different stages in the course of their journey, wasps and bees select and use different strategies, based on different types of landmark information. The authors discuss the role of image-matching mechanisms, as well as the significance of cues derived from self-generated image motion. In all cases, oriented locomotion that relies on the use of landmarks is based on learning processes.

In the next chapter ("Course control and tracking: Orientation through image stabilization"), Kuno Kirschfeld describes hard-wired mechanisms of visual orientation, i.e., such that do not require the animal's previous experience with the particular situation at hand. In this chapter, results of behavioural studies on flies, and theoretical considerations, are reviewed jointly to arrive at inspiring models that might contribute much to our understanding of the principles by which cues derived from image motion act to enable course control and target detection.

Hard-wired mechanisms for motion perception are also dealt with in the chapter that follows ("Visual control of honeybee flight"). Here, however, Mandyam Srinivasan and Shaowu Zhang examine the performance of freely flying bees, rather than that of tethered flying flies. This chapter demonstrates the use of *self-generated* image motion for coping with several orientational tasks, such as avoiding collisions with obstacles, flying safely through narrow gaps, controlling flight speed and height, and performing smooth landings. Based on analyses of the spatial and temporal properties of various performances, the authors arrive at the conclusion that several distinct motion detection systems are involved in the control of flight.

Still focusing on visual orientation, my own contribution ("Honeybees' visual spatial orientation at the feeding site") is concerned with the spatial and spatio-temporal cues that honeybees use for localizing and recognizing their target, i.e., a food source they have previously been trained to visit. The results reviewed in this chapter reveal the insect's flexibility in selecting particular visual cues, depending on the experimental situation. Some of these must be learned before they can be used, whereas the use of others does not require learning.

The chapter by Rüdiger Wehner ("The ant's celestial compass system: spectral and polarization channels") is the last chapter concerned with purely visually-guided locomotion. This chapter deals with the capacity of ants and bees to use celestial cues for compass orientation. Based on a large body of experimental evidence, the author describes the mechanisms underlying the insect's polarization compass. It shows, in addition, that the sky provides the insect with directional information that is based not only on the sun's azimuthal position and, in the absence of the sun, on the sun-position-dependent polarization pattern of the sky, but also on the spectral and intensity distribution of skylight. The results further suggest that the animal is able to switch between one skylight compass mechanism and another, depending on the experimental situation, demonstrating, again, the flexibility of the insect's orientation performance.

Whereas the chapter by Rüdiger Wehner shows that there is more than just one compass in the sky, the next chapter, by Michael Walker ("Magnetic orientation and the magnetic sense in arthropods") shows that compass information is available not only from the sky. This chapter deals with oriented behaviours based on cues extracted from the earth's magnetic field. The author describes the parameters of the geomagnetic field and explains why they are suitable to serve as compass cues. Different parameters of the geomagnetic field are shown to be used not only for guiding active locomotion, but also in simpler types of behaviour, such as alignment in space, and, in social insects, building activities. The author discusses possible mechanisms of magnetic perception, and highlights experimental methods by which work on magnetic orientation in arthropods may lead to further insights in the future.

In the next chapter ("Chemo- and mechanosensory orientation by crustaceans in laminar and turbulent flows: From odour trails to vortex streets"), Marc Weissburg is concerned with mechano- and chemosensory orientation in aquatic crustaceans. Based on knowledge of the physical properties of fluid flow, the author emphasizes that, depending on whether the animal lives in laminar or turbulent flow conditions, different receptor properties and neural processing circuits are required for successful orientation. Based on the results of behavioural and electrophysiological experiments, the author demonstrates that the mechanisms underlying mechano- and chemosensory orientation in aquatic environments indeed mirror the animal's adaptation to the properties of the flow within which these orientational capacities are manifested.

The visual, magnetic, mechanical and chemical cues shown to be used in the orientational performances described in the first eight chapters constitute external signals provided by the environment. Although the animal is free to use them, it can take little influence on their presence or absence, or on their modality and intensity. The situation is different when signals are considered that are produced by the animal itself and function to elicit oriented behaviour on the part of a conspecific. The four chapters that follow are concerned with this type of oriented behaviour. They exemplify the close relation between orientation and communication. Even in this case, different sensory modalities are involved, depending on the ecological needs of the animal, on its ability to produce and perceive a particular signal, and on the signal-propagating properties of the medium.

The chapter by Friedrich Barth ("Vibratory communication in spiders: Adaptation and compromise at many levels") describes the use of self-produced substrate vibration by wandering spiders in the task of mate finding and sex recognition. The results of behavioural, anatomical and electrophysiological work described in this chapter are combined with analyses of the physical properties of the substrate. This chapter provides a fascinating example of the large number of questions that need to be considered before full understanding of effective communication and orientation is gained.

In the next chapter ("Acoustical communication in social insects"), Wolfgang Kirchner is concerned with communication in those groups of arthropods in which communication is most crucial, namely social insects. Here, airborne and substrate-borne sounds represent the relevant signals. This chapter describes various mechanisms of sound production and sound perception, as well as the physical properties of the various signals. Mainly, it documents the large variety of functions that acoustical communication plays in oriented (and other types of) behaviour in social insects.

Whereas, in the chapter by Wolfgang Kirchner, acoustical perception involves measuring substrate displacement, or air-particle movement, the sensory modality dealt with in the chapter by Dagmar von Helversen ("Acoustic orientation in the grasshopper *Chorthippus biguttulus*") is hearing of the type realized in vertebrates, namely by measurement of sound wave pressure. Here, acoustical orientation and communication is examined in a solitary, rather than a social insect. As in the chapter by Friedrich Barth, the function of the signal, although it is of a different modality, is sex recognition and mate finding. The author is mainly concerned with those properties of the auditory system that enable the animal to localize the sound source (i.e., the potential mate). She approaches the question from behavioural, anatomical, biophysical and electrophysiological points of view, and shows how all of these approaches add up to render an understanding of the mechanisms involved, as well as of the auditory pathway along which information processing occurs.

In the chapter that follows ("Pheromone-controlled anemotaxis in moths"), Karl-Ernst Kaissling is concerned with oriented behaviour of male moths elicited by the female sex pheromone. In intraspecific communication, pheromones represent one of the most impressive examples of effective signals. Based on behavioural and electrophysiological experimental evidence, the author concludes that the olfactory cue plays a role only in triggering and controlling the oriented behaviour, whereas locomotion itself is guided by information on the wind direction, mediated by mechanical and visual cues. An inspiring model is proposed to describe the possible neural substrate of the observed behaviour.

The last chapter, by Raghavendra Gadagkar ("The evolution of communication and the communication of evolution: The case of the honey bee queen pheromone"), discusses the role that the queen pheromone plays in regulating the worker bees' activities in the honeybee colony. The author discusses the observed phenomena in the light of several theories concerning the evolution of effective intraspecific signals. He proposes that the insights gained on the evolution of chemical signals in social insects may help towards a better understanding of the evolution of sociality itself. This chapter is exceptional in that it is the only one in this volume that is concerned with communication that is not directly related to oriented locomotion. I have chosen this chapter to be the last one in this volume because the ideas expressed in it may become a source of inspiration for further volumes.

We are well aware of the fact that this volume does not contain everything that has been learned in the course of many decades on orientation and communication in arthropods. Nonetheless, we hope to provide here a representative, updated collection of examples, including most recent results, ideas and theories, as well as a large number of references that may guide the interested reader to the work of further researchers.

I am very grateful to the publisher, Birkhäuser Verlag, and particularly to the Life Sciences Editor, Dr. Petra Gerlach, and her assistant Ms. Janine Kern, for helping realize this project. I greatly appreciate their efforts to publish this volume well within the time schedule that we had planned together.

Mainly, however, I wish to thank my colleagues, the authors, leading scientists from all around the world – Australia, Austria, China, England, France, Germany, India, New Zealand, Switzerland, and the USA. All of them have agreed spontaneously to contribute to this volume. Most of them have submitted their typescripts on time (special thanks for that!), and each of them has, to my mind, done an excellent job.

Miriam Lehrer
May 30th, 1997

Orientation and Communication in Arthropods
ed. by M. Lehrer
© 1997 Birkhäuser Verlag Basel/Switzerland

Tactic components in orientation

R. Campan

Laboratoire d'Ethologie et Psychologie Animale, UMR CNRS 5550, Université Paul Sabatier, 118 route de Narbonne, F-31062 Toulouse Cedex, France

Summary. In the first half of this century, taxes were considered the best models for working out the rules of stimulus-response systems. The interest for tactic behaviours suddenly disappeared in the mid-1960s, out of reasons specified in the present review. However, results of several recent studies reviewed in the present article suggest that tactic behaviours constitute, from an ontogenetic as well as phylogenetic point of view, a first step towards more complex oriented behaviours that have received much attention in recent years. The aim of this chapter is to update the implications of tactic responses in complex oriented behaviours. We argue that taxes are basic in the process of acquiring most, if not all of these behaviours, and that they often constitute the first steps in the ontogeny of orientation. Taxes are determined by a flexible balance between genetic and epigenetic factors. Their main function is to assist the ecological adaptation of the animal to the constraints of its environment. Finally, we plead for a revival of the studies of taxes in the light of a theory on the development of behaviour, based upon selforganization of autonomous living systems.

Introduction:
Historical vicissitudes of the studies of tactic behaviours

The study of elementary behavioural responses to elementary physical stimuli was milestoned at the beginning of our century by a succession of pioneer works. They were rooted in their time, at a crossroad between the positivism taught by Auguste Comte in his courses, which called for searching only for the rules underlying processes, with no reference to theology or metaphysics, and the post-Darwinian psychology from which emerged, in 1913, the "behaviourism manifesto" of Watson, rejecting any reference to intermediate variables between stimulus and response for explaining behaviour. Both currents inspired experiments on simple, heteronomous "input systems", such as lower animals, whose rather stereo-typed responses to well defined physical stimuli can easily be described quantitatively. The idea was that the rules underlying the response can be deduced from the relation between well defined stimulus parameters and the animal's measurable locomotory response, without opening the "black-box".

The first half of the 20th century, up to the 1960s, was the golden age of the study of taxes. The classical works of Jennings (1906) and Loeb (1918), among many others, led to the basic classification of taxes by Kühn (1919), later revised by Fraenkel and Gunn (1961) (Tab. 1). This revision is valuable not only for the sake of description and classification of taxes it offers,

Table 1. Designation and definitions of terms involving kineses, routes, and taxes to be considered in the present chapter. S: Stimulus

Kinesis	Routes	Name of the taxis	Tactic stimulus
orthokinesis	tropotaxis	phototaxis	light
		geotaxis	gravity
		chemotaxis	odour
klinokinesis	telotaxis	anemotaxis	wind
		thigmotaxis	contact
		rheotaxis	water current
		astrotaxis	sun or moon
		polarotaxis	polarized light
	menotaxis	magnetotaxis	magnetic field
		scototaxis	low reflecting areas
		perigrammotaxis	vertical contrasted edges
		hypsotaxis	highest silhouettes
		photohorotaxis	contrasted lines underneath

but also, as Jander (1965) has put it, for relating the nature of oriented responses to the animal's sensory capacities, and for considerations involving the phylogeny of taxes.

The huge interest in tactic behaviours disappeared suddenly in the mid-1960s, like the dinosaurs at the end of the Cretaceous Period. An attempt to explain this sudden extinction can be found in Hinde (1965). Hinde's criticism of the earlier conclusions derived from the studies on taxes is essentially concerned with two points. Firstly, the division of taxes in various categories has mainly been inferred from laboratory work on relatively simple animals in simplified experimental conditions. Had the same animals been examined in more complex situations, or in the field, or had higher animals been studied, then classification would have been much more difficult, because different types of orientation may occur simultaneously, or in close succession. This argumentation, of course, does not invalidate the classification, but merely sets limits to its usefulness. Secondly, responses belonging to any one category do not necessarily share a common mechanism. Whereas assignment of an observed tactic behaviour to a particular category depends on the assessment of the receptor organs involved, it ignores all of the neural events that occur between stimulation and response. Therefore, the conclusions tend to be oversimplified. Hinde (1965) expressed a further reservation by saying: "Fraenkel and

Gunn restricted their classificatory system to cases which did not involve configurational stimuli". By now we have indeed already learnt how difficult it may be to identify the releasing stimulus for an instinctive response that is distinct from taxis, or even to distinguish between a behaviour that might be governed by taxis and one that might not. In addition, it is often not easy to identify a tactic component when it is masked by a more complex behaviour.

Despite these difficulties, several fundamental insights have emerged from the extensive work on taxes.

i) A distinction should be made between "agent stimuli" and "signal stimuli" (Viaud, 1951), or between orienting and releasing stimuli (Tinbergen, 1951). Agent stimuli influence behaviour in two ways. Depending on their intensity, they may induce a certain level of activation that becomes manifest through a general activity, termed "kinesis" (Fraenkel and Gunn, 1961). When the physical stimulus is localized at a fixed source and acts both by its intensity and its direction, it induces an oriented behaviour that is termed "taxis". Signal stimuli, on the other hand, induce "perceptive" responses (Viaud, 1951) that are triggered not only by intensity but also by the quality (configurational properties) of the stimulus.

ii) During phylogeny, as sensory capacities and neural integration improve and behavioural complexity increases, there is a progressive shift from tactic to perceptive responses. In the hierarchy of adaptive behaviour – reflex, taxis, instinct, learning and reasoning – the dominance of taxes decreases with phylogeny, more so in vertebrates than in invertebrates (Maier and Schneirla, 1935; Dethier and Stellar, 1961; Campan, 1980).

iii) The dominance of taxes also depends on the stage of ontogeny. Early in ontogeny, elementary behaviour consists of a repertoire of oriented responses governed by the intensity of the stimulus (Fig. 1 A). Low intensity induces an approach response, whereas high intensity induces a withdrawal response (Maier and Schneirla, 1964; Schneirla, 1965). In lower animals whose sensory capacities allow no more than intensity discrimination, ontogenetical adjustment only modifies the thresholds of the approach and withdrawal responses (Fig. 1 B), and taxes dominate the whole behaviour. Higher in phylogeny, however, responses become tuned to qualitative (configurational) properties of the stimulus (Fig. 1 C).

iv) Taxes are often part of more complex behaviours. They may be involved as tactic components of an innate releasing mechanism (Lorenz and Tinbergen, 1938; Tinbergen, 1951). Kineses and taxes may also play an appetitive function in many if not all behaviours, or be included in the perceptive responses to either innately attractive or learned signals (Rabaud, 1949; Tinbergen, 1951; Viaud, 1951; Médioni, 1967).

The aim of this chapter is to update the implications of the basic tactic components in complex oriented behaviours. In the first section we will

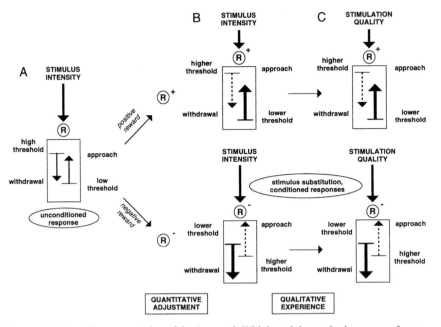

Figure 1. Schematic representation of the Approach-Withdrawal theory. In the course of onto-geny, from initial unconditioned responses to stimulus intensities (A), an animal will adjust its thresholds to stimulus intensity according to the sign of the reinforcement R⁺ and R⁻ (B). Later on, by stimulus substitution and control transfer, it will switch to conditioned responses to stimulation quality (C). After Schneirla (1965) and Suboski (1990).

evaluate the genetic and epigenetic determination of taxes. In a second part we shall provide evidence that taxes are elementary components of behaviour. The third part will be concerned with the role of kineses and taxes in the ontogeny of oriented behaviour. The fourth section will deal with the function of taxes in the animal's ecological adaptation. And final-ly we will discuss a theory of behavioural development based upon self-organization of the autonomous living system, and plead for a revival of the studies of taxes in the light of this theory.

Tactic behaviours: Innate but modifiable

In an early formulation of the objectivist theory of instinct, Lorenz and Tingergen (1938) and Lorenz (1956) stated that, to be considered innate, a behaviour must be present at birth and be expressed similarly by all members of the species; it must occur despite partial or total deprivation of corresponding sensory-motor experience, and it cannot be modified by learning. In this section we shall review data that provide evidence for the innateness of taxes, but also for their modifiability.

Innateness of taxes

In the first 50 years of this century, innateness of species-specific taxes was taken for granted. But it is only in the last decades that innateness of taxes has been demonstrated.

Phototaxis
When the young of the spider *Arctosa varianna*, born and reared in laboratory conditions and deprived of any orienting light, are released for the first time on a natural river bank, they escape consistently in a direction opposite to the sun azimuth, whatever the hour of day (Papi and Tongiorgi, 1963). Such an innate negative phototaxis was also described for the cockroach *Blaberus craniifer* (Bell et al., 1983) and the cricket *Nemobius sylvestris* (Campan and Médioni, 1963; Beugnon, 1984). Phototaxis is, however, positive in inexperienced, laboratory-born littoral isopods *Tylos europaeus* (Ugolini et al., 1995), in newly hatched larvae of the earwig *Labidura riparia* (Ugolini and Chiussi, 1996), of the water strider *Velia currens* (Birukow, 1956), and of the trichopteran *Hydropsyche cockerelli* (Coutant, 1982). In adults of the fly *Boettcherisca peregrina*, positive phototaxis is not impaired even after a 4-day period in total darkness (Mimura, 1986). An innate positive phototaxis with colour preferences (for example for UV) has been reported for the honeybee (Giurfa et al., 1995) and for larvae of *Hydropsyche cockerelli* (Lavoie-Dornik and Pilon, 1987).

Scototaxis
Innateness of scototaxis was demonstrated in the wood cricket *Nemobius sylvestris* (Campan et al., 1987), but the response is delayed in its development and it is in part irreversibly impaired by visual deprivation (Charii and Lambin, 1988; Meille et al., 1994). Thus, scototaxis (term coined by Alverdes, 1930) is a special case of elementary behaviour. Although a scototactic stimulus is a local orienting source, it does not act by its intensity and is thus distinct from negative phototaxis (Campan et al., 1987). A species may actually be at the same time scototactic and phototactic positive (Fraenkel and Gunn, 1961). Michieli (1959) examined scototaxis in many species, and found that attractiveness is mainly provided by the contrast perceived at the edges (perigrammotaxis). Still, because scototaxis (which is always positive) is not always directed towards edges, which constitute a configurational stimulus, a scototactic stimulus seems to be more than an "agent", but less than a "signal". Scototaxis thus seems to be intermediate between negative phototaxis and response to contrast (Bui Huy and Campan, 1982). As will be shown in the section on *From scototaxis to landmark orientation*, scototaxis is mainly effective in combination with further taxes.

Geotaxis
Based on observations on isopod species of the genus *Tylos*, Brown and Odendaal (1994) concluded that geotaxis is innate. It requires neither

learning nor compass calibration, as it is manifested on any sloped beach, irrespective of slope direction.

Preferred direction

An innate preferred direction of locomotion based on taxes has been found in several species of arthropods. Wolf spiders *Arctosa varianna*, reared in laboratory conditions, display negative phototaxis during the first days of their lives, but they assume a preferred direction towards north after about 30 days (Papi and Tongiorgi, 1963). The preferred direction is then time-compensated, as it remains the same regardless of the sun azimuth. A similar performance, but with somewhat different timing, was reported for the sandhopper *Talitrus saltator* (Pardi, 1960) and for the isopod *Tylos europaeus* (Ugolini et al., 1995). An innate orientation direction was also inferred for the water strider *Velia currens* (Birukow, 1956). The southwards and northwards migrations in autumn and spring, respectively, of monarch butterflies constitute another convincing example for innateness of the preferred direction. Both of these migrations, undertaken by different generations, take the animals to destinations where none of the individuals involved has ever been before (Brower, 1996).

Genetics of taxes

Studies on the genetics of taxes provide further evidence for the innateness of tactic behaviours. Erlenmeyer-Kimling and Hirsch (1961), Hirsch and Erlenmeyer-Kimling (1961) and Dobzhanski and Spasski (1967), using a multiple T-maze, were able to select strains of *Drosophila* according to their geotactic tendencies. An initially polymorphic population segregated into two strains, geotactic positive and negative, after only two to three generations of selection. Genes responsible for the sign of geotaxis were found to be located on each of the three major chromosomes of *Drosophila*: II and X for positive geotaxis, III for negative geotaxis. Using a similar method, Hirsch and Boudreau (1958) and Dobzhanski and Spasski (1967) selected two strains, positive and negative for phototaxis. The genetic determination seems, again, to be broadly polygenic. Benzer (1973), using a technique for fast fractioning of *Drosophila* populations, found many mutations with a pleiotropic effect on the phototactic responses. Variations of phototactic responses among mutants of *Drosophila* were found to be correlated with major alleles (Médioni, 1963). Light reactivity threshold, for example, is 100 times higher in the strain "bar" than in the "wild" type, whereas "yellow" is linked to a high light sensibility, and "tan" mutants are indifferent to light. Médioni (1963) found, in addition, a geographic cline in wild populations, with photonegativity increasing from east to west (from Japan to North America) and from north to south in the northern hemisphere.

In *Talitrus*, Pardi and Scapini (1983) and Scapini et al. (1985) demonstrated that the sun compass orientation for escape direction is genetically determined, with hereditary differences between populations living on differently oriented sea shores. Similar to findings in migratory birds (Helbig, 1996), an oligogenic mechanism was proposed for heredity of preferred direction in talitrids (two genes with two alleles, determining eight different directions, 45° apart) (Scapini and Buiatti, 1985; Scapini and Fasinella, 1990).

A study of progenies of mixed pairs of wild *Talitrus* from three different natural populations revealed differences of intra-population variability, probably depending on the stability of the shorelines (Scapini and Buiatti, 1985). More recently, Scapini et al. (1995) demonstrated a positive correlation within sandhoppers populations between heterozygosity and both coast stability and accuracy of the sun compass direction.

Thus, a large body of evidence supports the conclusion that basic tactic orientation mechanisms are genetically determined.

Variability of tactic behaviours

Although tactic behaviours are widely accepted to be innate and inheritable movements that are enforced on the animal, several observations show that they are variable. In the course of life, changes in stereotype tactic behaviour may occur due to several factors.

Age

Larvae of *Drosophila* are strongly photonegative, whereas adults display a polyphasic phototaxis (Médioni, 1963). Similarly, as they are ready to pupate, photonegativity in *Sarcophaga* fly maggots is weakened, and it sometimes disappears altogether (Zanforlin, 1969). Positive phototaxis of the trichopteran *Hydropsyche cockerelli* at hatching is rapidly reversed to negative phototaxis as the larvae get older (Coutant, 1982), while UV light sensitivity increases, reaching a peak at imaginal emergence (Lavoie-Dornik and Pilon, 1987).

Motivation and physiological state

In the coffee berry borer *Hypothenemus hampei*, the sign of phototaxis reverses from negative in virgin females to positive during oviposition (Mathieu, 1995). Honeybee workers and drones display negative geotaxis and positive phototaxis when they are about to leave the hive, but the sign of both of these taxes is reversed when they are returning to the hive (Jacobs-Jessen, 1959). Ebert (1980), studying the role of juvenile hormone in female honeybee larvae, found that queen and worker larvae (that are genetically identical) differ with respect to geotaxis as expressed in their orientation in the cells at the onset of metamorphosis. The queen larva, whose cell opens from below, is geotactic positive, i.e., the larva's head is oriented downwards. Worker larvae, whose cells are slightly slanted upwards, are not influenced by gravity; they pupate with head oriented either downwards or upwards. However, worker larvae treated with queen hormones (JHI or JH2) during the third day of development exhibit a positive geotaxis: they pupate head downwards (Ebert, 1980).

Environmental and physical factors
In the scolyte beetle *Blastophagus piniperda*, spring adults are photopositive at temperatures between 10° and 35°C, but photonegative at any other temperature (Perttunen, 1958, 1960). This range is restricted to 20°–30°C for autumn adults. Thus, for a 15°C temperature, individuals of the same species are photopositive in spring and photonegative in autumn. Honeybees and the ant *Camponotus* are photopositive at temperatures above 16°C and photonegative at lower temperatures (Müller, 1931). In water bugs *Naucoris*, photopositivity appears only when they start lacking oxygen for respiration. Otherwise, they exhibit negative phototaxis. In many aquatic arthropods, response to light changes with the concentration of carbon dioxide or the presence of other chemicals (potassium chloride, calcium and magnesium dichloride, etc.) in the water (Médioni, 1963). Even geotaxis is influenced by environmental factors: by gravity itself, as the sign of taxis depends on the slope, but also by light intensity, humidity and temperature. These factors modify the proportion of climbing or descending a slope or a vertical support (Horn, 1985). The tendency of zooplankton to come up to the water surface when carbon dioxide concentration increases (Médioni, 1963) is probably due to the fact that presence of carbon dioxide increases both negative geotaxis and positive phototaxis.

Innate rhythms
The sign of taxes may also change with the circadian rhythm. The apterigotan *Hercinothrips femoralis* (Thysanura), for example, is photopositive during day hours, when it usually migrates to the upper side of the bean leaves, and negative during night hours when it shelters below leaves (Koch, 1981). Similarly, honeybees and ants are photopositive and geonegative in the morning, but photonegative and geopositive in the evening (Jander, 1963). Rhythmic responsiveness to visual stimuli was also demonstrated in talitrids (Mezzetti et al., 1994).

Experience-based modifications of tactic tendencies

Several tactic behaviours have been shown to be modifiable through the animal's experience in its natural environment or through experimental manipulations.

Geotactic tendencies

Despite their fixed responses to gravity (positive, negative or transverse), most insects are able to change, spontaneously or after learning, their menogeotactic angle on an inclined surface. For example, the second larval instar of the green lacewing *Chrysopa carnea* consistently climbs up vertical rods for hunting aphids (Bond, 1983). However, if the experiment is repeated, without reinforcement, geotaxis habituates and becomes polyphasic. Social insects that have learnt a new geomenotactic angle reproduce it with only very small errors (Markl, 1964, 1966).

Scototactic tendencies

In *Nemobius sylvestris*, scototaxis may vanish when the animals are submitted to 10 daily releases during 40 days in an arena where the scototactic stimulus offers no reward (Campan et al., 1987). In a training experiment, when a reward of food, shelter and water is associated with a white stripe on a black background, as opposed to a non-rewarded black stripe on a white background, Campan and Lacoste (1971) were able to reverse the cricket's scototactic tendency.

Sun compass direction

Talitrus, whose escape from a beach in a landwards direction is innate and inherited, is able to adjust its sun compass direction to a quite differently oriented beach when it has been displaced (Scapini, 1995).

Taxes as early steps in the ontogeny of oriented behaviours

Taxes are elementary components of behaviour, because they are the simplest ones and because they appear earliest in life (Smith, 1993; Menzel et al., 1993). We will now provide evidence showing that these elementary behaviours are involved in the ontogeny of oriented behaviour, and that they are basic for acquiring more complex orientational cues, as well as for developing signal selectivity that represents by far more than simple stereotyped behaviour.

From geotaxis to slope

Jander (1963) distinguishes between i) progeotaxis, when menotactic angle varies with intensity of gravity stimulation, i.e., with steepness of slope, and ii) metageotaxis, when the menotactic angle does not change with the slope (Jander, 1963). Progeotaxis is more thoroughly a taxis, i.e., a response to an agent stimulus. It is mainly negative (ascending) and seems to be very widespread in arthropod species (Rabaud, 1949). Metageotaxis, on the other hand, is more complex, coming closer to some kind of "slope compass".

In the sandhopper *Taliturs*, Scapini et al. (1993) and Scapini (1995) compared inexperienced and experienced animals, both originating from the same population, with respect to their response to slope. In both groups, geotaxis was positive on a dry substrate, but negative on a wet one, as is the case on a natural beach. When a large black stripe occluding the view of the horizon, thus simulating the dune crowned with its vegetation outline, provided scototactic stimulation, half of the inexperienced animals reared in the laboratory in horizontal tanks with a black stripe followed the visual cue, and the other half the slope. Same age animals from a wild population, however, responded exclusively to the black stripe, and not to the slope. Animals reared in a slanting tank (where humidity is in the lower part) with a dark horizon in the upper part of the tank, responded positively to gravity and oriented down the slope. Thus, the hierarchy of cues that govern orientation is calibrated by experience.

In *Tylos europaeus*, terrestrial cues (hypsotaxis) were shown to dominate over geotaxis (Mead and Mead, 1974). The sign of geotaxis is determined by the degree of humidity of the sand, which dominates over all other possible orienting factors. When moving on wet sand, this isopod climbs up the beach slope, but it does the opposite on dry sand (Brown and Odendaal, 1994). In *T. punctatus*, $1°15'$ of declivity is sufficient to elicit a response.

Mieulet (1980) and Campan et al. (1987) described an experiment on the wood-cricket *Nemobius sylvestris* living in a forest on a slope. When the experimental platform used to measure the escape direction is set horizontally, crickets of all ages escape towards the darkest part of the panorama. When the visual cues are screened, they orient at random. When the platform is inclined by about $40°$, with all cues visible, they flee towards the forest, whatever the direction of the slope. However, when the platform is inclined and the visual cues are occluded, they step down the slope, even when this brings them to the open, where they would otherwise not go (except at night). The accuracy of orientation to slope improves with age. Tested at the same site, 7–9-month-old crickets captured in a distant flat forest did not display geotaxis.

These data suggest that fundamental geotaxis is calibrated during ontogeny to cope with the local conditions.

From scototaxis to landmark orientation

Fundamental scototaxis (Alverdes, 1930) may involve several derived behaviours: i) orientation towards contrasting edges (perigrammotaxis) (Michieli, 1959); ii) tendency to follow continuous contrasting edges (photohorotaxis) (Kalmus, 1937; Lehrer et al., 1985), and iii) attraction to the highest outlines at the horizon (hypsotaxis) (Schneider, 1952; Couturier and Robert, 1958). Although little is known so far about the role of these specific behaviours in the development of complex oriented behaviour, scototaxis in its general sense has been shown several times to be a basic factor in the ontogeny of advanced orientational performances.

In the wood cricket *Nemobius sylvestris*, Campan et al. (1987) have demonstrated that the initial innate scototactic tendency is being replaced, during the first 2–3 weeks after hatching, by a tendency to use more specific terrestrial cues, such as edges (the forest outlines, as well as those of tree trunks flanking the forest trails), for orienting its daily migrations away from or towards the forest ecotone, to find food or shelter, respectively. Later in ontogeny, after 6–10 weeks, it will have associated the familiar routes with celestial directional cues (Campan et al., 1987).

In the wolf spider *Arctosa varianna*, scototaxis is directed towards vegetation outlines along banks in the natural habitat (Papi and Tongiorgi, 1963). During the phase in which the spider is guided by this innate scoto-taxis, it learns to associate the dark signal of the bank with a sun compass direction and to use it for returning to a familiar bank. A similar onto-genetical development was found in riverine crickets *Pteronemobius heydeni* and *P. lineolatus* (Beugnon, 1985, 1986), and in the sandhopper *Talitrus* (Scapini, 1995).

These results render scototaxis a good candidate for constituting the first step in ontogeny towards the use of terrestrial cues, being later associated with celestial cues, in more complex orientation tasks. It seems very likely that similar developmental processes occur in further arthropod species, mainly those living in ecotone systems (beaches, river or lake banks, trail and road sides, etc.) in which regular locomotor activity is associated with terrestrial cues located in a particular compass direction. This may be the case, for example, in the pronounced directional preferences observed in riverine Carabidae (Papi, 1995; Colombini et al., 1994) and earwigs *Labidura riparia* (Ugolini and Chiussi, 1996), whose ontogenetic develop-ment, however, has not been examined so far.

An initial scototactic tendency may also lead to the development of hypsotaxis. For example, may beetles fly towards the forest outlines to forage (Couturier and Robert, 1958); ladybirds gather at the highest summit of the area where they stay quiescent over the summer under rocks and within crevices (Iperti, 1966), and winged ants *Formica subnuda* and *Leptothorax muscorum* will meet for mating at the highest hilltop (Chapman, 1969). Similarly, photohorotaxis, also emerging from initial scototaxis, could explain the tendencies of honeybees (Lindauer, 1969) and digger wasps *Bembix rostrata* to fly along uninterrupted lines provided by underneath topographic features such as roads, forest edges, rivers, or lake banks (Chmurzynski, 1964). Of course, these behaviours would not require a compass and need not develop further.

From phototaxis and polarotaxis to sun compass orientation

Jander (1963) has proposed a hierarchy of various forms of phototactic behaviours which he believes to correspond to their chronological succession, namely: i) archeophototaxis, the most primitive form, similar to photokinesis in as far as the response is only intensity dependent; ii) prophototaxis, which is more advanced, because the response depends on both intensity and direction; iii) metaphototaxis, the most advanced form, in which the response is based on the direction alone. In the course of phylogeny, as well as of ontogeny, each form might have emerged from the previous one. Astrotaxis, the most advanced phototactic behaviour, would then be built on photomenotaxis by including a time-compensating mechanism.

In laboratory-born young wolf spiders *Arctosa varianna*, negative phototaxis is already present at birth (Papi and Tongiorgi, 1963). During maturation, they progressively develop a northwards orientation, while photonegativity still persists. After 30 days, however, photonegativity disappears and the tendency to orient northwards prevails. Animals of the same age from the wild consistently preferred a northwards direction when they were tested for the first time. When animals were time-shifted prior to testing them by manipulating the photoperiod, their escape direction shifted accordingly, showing that it is time-compensated. In the wild, however, learning a functionally meaningful direction occurs already during the first days of the spiders' life. Even at an age of only 2 days, they orient towards their native shoreline. Orientation accuracy improves further in the course of ontogeny.

In another spider species, *Arctosa cinerea*, one that hardly ever seeks contact with water, even very young animals escape landwards when they are brought in contact with water for the first time. Papi and Tongiorgi (1963) proposed that the spiders determine land direction by monitoring the darkest area of the landscape, but also humidity gradients and slope may play a role. Indeed, 3 days of experience on a simulated slope in laboratory conditions were sufficient to make the animals adopt a new escape direction. The authors suggested that a learned association may occur between a scototactic signal and a local cue, thus calibrating menotaxis with respect to celestial cues. However, the preference for the learned direction may disappear if it is not regularly exercised, for example after hibernation (Papi and Tongiorgi, 1963).

In the wood-cricket *Nemobius sylvestris*, newly hatched larvae are consistently photonegative (Campan and Médioni, 1963; Beugnon, 1984). They escape westwards in the morning, i.e., in a direction opposite to the sun's azimuth, whether or not terrestrial cues are visible. By using a mirror to simulate a wrong position of the sun, Beugnon et al. (1983) were able to manipulate the insect's orientation. In the wild, crickets of up to 6 weeks of age showed a consistent photomenotaxis whose angle with respect to the light source varied from one animal to another. In older animals, however, the menotactic angle was common to all individuals of the same area. Laboratory-reared crickets of the same age showed no preferred escape direction. These findings again support the idea of a basic innate negative phototaxis that shifts, during ontogeny, first to a photomenotactic response, and later to astrotaxis, a time-compensated sun compass orientation. This idea has already been hinted at by Viaud (1951) who wrote: *"Dans un premier temps, c'est la phototaxie qui guide les insectes; ensuite ils utilisent la lumière pour se guider".*

In analogy to the findings reviewed above, Jander (1957) proposed that, in the ant *Formica rufa*, photomenotaxis is an experience-based improvement of phototropotaxis. Menotactic orientation is learned during repeated foraging trips to a particular food source. Phototropotaxis should assist the insect in compensating for errors during the period of acquisition of menotaxis. Once menotaxis is acquired, it becomes largely independent of phototropotaxis. Indeed, when the sign of phototaxis changes from positive to negative (when the ant returns to the nest), menotaxis is not impaired. Thus, during ontogeny, photomenotaxis is superimposed on the initial phototropotaxis and finally replaces it. Sun compass orientation is then calibrated with menotaxis by a time compensation mechanism (Jander, 1957).

In bees, again, photomenotaxis precedes sun compass orientation. Lindauer (1959) suggested that young honeybee foragers learn to use the sun compass on the basis of an initial innate photomenotaxis. A similar conclusion was drawn by Dyer and Gould (1981) and Gould (1982). The switch from photomenotaxis to sun compass orientation requires knowledge on the apparent movement of the sun in the course of the day. When conspicuous visual marks are present, bees may memorize the sun's daily trajectory relative to these marks (Gould, 1993). However, they need not see the complete trajectory in order to learn it. Bees that had seen the sun exclusively in the morning were able to extrapolate its afternoon position, as revealed by their dances (Dyer, 1996).

Adults of the earwig *Labidura riparia*, tested on the beach where they live, escape landwards, whatever the hour of day. They do so even when terrestrial cues and the sun are hidden (Ugolini and Chiussi, 1996), suggesting that they use a time-compensated compass mechanism. When a shifted sun position is simulated using a mirror, the animals' escape direction is shifted by the appropriate angle. At night, they assume an average menotaxis to the moon which allows them a rather correct orientation. It is then possible to shift the escape direction by hiding the moon and guiding the insects with a flash light. After a week of captivity in the lab, however, the preference for the compass direction disappears and the animals only respond phototactically. When earwigs of stage III and older, appropriately oriented on their original beach, are released on a shoreline whose orientation is reversed or at a right angle to the original one, they adopt, after a while, an escape direction that agrees with the new local conditions (Ugolini and Chiussi,

1996). Animals younger than that, however, are just photonegative and escape in the opposite direction to the solar azimuth, in the same way as do laboratory-born animals of the same age. Thus, phototaxis, although innate, allows, in the course of ontogeny, calibration of the sun compass. In *Talitrus*, such a calibration also occurs with respect to the moon (Papi, 1960). In the case of the moon compass of the earwig *Labidura riparia* mentioned above (Ugolini and Chiussi, 1996), it remains to investigate whether or not it is time compensated.

For sun compass orientation, free view of the sun is a sufficient, but not necessary condition. In many species (honeybee, ants, crickets, sand-hoppers), sun compass orientation was shown to occur even when the sun is not visible, by using the sky pattern of polarized light. Because the distribution of e-vector directions in the sky varies systematically with the sun's position, the sky pattern can inform the animal about the sun's position at any time of the day.

Polarotaxis, a photomenotactic response to e-vector direction, might constitute an early but essential step towards utilization of the sky pattern compass. Polarotactic responses are widespread among arthropods (Wehner, 1981). They have been observed in mosquitoes and moths (Kovorov and Monchadskiy, 1963; Danthanarayana and Dashper, 1986), as well as in aquatic insects searching for ecologically significant sites, such as ponds (Schwind, 1989, 1995; Horvath and Zeil, 1996). In laboratory experiments, Ruiz (1991) was able to train crickets *Gryllus bimaculatus* to associate an e-vector direction with localized black stripes rewarded by food and shelter. E-vector information alone proved to allow the trained crickets, even in the absence of the stripes, to find the place where they have previously been rewarded. The animal's capacity to orient to e-vector direction is clearly a prerequisite for establishing a compass based on the celestial e-vector pattern.

Arthropods that repeat regular routes in a specific environment may possess an inherited compass direction, as we have seen in many examples cited above (see Section on *Innateness of taxes/Preferred direction*). However, despite this genetic determination, a large amount of flexibility is needed not only for adjusting locomotory directions to new ecological requirements, but also for switching from one compass mechanism to another when circumstances require such a measure. In the time compensated menotaxis of *Talitrus*, direction preference is innate and inherited (Pardi, 1960; Pardi and Scapini, 1983; Scapini et al., 1985; Scapini and Buiatti, 1985). Still, the sandhopper is perfectly able to adjust the various compasses (sun, moon, slope, polarized light) to the actual ecological conditions (Ugolini and Macchi, 1988; Ugolini et al., 1988, 1991; Scapini, 1995). In many other arthropods, compass direction is not innate and must therefore be learned first. But even in these cases, flexibility is largely retained, to enable the animal to cope with unexpected new situations, as has been demonstrated in the wolf spider *Arctosa varianna* (Tongiorgi,

1962; Papi and Tongiorgi, 1963), the crab *Goniopsis* (Schöne, 1965), the riverine cricket *Pteronemobius lineolatus* (Beugnon, 1986), the mole cricket *Gryllotalpa gryllotalpa* (Felicioni and Ugolini, 1991), the shrimp *Palaemonetes antennarius* (Ugolini et al., 1989), the isopod *Idotea baltica* (Ugolini and Pezzani, 1993), the earwig *Labidura riparia* (Ugolini and Chiussi, 1996), and the desert ant *Cataglyphis* (R. Wehner, this volume).

Flexibility is particularly important in tasks the require improvisation on the part of the animal. For example, scouts of foraging honeybees and ants searching for a novel feeding site do not have *a priori* a particular goal and therefore also no particular preferred direction. Still, they must keep a record of the various directions (and distances) they have taken in the course of the outbound journey, in order to compute, by path-integration, the direction (and distance) that will bring them on a straight route back home from any location along the foraging route (Wehner, 1992; Wehner et al., 1996). For updating its home vector, the animal may use any type of directional information available to it. The flexibility required in this task may make use of various forms of taxis as primary elementary tendencies to support the organization of a complex orientation system.

From frequency response to acoustic signal selectivity

Cricket and grasshopper females respond to the calling songs of con-specific males by walking or flying in the direction of the sound source. It has been demonstrated many times that this response, termed "phono-taxis", is very species-specific. Adult females are highly selective in re-sponding to various song parameters, specified below. Song selectivity, however, represents a perceptive response to a configurational signal, and thus it falls in the uncertainty range where true tactic behaviour and per-ceptive response overlap. We here review experimental results from which we conclude that only the ontogenetically initial part of phonotactic behaviour can actually be termed a taxis in the sense put forward in the *Introduction*, but that it constitutes the first step towards a more complex response to a signal stimulus.

Popov and Shuvalov (1977) and Shuvalov and Popov (1984) demonstrat-ed, in dual-choice experiments, that discrimination of *Gryllus bimaculatus* females between the calling song of their conspecific males and that of the closely related species *G. campestris* is based on the basic temporal characteristics of the song (carrier frequency), on syllable (= pulse) rate, and on chirp frequency and duration. Investigating responses of mature female *G. bimaculatus* to artificially manipulated conspecific songs (Fig. 2), Shuvalov and Popov (1873, 1984) found that a trill (i.e., a con-tinuous train of pulses, Fig. 2 A, Stimulus II) does not induce a phonotactic response (Fig. 2 B, II, hatched bar), except when the trill is the first signal in the evening after the daily silence (Thorson et al., 1982). However,

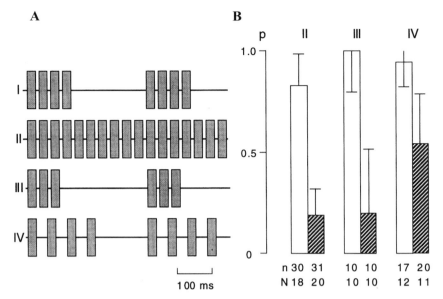

Figure 2. Experiments on recognition of male calling songs by conspecific females *Gryllus bimaculatus*. (A) Artificially produced calling songs. Each bar represents a sound pulse (syllable); a group of pulses separated from the next one by a time interval is termed a "chirp". (I) Natural calling song of *Gryllus bimaculatus* (chirps of four pulses, pulse repetition period 36 ms, chirp repetition frequency 3 Hz). (II–IV) Modifications of natural song. (B) Normalized proportion of positive responses of experienced and inexperienced (hatched and white bars, respectively) *G. bimaculatus* females to the source of the acoustic signals II–IV shown in (A). Vertical lines denote standard errors. n: number of responses; N: number of animals tested. Redrawn after Shuvalov and Popov (1984).

females that have been reared in isolation do respond to trill (Fig. 2B, II, white bar) for as long a pulse frequency agrees with that of the natural song (Stimulus I in Fig. 2A). They respond to a calling song even if the interchirp interval is longer than the natural one (Stimulus III in Fig. 2A), as long as pulse and chirp frequencies are in accordance with those of the natural song (Fig. 2B, III, white bar). They respond even when pulse frequency within chirps is decreased (Stimulus IV in Fig. 2A), if the chirp frequency of the natural song is retained (Fig. 2B, IV, white bar). Thus, inexperienced females, as opposed to experienced ones (Fig. 2B, II–IV, hatched bars) respond to calling songs without much selectivity, but within the limited range of the natural pulse and chirp frequencies, which are believed to be genetically determined (Shuvalov et al., 1990). Adult females of *Gryllus campestris* respond to calling songs even when the number of syllables is doubled, and some respond to trills as well, for as long as the carried frequency is in accordance with the natural one (Weber and Thorson, 1989). Thus, innate phonotaxis is indifferent to the specific structure of the natural song, it is merely frequency-dependent.

Several workers (e.g., Alexander, 1962; Huber, 1962) proposed that phylogeny of cricket song has taken place in three steps: i) discrete unpatterned pulses; ii) trill with a steady repetition of syllables, the frequency of which is close to wing beat frequency; iii) a particular, species-specific chirp pattern. This conclusion is based on the finding that nerve structures that control the song commands are organized hierarchically, the more primitive commands being embedded in more advanced neural structures (Sergejeva et al., 1993). Results of experiments on females that had been sound-deprived at different ages during their larval development and tested as adults show that ontogeny of phonotaxis, measured as positive orientation during tethered flight (Shuvalov and Popov, 1984; Shuvalov et al., 1990; Sergejeva et al., 1993), runs through three successive stages: i) True phonotactic response to carrier frequency, without much selectivity; ii) highly selective response to species-specific calling song; iii) decrease of selectivity (non-selective phase). Thus, the ontogeny of *G. bimaculatus* phonotactic responses is a rough repetition of the phylogenetic steps, except for the final decrease of selectivity that was described in walking females as well (Walikonis et al., 1991). These results support the hypothesis that pulse rate is effective earlier than is chirp frequency, but that both of these parameters are effective in adult crickets.

By subjecting previously sound-isolated females to a song prior to testing them, Shuvalov et al. (1990) demonstrated that the ontogeny of song selectivity requires at least a short auditory experience. The authors concluded that the genotype roughly controls the species-specific song preference, experience acting to precisely calibrate song features by learning and imprinting. Learning during development is clearly predetermined, because only the species-specific song is learnt. A similar conclusion was drawn for bird song learning (e.g., Gould and Marler, 1991).

The role of learning in song selectivity was confirmed in several field studies (Shuvalov and Popov, 1984; Shuvalov, 1985; Shuvalov et al., 1990; Sergejeva et al., 1993). Crickets live in dense populations and males begin to sing 1 or 2 days before female positive phonotaxis appears. Females thus experience conspecific male songs before having established the final representation of the acoustic signal. When they first fly or walk towards a choir of males, they initially respond to syllable (pulse) frequency. When they land in the choir area, they express selectivity towards chirp frequency. Later on, when females adopt a final orientation toward a particular male, chirp selectivity decreases. Chirp selectivity also decreases when females move from one sub-population to another that displays slightly different characteristics of the song. This tolerance has a biological significance, as it may allow mature females to migrate to neighbouring populations (which actually occurs) and to easily become tuned to interpopulation variations of song structure.

The main conclusion to be drawn in the present context is that the only true phonotactic element of phonotaxis is the response to a carrier frequency within a genetically fixed range of syllable and chirp frequencies. As in the previous examples, this tactic behaviour is a first step, both phylogenetically and ontogenetically, towards a more complex response to a signal stimulus, culminating in high selectivity towards the stimulus.

From anemochemotaxis to food and host-plant selection

Anemochemotaxis, i.e., the innate attraction towards a particular chemical signal carried by air, as well as the innate rejection of such, is very pronounced in all insects that use a particular host (plant or animal) for feeding or laying eggs. A specific attractive chemical stimulus, signalling food, or a host, triggers a response globally directed at the source of stimulation.

However, anemochemotaxis may be only one more step towards the development of selectivity, because the final preference will depend on the insect's experience with the source of stimulation once it has arrived there, whatever innate search image of the target the insect might have had initially. A reward will result in a preference, no reward (or punishment) in avoidance of that particular target (Bernays, 1993; Mathieu, 1995). In the course of this process, new associations are developed. Avoidance of toxic prey, for example, is often based on a conditioned association between aversive experience and the visual appearance of the prey (Bernays, 1993).

Also, variations in host plants colours, linked to the natural vegetative cycle, can modify the tactic response through learning, for example if the change of colour indicates that the plant is deficient in some nutrient and should therefore be avoided. Thus, the initially unconditioned tactic components of the original searching behaviour will become, through experience, part of the signals stimulus. Generalist insects are much more likely to develop preferences and aversions based on experience than are specialists, whose selectivity is usually predetermined (Bernays, 1993).

Young laboratory-reared honeybees, on their very first foraging trip, do not land on flowers that lack scent, regardless of their colour (Giurfa et al., 1995). Thus, initially, they are innately attracted to natural flower scent, but not to colour. However, after only one reward on a scented flower, their colour preferences towards differently coloured flowers, all equally scented, can easily be examined. They then exhibit an innate preference for UV and blue-green over other colours (Giurfa et al., 1995). Through further foraging experience, however, the initial preference may change and become directed towards other colours. Innate preferences for a particular colour thus leads to a capacity of learning further colours. Bees actually learn any colour at which they are being rewarded (e.g., Menzel, 1967), and they easily reverse their preferences when they are later rewarded at another colour (Menzel, 1969). The same holds true in the case of shape learning: although bees express pronounced innate preferences towards flower-like shape parameters (Lehrer et al., 1995), they can easily be trained to shapes that do not resemble flowers at all (e.g., Hertz 1933; Lehrer et al., 1995). Thus, by operant conditioning, the insect may associate the orienting stimulus with the reward, and the latter, in a further step, with additional stimuli also present at the target. This learning process will result in a new selective behaviour (Bernays, 1993).

Orientation to naturally relevant and attractive chemical signals (chemo-taxis) may play an important role in the development of complex behaviours not only by constituting the first step in the ontogeny of such behaviours, but, in addition, by containing already initially further stimu-latory components. Chemotaxis may be associated not only with anemo-taxis (see K.-E. Kaissling, this volume), but also with phototaxis, scoto-taxis, and geotaxis. These additional tactic components may later partici-pate in the updating of search images, leading to new preferences.

Possible role of magnetotaxis in ontogeny of orientation

Earth magnetic field is an omnipresent, reliable source of orientational information. It has been suggested for several bird species that magneto-taxis is basic and that all other compasses (sun, moon, stars) are calibrated by it (Wiltschko and Wiltschko, 1975a, b; Able, 1991; Able and Able, 1996). The hypothesis that the magnetic field may, even in arthropods, be used for calibrating or improving the accuracy of astronomical compass mechanisms was proposed by Martin and Lindauer (1973), Towne and Gould (1985), and Leucht and Martin (1990) for honeybees, by Baker (1987) for moths, and by Pardi et al. (1988) for sandhoppers. (See also M.M. Walker, this volume.)

In experiments on an equatorial adult *Talorchestia martensii* population, Pardi et al. (1988) demonstrated that the sandhopper uses a magnetic com-pass. When magnetic and sun compass are set in competition, the magnetic compass dominates, sun compass being mainly used at dawn and at dusk. *Talitrus* uses a magnetic compass for inferring a sea-land escape direction which agrees with the one of the beach where it lives, when no other cues are available (Arendse and Kruyswijk, 1981; Pardi et al., 1984, 1988). Ugolini and Pardi (1992) suggested that magnetic compass provides the sandhopper with the general sea-land orientation axis, whereas the sun compass would provide the necessary information about land or sea direc-tion. In Italian populations, however, sea-land orientation behaviour seems to be governed exclusively by the sun compass (Scapini and Quochi, 1992).

Ugolini and Pezzani (1995) studied the role of magnetic cues in learning the migration direction beach-sea in the marine isopod *Idotea baltica basteri*. They demonstrated the animal's capacity to use a magnetic com-pass in the laboratory in the dark, whereas in the field the animal adjusted itself to other local cues. This isopod also seems to use the sun (Pardi, 1963; Ugolini and Messana, 1988), and it exhibits a remarkable plasticity in changing its escape direction when required to do so by environmental conditions (Ugolini and Pezzani, 1993). *Idotea baltica* lives in conditions of low visibility, among floating banks of seaweed and *Posidonia*. Animals tested in the absence of visual cues display a bimodal distribution of escape directions along the theoretical sea-land direction of the beach where they have been captured. When magnetic north is artificially shifted by 90 degrees, bimodality axis shifts accordingly, but if the horizontal component of the magnetic field is compensated to zero, the animals are disoriented. In a learning experiment, when a slope at a right angle to the natural one is offered, the animals shift the escape direction by 90 degrees. It is concluded that these isopods have a basic capacity to orient to the magnetic field, but that orientation is functionally more meaningful when it is based on slope direction (or other external cues available, mainly the sun) (Ugolini and Pezzani, 1993).

Monarch butterflies possess an innate migratory direction that does not require experience (Brower, 1996). Gould (1990) suggested that they might determine this direction by learning the angle between the rotational axis of the terrestrial cues and the direction of magnetic North. Magnetic orientation is also involved in other orientational tasks. For example, systematic directional errors observed in honeybee waggle dances seem to be caused by earth magnetic field (von Frisch, 1965). The error values are linked to the orientation of the comb on which the bee is dancing, and can be predicted from the local characteristics of the earth magnetic field. Martin and Lindauer (1977) suggested that magnetotaxis might stabilize the dance against disturbances, such as light.

Taxes involved in more complex oriented behaviours

It is generally admitted that taxes are involved in more complex oriented behaviours (e.g., Fraenkel and Gunn, 1961; Médioni, 1963; Hinde, 1965). Lorenz has, already in 1938, described tactic components of instinct and emphasized the overlap between tactic and instinctive behaviours. The distinction between agent and signal stimuli mentioned in the Introduction may help distinguish between the two types of response.

Appetitive functions of kineses and taxes

Although kineses do not guide the animal to its target, they play an important role in all types of tactic behaviour due to their appetitive function (Cardé, 1984), resulting, in most cases, in locomotory activity. Kineses are phylogenetically more primitive than taxes, because they do not require sense organs to localize a stimulation source. The intensity of the ortho-kinetic component (i.e., straightforward movement), as well as that of the klino-kinetic component (i.e., frequency of direction changes) (see Tab. 1), increases with increasing intensity of the stimulus and with increasing responsiveness of the animal (the latter being determined by physiological and environmental factors), thus increasing the probability that a tactic behaviour will be initiated, resulting in consumatory actions. The role of kineses in initiating tactic behaviour has been proposed by Akers and Wood (1989) for anemotaxis, by Bell and Kramer (1979) and Mathieu (1995) for anemochemotaxis, and by Médioni (1963) for phototaxis. The same appetitive function can be extended to all taxes. With respect to functional efficiency of appetitive tactic behaviours, polyphasic taxes, as described for phototaxis (Médioni, 1963) and geotaxis (Jander, 1965; Bond, 1983), are certainly optimal.

Thus, both kineses and taxes increase the animal's probability to explore its world (Bernays, 1993). Far beyond the innate responses to signal stimuli, the animal learns the functional meaning of new signals and enlarges or updates its selectivity for food or prey, as it also learns to avoid aversive stimuli.

Combination, dominance, and succession of taxes

Taxes are often combined for monitoring a given oriented response. They often appear in a particular succession in the course of ontogeny, and often one type of taxis is dominant over others that are present simultaneously.

The set of innate tactic tendencies in *Talitrus* (Pardi and Scapini, 1987) are differently balanced in various populations according to the local conditions, but they will all be synergic to allow a biologically meaningful orientation. When the beach where *Talitrus* lives is sloped and the outline of the land vegetation inconspicuous, geotaxis and astrotaxis dominate over phototaxis and scototaxis (Hartwick, 1976; Ugolini et al., 1986). In *Nemobius*, phototaxis and scototaxis are less pronounced in populations living in deep wood, compared to those inhabiting the ecotone between wood and meadow (Campan et al., 1987).

Larvae of the fly *Sarcophaga* are consistently photonegative, but at the time immediately prior to pupation, this tendency decreases, which causes the maggot to desert natural or artificial traps such as holes, crevices or tubes which are not suitable for pupation, by moving towards the light (Zanforlin, 1969). This behaviour disappears when the maggot has found a shelter where the thigmotactic stimuli promise convenient conditions. Then the maggot settles itself facing the light, stops exploring and may start pupation. Thus, phototaxis and thigmotaxis act together to enable the maggot to choose the suitable place for pupation.

In *Carpophilus hemipterus*, a parasite of apple trees, an initial positive phototactic response that induces take-off is later inhibited by a food signal that causes the insect to land (Blackmer and Phelan, 1991). It resumes its appetitive flight if the food odour is removed. A similar competition between two types of stimulus was also described by Graham (1959) in the scolyte *Trypodendron*: Wood odour inhibits the initial phototaxis, i.e., phototaxis helps the insect to detect the chemical signal of the host.

Larvae of *Ecdyonurus venosus* (Ephemeroptera) observed in a fluvarium (Butz, 1975) exhibit, during daytime, positive rheotaxis, swimming against the water current, thus compensating for the nocturnal drift. At the same time, they are guided by negative phototaxis and possibly also by scototaxis for finding a shelter such as crevice, crack or slit in a rock, to protect them from diurnal predators. Thigmotaxis and negative phototaxis then act synergically to keep the larvae within the shelter. In this phase, thigmotaxis is dominant (Butz, 1975).

Another hierarchical combination of various elementary tendencies governs the aggregation of the rice grasshopper *Heiroglyphus banian* (Nayak et al., 1990) in areas of favourable abiotic conditions. Humidity preference and photopositivity play a more important role than do thigmotactic stimuli provided by the substrate. In the aggregation, males exhibit negative geotaxis, whereas females display positive geotaxis. This combination of taxes allows the insects to select favourable abiotic conditions for performing their species-specific behaviour. The negative sign of geotaxis in males is in agreement with their feeding habits on the upper part of the foliage, whereas the positive geotaxis in females is linked to egg laying in the soil underneath host plants.

In several cases, a relationship between the menotactic angle with respect to one type of stimulus and that with respect to another has been reported. For example, in Trichoptera, geomenotactic and photomenotactic angles are linearly proportional (Jander, 1960). When the two stimuli are set in competition with each other, the resulting orientation is intermediate. A similar photogeomenotactic transposition has been described in the beetle *Geotrupes* (Birukow, 1954), the ant *Myrmica* (Vowles, 1954), and the water strider *Velia* (Birukow and Oberdorfer, 1959).

Another frequent combination of taxes is involved in scototaxis. In many cases, it is perigrammotaxis (orientation to edges) that renders the signal efficient (Michieli, 1959). In the mealworm beetle *Tenebrio molitor*, for example, scototaxis is accompanied by edge fixation (Varjú, 1987). Walking moths *Lymantria dispar* are guided by contrasting edges as well as by negative phototaxis (Preiss and Kramer, 1984). Scototaxis in the wood cricket *Nemobius* is accompanied by both perigrammotaxis and photonegativity (Campan and Médioni, 1963). The response increases with contrast and is mainly guided by the albedo (ratio between incident and reflected light) of the signal, as a black stripe remains attractive even when it reflects more light than does the white background (Campan et al., 1987). Similarly, phototaxis and scototaxis seem to combine in the orientation of talitrids from Atlantic beaches to allow emergence and burrowing at the right time of the day, whereas in Mediterranean *Talitrus* these two taxes act independently from each other (Mezzetti et al., 1994).

Finally, one of the best examples of taxes combination is involved in honey bee language. To indicate the direction of a feeding place, the worker bee performs the waggle dance on the vertical comb inside the hive in such a way that the angle between the straight part of the dance and the force of gravity equals the angle between the food source direction and the sun azimuth (von Frisch, 1965). This is a functional case of photogeomenotactic transposition in which astrotaxis (or polarotaxis) acts together with geotaxis and magnetotaxis.

Taxes participate in complex oriented behaviours: The special case of anemochemotaxis

Selection of a host, be it a plant or an animal, as well as mate finding, whenever these behaviours are guided mainly by odour cues or by pheromones, provide good examples for the participation of taxes in more complex oriented behaviours. However, because odours (particularly pheromones) are innately preferred configurational signals acting by their quality, anemochemotaxis, similar to phonotaxis (see Section on *From frequency response to acoustic signal selectivity*), cannot be regarded as a true taxis. However, it actually involves several intermingled tactic components that cannot be held apart.

Flight, initiated by chemo- or anemokinesis, takes the insect either upwind, downwind or transverse with respect to wind direction. Along the route, it may be guided by a combination of chemotaxis, anemotaxis, phototaxis, and scototaxis, acting either synergically or sequentially, as well as by a variety of configurational visual cues. For example, *Rhagoletis pomonella*, a parasite of apple trees, is attracted, at a distance, by the odour of the host tree, but also by its outline, colour, and size (Moericke et al., 1975; Owens and Prokopy, 1984; Duan and Prokopy, 1992). The visual cues will keep the insect within the odour plume, maintaining a constant angle with respect to the source of the stimulus. When the target provides no visual cues, an insect flying in an odour plume cannot maintain a straight flight course (Kennedy, 1983; Kaissling and Kramer, 1990). This type of behaviour may also apply to rheochemotaxis, i.e., the oriented responses in a water current carrying chemical information (Bell and Tobin, 1982; see also M.J. Weissburg, this volume).

The odour plume is roughly a cone, consisting of moving air carrying the chemical substances provided by the source (Bossert and Wilson, 1963). The general structure of the odour plume is discontinuous (Murlis and Jones, 1981). It consists of a succession of odour puffs (Fig. 3), variable in size and concentration, distributed periodically in space and time. The odour plume needs to be patterned in order to trigger oriented behaviour

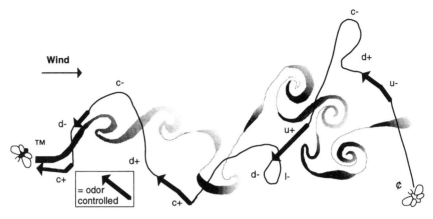

Figure 3. Schematized odour plume in turbulent air (dotted) originating from a point source (female gland), and imaginary flight track of a male moth, composed of several flight sequences leading the male on a zig-zag trajectory (continuous line) to the female. u: upwind turn; d: downwind turn; c: counter-turn; 1: loop; +: right; −: left. Redrawn from Kaissling and Kramer (1990).

(Kramer, 1992), a conclusion supported by findings on the properties of pheromone receptor cells (Kaissling, 1986; Kaissling and Kramer, 1990). The period between puffs increases with the distance from the source (Murlis, 1986). Within an odour plume, however, there is no clear gradient, i.e., the plume structure does not provide reliable information about the distance of the source (Masson and Mustaparta, 1990), although flow and dose are roughly correlated with distance (Murlis et al., 1992). The size of the source, wind discontinuities and turbulences, as well as the opening of the habitat, also influence the shape of the plume cone. In such a complex stimulus situation, the oriented behaviour has to be klinokinetic and klinotactic, i.e., with frequent changes of flight direction, which enhance sampling efficiency by a kind of casting behaviour (Kaissling and Kramer, 1990). In the whole orientation process, idiothetic feed-back contributes to self steering of flight direction. As a result, flight appears zig-zagging, as if the insect were trapped within the invisible cone, with the odour source as the target (see Fig. 3). In the final approach, the source is identified and acts as a signal stimulus that need no longer be exclusively olfactory. In the fly *Rhagoletis pomonella*, for example, the final approach to the fruit target is triggered by the colour contrast between fruit and foliage (Moericke et al., 1975; Owens and Prokopy, 1984; Duan and Prokopy, 1992). The scolyte *Hypothenemus hampei* is attracted to tree silhouettes or to coffee berries by a hypsotactic and a scototactic stimulus, respectively (Mathieu, 1995). The visual signal acts synergically with odour, the two types of information being effective according to the physiological condition of the female.

Another example for the role of taxes in more advanced oriented behaviours is host finding in female mosquitoes, which again represents a complex combination of responses to visual and olfactory stimuli (Bidlingmayer, 1994). Appetitive flights (kinesis) of the females take place mainly at dusk and at night, according to their activity rhythm. The first step of orientation is guided by hypsotaxis to the highest silhouettes in the panorama. Elevation above ground is then controlled by optic flow (opto-motor anemotaxis) (Kennedy, 1939). At the same time, the insect responds positively to wind by flying against the draft direction (positive anemo-menotaxis), which increases the probability to encounter chemical signals originating from a potential host. Flight control (speed, direction, etc.) is mediated by internal (idiothetic) self-steering mechanisms. It results, again, in a characteristic zig-zagging flight, limited by the plume cone. Close to the host, just before landing, a different hierarchy of stimuli will guide the insect's approach (Bidlingmayer, 1994).

Even in walking insects, although we do not know exactly how close the odour plume is to the ground, with many obstacles that might increase turbulences, the two-dimensional structure of approach is comparable to that of flying insects. A study on the bark beetle *Ips paraconfusus* (Akers, 1989) shows that orientation of the approach track is roughly towards the target, however, again, on a zig-zagging route. Path corrections are weak and less frequent at a distance, but klinotactic tendencies increase with increasing odour concentration as the insect gets closer to the source and the plume cone narrows. Of course, in a walking insect it is very likely that geotaxis, thigmotaxis, or contact chemotaxis, play a role in the host finding behaviour, particularly for selecting the appropriate site on the host (Bond, 1983).

All of these examples suggest that oriented behaviours are multi-channelled, with a variable and complex ratio, or succession, among various orienting stimuli. In some cases, synergic stimuli may enhance each other, as reported by Bell and Tobin (1982). For example, detection of conspecific pheromone by aphids enhances visual orientation to a con-specific or to a visual stimulus marking the area where the probability to find a conspecific is high. The bark beetle *Dendroctonus ponderosae* ex-hibits increased scototactic tendency when it is exposed to aggregation pheromone. Carbon dioxide concentration renders black flies (Simuliidae) responsive to coloured outlines, and sexual pheromones stimulate positive geotaxis in male cockroaches (Bell and Tobin, 1982).

Ecological adjustment: Developmental models

Although certainly not exhaustive, the previous sections provide evidence that arthropods develop, during ontogeny, multisensory orientation mecha-nisms that are based on innate tendencies. The interactions with environ-

mental factors shape the individual behaviour to render the animal opti-
mally adapted to the unique ecological environment of its home range. The
basic tendencies on which oriented behaviours are built persist and are
included, later in life, in most of the instinctive behaviours. They play a
role in updating behavioural capacities and shaping the animal's *Umwelt*
(von Uexküll, 1934).

    Scapini (1995) distinguishes between two types of developmental model.
According to the first, termed "instructive model", only simple behaviours
allowing no more than rough orientation would be inherited. Environmental factors would then be integrated during ontogeny to shape and
refine the behaviour, adjusting the individual to the local environment.
According to the alternative model, termed "selective", a large variety of
adaptive behaviours are present at birth. In the course of ontogeny, some of
them will be selected by experience, whereas others will be switched off.
The two models are obviously not mutually exclusive. In either case, the
result would be an optimal adjustment to the local environment. In the
following we shall review some of the results described above in the light
of these two models.

Instructive model: The case study of the wood-cricket Nemobius sylvestris

The life cycle of *Nemobius sylvestris* is a very long one. The eggs are laid
in summer and autumn, but the larvae do not hatch before June–July of the
following year. The larval development will take another year and the
adults emerge in June of the second year. Most of the adults die during the
following winter, but, at least in Southern France, a certain number of them
survive. They may be found in June–July of the third year amongst the new
adult cohort.

    Nemobius lives under the litter of European oak forest, sometimes mixed
with pine trees, maples and other plant species. More and more it has been
colonizing hedges and embankments where the litter of dead leaves is as
thick as that in the forest. Population density is particularly high in ecotones between wood and meadow, on trail sides and in clearings, and lower
in deep wood. Along ecotone lines, crickets have been observed migrating
regularly each day, away from the trees in the late afternoon, and back to the
trees in the morning. Movements to the open are associated with feeding
and reproduction occurring after nightfall, whereas woodwards orientation
is linked to search of a humid shelter under trees and litter to avoid desiccation during the hot hours of the day (Campan et al., 1987). Similar migrations can be observed along ecotones of forests covering hill slopes.

    Despite these migrations, the animals are very faithful to the sites where
they live. By marking crickets and recapturing them several months later,
it was shown that 98% of the animals do not move more than 10 m away
from their first capture point (Campan et al., 1987). Spontaneously, they

never cross a litter-free forest trail, but they do cross it if they are passively transported from one side to the other. Therefore, it is likely that, when they actively move, they will later return to their starting point. Daily migrations are guided by visual terrestrial cues provided by trees and hedges, but also by the sky, whether or not the sun is directly visible (Campan et al., 1987). When celestial and terrestrial cues are experimentally manipulated to provide contradicting information, the latter always dominate.

As has been mentioned in the section on *From phototaxis and polarotaxis to sun compass orientation, Nemobius* exhibits innate photonegativity and scototaxis, and probably also geotaxis. At hatching, young larvae living at a forest edge are already able to escape woodwards, although their paths are anything else but straight. Straightness, however, improves with age. At an age of 2–3 weeks, the crickets have become familiar with the forest edge side where they live, as well as with their escape direction, that is now strictly correlated with the darkest zones of the surrounding. In the same situation, laboratory-born 5-month-old larvae, or adults collected as larvae in the field but kept for 6 months in the lab, display a rather scattered distribution of escape directions. Scototactic tendency, which does not show any daily modulation in first instar larvae, becomes tuned to the migration rhythm after 4 to 8 months, just as in the adults, with a peak in the morning and a minimum at night (Campan et al., 1987). Thus, innate scototaxis is progressively adjusted to the functional orientation towards or away from the forest edge.

Phototactic tendencies undergo a similar development as does scototaxis. Very young larvae are photonegative, escaping in the direction opposite to the sun azimuth. However, after 6 weeks of life, sometimes earlier, the same larvae are menotactic with respect to the sun, each individual adopting its own angle to the light source (Beugnon et al., 1983). Later in ontogeny, one direction becomes common to all individuals and is time compensated. Inexperienced crickets of the same age do not show any preferred escape direction. This ontogenetical sequence suggests that photonegativity can shift to menotaxis and later on to astro-orientation, using either the sun's position or polarized skylight, with time compensation. The whole ontogeny reflects a progressive association between terrestrial and celestial cues. Although native 2–3-week-old larvae are already capable of directing themselves towards dark zones, the capacity to use the astronomical component appears only between 6 to 10 weeks of age, when the adjustment to local environmental conditions has been accomplished.

Campan and Gautier (1975) studied the ontogeny of visual orientation in another natural habitat, along a N/S forest trail. Two-week-old larvae, despite a strong scototactic tendency, show a very diffuse escape orientation towards the trail sides. A preferred direction, however, already becomes significant in 3-week-old larvae. Path straightness improves accordingly. Futhermore, in the course of ontogeny, the crickets learn to associate a particular astronomical direction with the side of the trail on which they live. A significant sun compass orientation towards the familiar site, similar to that of adults, appears between 6 and 10 weeks of age. At this age, the crickets already seem to be familiar with the corresponding tree trunks pattern. When adult crickets living at the edge of a straight forest trail are captured at their natural site and released 80 m farther up the same forest trail, they display an orientation towards the closest trees, selecting directions that are significantly less accurate than those observed at their native site.

This type of ontogenetic process could also explain several observations made in other parts of the forest (Campan et al., 1987). In forest clearings, the cricket's escape directions again change from young larvae to adults according to the heterogeneity of the panorama. For each clearing, there is a particular preferred escape direction which suggests a relationship between that direction and some local ecological characteristics. Deep in the wood, the insects do not exhibit any preferred escape direction. When the forest edge is situated on a slope, slope direction may be used as a cue for guiding the insects down the slope, out of the wood, when visual information is not available (Campan et al., 1987).

All of these data support the conclusion that, based on innate tendencies, such as phototaxis, scototaxis, and possibly also geotaxis, the crickets become, in the course of ontogeny, progressively adjusted to their local environment, local signals being associated with rewards provided by the satisfaction of the animal's ecological needs and their attachment to a familiar area.

Selective model: The case study of the sandhopper Talitrus saltator

Sandhoppers come armed with a diversity of innate orientation mechanisms (Fig. 4), i.e., phototaxis, geotaxis, scototaxis, polarotaxis, astrotaxis to the sun and the moon, and magnetotaxis (Pardi and Scapini, 1987). They also possess the capacity to use local cues such as landmarks, wave activity, substrate chemicals, slope, humidity and sand grain (Scapini et al., 1992). They use these cues hierarchically according to the local ecological conditions. On a sloped beach, geotaxis predominates, similar to a slope compass, whereas on a beach with a stationary conspicuous dune, scototaxis dominates. Orientation based on a solar, polarized light or moon compass is generally used on flat beaches where conspicuous terrestrial cues are rare. In equatorial areas, sun compass is used only at dusk and at dawn, when the apparent sun azimuth provides reliable cues, whereas the magnetic compass dominates during the rest of the day (Pardi et al., 1988).

All adjustments to local ecological conditions take place early in ontogeny. Laboratory-born, inexperienced animals exhibit a much greater behavioural flexibility than do experienced animals (Scapini, 1995). Despite their inherited compass direction, they adjust much faster to a new sea-land

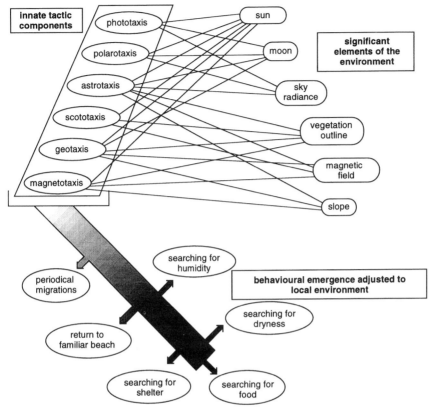

Figure 4. Diagram illustrating relationships among various innate tactic components of behaviour, significant cues of the environment, and ontogenetical development of orientation involved in ecological adjustment of the amphipod *Talitrus* to its natural environment.

direction, different from that of their parents, than do young individuals captured in the wild and released on a foreign beach. Similarly, animals living on shorelines highly variable in both space and time quickly learn new directions (Ugolini and Scapini, 1988).

In Italy, *Talitrus* mostly uses sun and moon compasses (Papi and Pardi, 1953), slope (Scapini et al., 1993), and visual heterogeneities at the horizon (Ugolini et al., 1986). Composition of sun light at the seaside may provide further local cues (Ercolini et al., 1983): sky above sea is richer in short wavelengths (400–490 nm) than is sky above land (Fiocco et al., 1983). Thus, chromatic gradients may participate in the animal's orientation along the sea-land axis (Ercolini and Scapini, 1976; Ercolini et al., 1983; Mezzetti and Scapini, 1995).

Scapini et al. (1992) examined the annual and daily patterns of activity, as well as zonation, direction, and extent of movements in the sandhopper

Talitrus saltator along stable beaches of the Italian Mediterranean coast. *Talitrus saltator* is usually restricted to the eu- and supralittoral part of the beach, but zonation can vary with climatic conditions. Activity patterns, as well as zonation, change in the course of ontogeny. An annual bimodal activity pattern is established with a first peak in the late spring and another in autumn. This bimodality is sharper in the juveniles, the two peaks corresponding to two juvenile hatching periods, i.e., to two generations. In summer (August–October), when air and sand temperature is high, the animals stay close to the sea, whereas for the rest of the year they are zoned more landwards. Daily activity shows two peaks, one at dusk, a second one 2–3 h after midnight or immediately after sunrise. These peaks were found to be correlated with atmospheric as well as substrate humidity (Scapini et al., 1992). Temperature might also be a limiting factor, because locomotory activity only occurs between 10° and 28.8°C. *Talitrus* copes with dehydration problems by moving along the sea-land axis, with an alternation of landwards and seawards migrations. Landwards migration usually corresponds to the first activity peak at dusk which is considered to be a foraging trip, whereas seawards migration usually coincides with a morning activity peak, ensuring that humid territory is reached before risk of dehydration must be taken.

Scapini (1995) compared the path directions of experienced *Talitrus*, collected in the wild, with those of inexperienced, laboratory-born animals of the same age, from the same population. In a situation where only celestial information was available, the path directions of the experienced animals were much more direct, and agreed much better with the sea-land direction in the natural habitat than did those of the inexperienced ones. When experienced animals were released on a beach oriented differently than the original one, they escaped on a straight route, but in a wrong direction, neglecting lcoal terrestrial cues. These animals will need some time to adapt to a new situation. Inexperienced animals, on the other hand, were found to make a much better use of terrestrial cues. They will get adjusted faster to the new beach.

Thus, ontogenetic canalization (Waddington, 1975) reduces the flexibility of behaviour. It renders the animal more adapted, but less adaptable. The hierarchy in the use of the various innate orientation mechanisms is the result of individual experience in the natural environment. This learning process consists of selection and modulation of innate mechanisms to match the ecological requirements.

A plea for studies of taxes in self-organizing ontogenies

In this chapter, we have considered orientation behaviours in the light of the dynamics of their construction and of the consequences of their plasticity for any single individual. We described experimental results showing that an individual has, at the beginning of its life, an innate repertoire of elementary responses to intensities of the physical agents of its environment. These innate tactic tendencies constitute a set of unconditioned responses, i.e., approach at low intensities and withdrawal at high intensities. Positive or negative reinforcements associated with these reactions have two effects. Firstly, they adjust the response threshold of the taxis, enhancing either approach or withdrawal, as has been proposed by Maier and Schneirla (1964) and Schneirla (1965). Secondly, still within the frame of Schneirla's theory, tactic unconditioned responses, when they are reinforced, are associated with further signals which become effective by a "releaser-induced recognition learning" (Suboski, 1990) through a Pavlovian stimulus substitution process. In other words, the major mechanism involved in the ontogeny of oriented behaviour is learning, defined as a property of the nervous system to change the informational status of stimuli as a consequence of the animal's being passively or actively exposed to stimuli and their combinations (Menzel et al., 1993). Tactic stimulation efficiency is transferred to a formerly neutral signal, a visual, acoustic or olfactory configuration, which will be discriminated from others for guiding the animal's movements according to the biological functions that must be fulfilled.

Ontogeny of behaviour is the result of intermingled processes of maturation and experience based on genetic and epigenetic factors with all of the functional traces which persist along the ontogenetical trajectory. This holds true not only in the case of oriented locomotion. Through these processes, an animal will assign significances (Bedeutung) to some configurations which acquire a functional value. It constructs its own world (Umwelt) by associating perceived signals (Merkwelt), motor responses (Wirkwelt) and the corresponding search images (internal world) within the so-called "functional loops" (von Uexküll, 1934) (Fig. 5). Ontogeny of behaviour is thus a unique historical process of constructing the Umwelt of an individual. The ontogeny of oriented locomotion follows this frame, beginning with the inherited elementary tactic tendencies, later improved by all epigenetic traces of experience. Taxes are involved in the functional loops (Fig. 5) as early responses to elementary stimuli, later replaced by more complex signals at various stages of complex oriented behaviours serving the fundamental biological functions. The behaviour retains plasticity, within species specific limits, and may be modified at any time according to the reinforcements experienced by the animal as a consequence of its own actions.

In a more modern view, the ontogeny of oriented behaviour is in accordance with the idea of self-organization by an individual of its own world and may be understood within the frame of the theory of autopoiesis of the autonomous living systems (as opposed to input systems) developed by Maturana and Varela (1992). According to this theory, an animal is considered to be a functionally closed (autonomous) system (a set of inter-

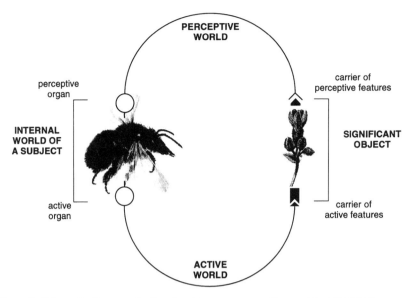

Figure 5. Schematic diagram of a functional loop as proposed by von Uexküll (1934).

acting elements) surrounded by an environment offering many potentially relevant stimuli. Some of them are detected, others not, but only a few are able to modify the system. When one of the environmental factors is associated with a positive or a negative reward, then the autopoietic system will perform a structural coupling in its neural network, assigning a meaning to that factor (Fig. 6). The system is considered active within the process and not reactive to an input. Within this frame, autopoietic ontogeny is the history of all the functional couplings occurring in the network in the course of the animal's life. At the beginning of the ontogeny of orientation behaviours, the autopoietic system assumes an initial state provided by species specific capacities and the ones directly inherited from its parents. We have demonstrated that these initial components are the set of innate taxes available at the beginning of ontogeny. Then, within inherited limits, the autopoietic living system will construct its behavioural phenotype through an epigenetic canalization as defined by Waddington (1975), and will optimize its individual adjustment to the local environment, while increasing its fitness.

Any experimenter studying, *hic et nunc*, orientation behaviour of an animal, usually catches only one or a few snapshots of the ontogenetical history and not its whole dynamics. At present, it appears justified to reconsider taxes (innate but modifiable) as elementary components of oriented behaviour within the frame of their involvement in the autopoietic ontogeny of the individual.

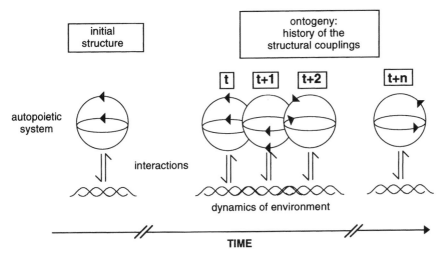

Figure 6. Diagram representing the ontogeny of behaviour as the history of structural couplings between an autopoietic (autonomous) system provided with an initial structure (innate
components, left-hand panel) and its environment in the course of ontogeny. After Maturana and
Varela (1992).

Acknowledgements
I am much indebted to Miriam Lehrer for fruitful discussions and helpful suggestions, and to
Felicita Scapini for her comments. Michelle Corbière and Anne Grimal also deserve acknowl-
edgement for their invaluable help in collecting the bibliography. I thank Sophie Semenoff
Tian-Chanski for translations from Russian, and my colleague Guy Thétaulaz for helping preparing the figures. Finally, I am grateful to Professor Floriano Papi for reading the manuscript
and for his encouraging remarks.

References

Able, K.P. (1991) The development of migratory orientation mechanisms. *In*: P. Berthold (ed.):
    Orientation in Birds. Birkhäuser Verlag, Basel, pp 166–179.
Able, K.P. and Able, M.A. (1996) The flexible migratory orientation system of the Savannah
    sparrow (*Passerculus sandwichensis*). *J. Exp. Biol.* 199:3–8.
Akers, R.P. (1989) Counterturns initiated by decrease in rate of increase concentration. Possible
    mechanism of chemotaxis by walking female *Ips paraconfusus* bark beetles. *J. Chem. Ecol.*
    15:183–208.
Akers, R.P. and Wood, D.L. (1989) Olfactory orientation responses by walking female *Ips para-
    confusus* bark beetles. II. An anemotaxis assay. *J. Chem. Ecol.* 15:1147–1159.
Alexander, R.D. (1962) Evolutionary change in cricket acoustical communication. *Evolution*
    16:443–467.
Alverdes, F. (1930) Die lokomotorischen Reaktionen von decapoden Krebsen auf Helligkeit
    und Dunkelheit. *Z. wiss. Zool.* 137:403–475.
Arendse, M.C. and Kruyswijk, C.J. (1981) Orientation of *Talitrus saltator* to magnetic fields.
    Netherl. J. Sea Res. 15:23–32.
Baker, R.R. (1987) Integrated use of moon and magnetic compass by the heart-and-dart moth,
    Agrotis exclamationis. Anim. Behav. 35:94–101.
Bell, W.J. and Kramer, E. (1979) Search and chemotactic orientation of cockroaches. *J. Insect
    Physiol.* 25:631–640.

34   R. Campan

Bell, W.J. and Tobin, T.R. (1982) Chemo-orientation. *Biol. Rev.* 57:219–260.
Bell, W.J., Tobin T.R., Vogel, G. and Surber, J.L. (1983) Visual course control of escape responses in the cockroach *Blaberus craniifer*: role of internal and external orientation information. *Physiol. Entomol.* 8:12–132.
Benzer, S. (1973) Genetic dissection of behavior. *Sci. Amer.* 229:24–37.
Bernays, E.A. (1993) Aversion learning and feeding. *In*: D.R. Papaj and A.C. Lewis (eds): *Insect Learning*. Chapman and Hall, New York, London, pp 1–17.
Beugnon, G. (1984) De la photonégativité à l'astroorientation chez le grillon des bois, *Nemobius sylvestris* (Bosc). *Monit. Zool. Ital.* 18:185–197.
Beugnon, G. (1985) Orientation of evasive swimming in *Pteronemobius heydeni* (Orthoptera: Gryllidae, Nemobiinae). *Acta Oecol. Oecol. Gen.* 6:235–242.
Beugnon, G. (1986) Learned orientation in landward swimming in the cricket *Pteronemobius lineolatus*. *Behav. Proc.* 12:215–226.
Beugnon, G., Mieulet, F. and Campan, R. (1983) Ontogenèse de certains aspects de l'orientation du grillon des bois, *Nemobius sylvestris* (Bosc), dans son milieu naturel. *Behav. Proc.* 8:73–86.
Bidlingmayer, W.L. (1994) How mosquitoes see traps: role of visual responses. *J. Amer. Mosq. Cont. Ass.* 10:272–279.
Birukow, G. (1954) Photo-Geomenotaxis bei *Geotrupes silvaticus* Panz. und ihre zentralnervöse Koordination. *Z. vergl. Physiol.* 36:176–211.
Birukow, G. (1956) Lichtkompassorientierung beim Wasserläufer *Velia currens* F. (Heteroptera) am Tage und zur Nachtzeit. I. Herbst- und Winterversuche. *Z. Tierpsychol.* 13:463–484.
Birukow, G. and Oberdorfer, H. (1959) Schwerekraftorientierung beim Wasserläufer *Velia currens* F. (Heteroptera) am Tage und zur Nachtzeit. *Z. Tierpsychol.* 16:693–705.
Blackmer, J.L. and Phelan, P.L. (1991) Behavior of *Carpophilus hemipterus* in a vertical flight chamber: transition from phototactic to vegetative orientation. *Entomol. Exp. Appl.* 58:137–148.
Bond, A.B. (1983) The foraging behaviour of lacewing larvae on vertical rods. *Anim. Behav.* 31:990–1004.
Bossert, W.H. and Wilson, E.O. (1963) The analysis of olfactory communication among animals. *J. Theoret. Biol.* 5:443–469.
Brower, L-P. (1996) Monarch butterfly orientation: missing pieces of a magnificent puzzle. *J. Exp. Biol.* 199:93–103.
Brown, A.C. and Odendaal, F.J. (1994) The biology of oniscid isopoda of the genus *Tylos*. *Adv. Mar. Biol.* 30:89–153.
Bui Huy, B. and Campan, R. (1982) Etude comparative de quelques aspects de la vision des formes chez deux espèces de chenilles de Lépidoptères: *Arctia caja et Galleria melonella*. *Bull. Soc. Hist. Nat., Toulouse* 118:199–222.
Butz, I. (1975) Strömungsverhalten von *Ecdyonurus venosus* (Fabr.) (Ephemeroptera). *In*: K. Pasternak and R. Sowa (eds): *Proceedings of the Second International Conference on Ephemeroptera*, Warsaw, pp 199–212.
Campan, R. (1980) *L'animal et son univers*. Privat, Toulouse.
Campan, R. and Gautier, J.Y. (1975) Orientation of the cricket *Nemobius sylvestris* (Bosc) towards forest trees. Daily variations and ontogenetic development. *Anim. Behav.* 23:640–649.
Campan, R. and Lacoste, G. (1971) Les préférences visuelles spontanées chez le grillon *Nemobius sylvestris* et leurs modifications sous l'effect de l'expérience. *96ème Congrès National des Sociétés Savantes*, Toulouse 3:465–483.
Campan, R. and Médioni, J. (1963) Sur le comportement "scototactique" du grillon *Nemobius sylvestris* (Bosc). *C.R. Soc. Biol.* 157:1690–1695.
Campan, R., Beugnon, G. and Lambin, M. (1987) Ontogenetic development of behavior: the cricket visual world. *Adv. Stud. Behav.* 17:165–212.
Cardé, R.T. (1984) Chemo-orientation in flying insects. *In*: W.J. Bell and R.T. Cardé (eds): *Chemical Ecology of Insects*. Chapman and Hall, New York, London, pp 11–124.
Chapman, J.A. (1969) Winged ants return after removal from a summit swarming site. *Ann. Ent. Soc. Amer.* 62:1256–1259.
Charii, F. and Lambin, M. (1988) Ontogenèse du contrôle visuel de la marche orientée chez le grillon *Gryllus bimaculatus*. *Biol. Behav.* 13:49–58.

Chmurzynski, J.A. (1964) Studies on the stages of spatial orientation in females *Bembex rostrata* (Linné, 1758) returning to their nests (Hymenoptera, Sphegidae). *Acta Biol. Exp.* 24:103–132.

Colombini, I., Chelazzi, L. and Scapini, F. (1994) Solar and landscape cues as orientation mechanisms in the beach-dwelling beetle *Eurynebria complanata* (Coleoptera, Carabidae). *Mar. Biol.* 118:425–432.

Coutant, C.C. (1982) Positive phototaxis in the first instar *Hydropsyche cockerelli* Banks (Trichptera). *Aquat. Insect.* 4:55–59.

Couturier, A. and Robert, P. (1958) Recherches sur les migrations du hanneton commun (*Melolontha melolontha* L.). *Annales Epiphyties* 3:257–329.

Danthanarayana, W. and Dashper, S. (1986) Response of some night flying insects to polarized light. *In*: W. Danthanarayana (ed.): *Dispersal and Migration*. Springer-Verlag, Berlin, Heidelberg, pp 120–127.

Dethier, V.G. and Stellar, E. (1961) *Animal Behavior*. Prentice-Hall. Englewood Cliffs.

Dobzhanski, Th. and Spasski, B. (1967) Effect of selection and migration on geotactic and phototactic behavior of *Drosophila*. III. *Proc. R. Soc.* B 168:27–47.

Duan, J.J. and Prokopy, R.J. (1992) Visual and odour stimuli influencing effectiveness of sticky spheres for trapping apple maggot flies *Rhagoletis pomonella* (Walsh) (Diptera, Tephritidae). *J. App. Entomol.* 113:271–279.

Dyer, F.C. (1996) Spatial memory and navigation by honeybess on the scale of foraging range. *J. Exp. Biol.* 199:147–154.

Dyer, F.C. and Gould, J.L. (1981) Honeybee orientation: a backup system for cloudy days. *Science* 214:1041–1042.

Ebert, R. (1980) Influence of juvenile hormone on gravity orientation in the female honeybee larva (*Apis mellifera* L.). *J. Comp. Physiol.* 137:7–16.

Ercolini, A. and Scapini, F. (1976) Sensitivity and response to light in the laboratory of the littoral amphipod *Talitrus saltator* Montagu. *Monit. Zool. Ital.* (NS), 10:293–309.

Ercolini, A., Pardi, L. and Scapini, F. (1983) An optical direction factor in the sky might improve the direction finding of sandhoppers on the seashore. *Monit. Zool. Ital.* (NS) 17:313–327.

Erlenmeyer-Kimling, L. and Hirsch, J. (1961) Measurement of the relations between chromosomes and behavior. *Science* 134:1068.

Felicioni, S. and Ugolini, A. (1991) Learning and solar orientation in the mole cricket *Gryllotalpa gryllotalpa* (Orthoptera: Gryllidae). *J. Insect. Behav.* 4:431–439.

Fiocco, G., Guerrini, A. and Pardi, L. (1983) Spectral differences in sky radiance over land and sea and the orientation of the littoral amphipod *Talitrus saltator* Montagu. *Atti Accad. naz. Lincei Rc.* 74:25–33.

Fraenkel, G.S. and Gunn, D.L. (1961) *The Orientation of Animals*. Dover, New York.

Frisch, K. von (1965) Tanzsprache und Orientierung der Bienen. Springer-Verlag, Berlin, Heidelberg, New York.

Giurfa, M., Nunez, J., Chittka, L. and Menzel, R. (1995) Colour preferences of flower-naive honeybess. *J. Comp. Physiol.* A, 177:247–259.

Gould, J.L. (1982) *Ethology: the mechanisms and evolution of behavior*. W.W. Norton, New York.

Gould, J.L. (1990) Honey bee cognition. *Cognition* 37:83–103.

Gould, J.L. (1993) Ethological and comparative perceptives on honeybee learning. *In*: D.R. Papaj and A.C. Lewis (eds): *Insect Learning*. Chapman and Hall, New York, London, pp 18–50.

Gould, J.L. and Marler, P. (1991) Learning by instinct. *In*: W.S.Y. Wang (ed.): *The emergency of language: Development and evolution*. Freeman and Co., New York, pp 88–103.

Graham, K. (1959) Release by flight exercise of a chemotropic response from photopositive domination in a scolytid beetle. *Nature* 184:283–284.

Hartwick, R.F. (1976) Beach orientation in talitrid amphipods: capacities and strategies. *Behav. Ecol. Sociobiol.* 1:447–458.

Helbig, A.J. (1996) Genetic basis, mode of inheritance and evolutionary changes of migratory directions in palearctic warblers (Aves: Sylvidae). *J. Exp. Biol.* 199:49–55.

Hertz, M. (1993) Über figurale Intensitäten und Qualitäten in der optischen Wahrnehmung der Biene. *Biol. Zbl.* 53:10–40.

Hinde, R.A. (1965) *Animal Behaviour. A Synthesis of Ethology and Comparative Psychology*. MacGraw Hill Kogakusha, Tokyo.

Hirsch, J. and Boudreau, J. (1958) Studies in experimental behavior genetics I. The heritability of phototaxis in a population of *Drosophila melanogaster. J. Comp. Physiol. Psychol.* 51:647–651.

Hirsch, J. and Erlenmeyer-Kimling, L. (1961) Sign of taxis as a property of the genotype. *Science* 134:835.

Horn, E. (1985) Gravity. *In*: G.A. Kerkut and L.I. Gilbert (eds): *Comprehensive insect physiology, biochemistry and pharmacology.* Pergamon Press, Oxford, New York, pp 557–576.

Horvath, G. and Zeil, J. (1996) Kuwait oil lakes as insect traps. *Nature* 379:303–304.

Huber, F. (1962) Central nervous control of sound production in crickets and some speculation on its evolution. *Evolution* 16:429–442.

Iperti, G. (1966) Migration of *Adonia undecimnotata* in south-eastern France. *Proc. Symp. of Czechoslovak Acad. Sci.*, Prague, pp 137–138.

Jacob-Jessen, U. (1959) Zur Orientierung der Hummel und einiger anderer Hymenopteren. *Z. vergl. Physiol.* 41:597–641.

Jander, R. (1957) Die optische Richtungsorientierung der Roten Waldameise (*Formica rufa* L.). *Z. vergl. Physiol.* 40:162–238.

Jander, R. (1960) Menotaxis and Winkeltransponieren bei Köcherfliegen (Trichoptera). *Z. vergl. Physiol.* 43:680–686.

Jander, R. (1963) Insect orientation. *Ann. Rev. Entomol.* 8:95–114.

Jander, R. (1965) Die Phylogenie von Orientierungsmechanismen der Arthropoden. *Verh. Dtsch. Zool. Gesell.* Jena.

Jennings, H.S. (1906) *The Behavior of Lower Organisms.* Columbia University Press, New-York.

Kaissling, K.E. (1986) Peripheral mechanisms of pheromone reception in moths. *Chem. Senses* 21:257–268.

Kaissling, K.E. and Kramer, E. (1990) Sensory basis of phermomone-mediated orientation in moths. *Verh. Dtsch. Zool. Ges.* 83:109–131.

Kalmus, H. (1937) Photohorotaxis, eine neue Reaktionsart, gefunden an den Eilarven von *Dixippus. Z. vergl. Physiol.* 24:644–655.

Kennedy, J.S. (1939) The visual responses of flying mosquitoes. *Proc. Zool. Soc. Lond.* A 109:221–242.

Kennedy, J.S. (1983) Zig-zagging and casting as a programmed response to wind-borne odour: a review. *Physiol. Entomol.* 8:109–120.

Koch, F. (1981) Das circadiane phototaktische Verhalten von *Hercinothrips femoralis* (O.M. Reuter) (Thysanoptera, Insecta). *Zool. Jb. Physiol.* 85:312–315.

Kovorov, B.G. and Monchadskiy, A.S. (1963) The possibility of using polarized light to attract insects. *Entomol. Rev.* 42:25–28.

Kramer, E. (1992) Attractivity of pheromone surpassed by time patterned application of two nonpheromone compounds. *J. Ins. Behav.* 5:83–97.

Kühn, A. (1919) *Die Orientierung der Tiere im Raum.* Fisher, Jena.

Lavoie-Dornik, J. and Pilon, J.G. (1987) Rôle probable des rayons ultraviolets lors de l'émergence des Zygoptères Coenagrionides. *Odonatologica* 16:185–191.

Lehrer, M., Wehner, R. and Srinivasan, M.V. (1985) Visual scanning behaviour in honeybees. *J. Comp. Physiol.* A 157:405–415.

Lehrer, M., Horridge, G.A., Zhang, S.W. and Gadagkar, R. (1995) Shape vision in bees: innate preference for flower-like patterns. *Phil. Trans. R. Soc. London* B 347:123–137.

Leucht, T. and Martin, H. (1990) Interactions between e-vector orientation and weak, steady magnetic fields in the honeybee, *Apis mellifica. Naturwiss.* 77:130–133.

Lindauer, M. (1959) Angeborene und erlernte Komponente in der Sonnenorientierung der Bienen. Bemerkungen und Versuche zu einer Mitteilung von Kalmus. *Z. vergl. Physiol.* 45:590–604.

Lindauer, M. (1969) Behavior of bees under optical learning conditions. *In*: W. Reichardt (ed.): *Processing of Optical Data by Organisms and by Machines.* Academic Press, New York, London, pp 527–543.

Loeb, J. (1918) *Forced Movements, Tropisms and Animal Conducts.* Lippincott, Philadelphia, London.

Lorenz, K. (1956) Play and vacuum activity in animals. *In*: P.P. Grassé (ed.): *Symposium Singer-Polignac "L'Instinct dans le comportement des animaux et de l'homme".* Masson, Paris, pp 633–645.

Lorenz, K. and Tinbergen, N. (1938) Taxis and Instinkthandlung in der Eirollbewegung der Graugans. *Z. Tierpsychol.* 2, 1–29.

Maier, N.R.F. and Schneirla, T.C. (1935) *Principles of Animal Psychology.* MacGraw-Hill, New York, London. (Reissue 1964, Dover, New York).

Markl, H. (1964) Geomenotaktische Fehlorientierung bei *Formica polyctena* Förster. *Z. vergl. Physiol.* 48:552–586.

Markl, H. (1966) Schwerekraftdressuren an Honigbienen. II. Die Rolle der Schwererezeptorischen Borstenfelder verschiedener Gelenke für die Schwerekompassorientierung. *Z. vergl. Physiol.* 53:353–371.

Martin, H. and Lindauer, M. (1973) Orientierung im Erdmagnetfeld. *Fortschr. Zool.* 21:211–228.

Martin, H. and Lindauer, M. (1977) Der Einfluss des Erdmagnetfeldes auf die Schwereorientierung der Honigbiene (*Apis mellifica*). *J. Comp. Physiol.* 122:145–187.

Masson, C. and Mustaparta, H. (1990) Chemical information processing in the olfactory system of insects. *Physiol. Rev.* 70:199–245.

Mathieu, F. (1995) *Mecanismes de la colonisation de l'hôte chez le scolyte du café* Hypothenemus hampei *(Ferr.) (Coleoptera: Scolytidae).* PhD Thesis, Paris.

Maturana, H.R. and Varela, F.J. (1992) *The Tree of Knowledge: The Biological Roots of Human Understanding.* Scherz Verlag, München.

Mead, M. and Mead, F. (1974) Etude de l'orientation chez l'isopode terrestre *Tylos latreillei s.sp sardous. Vie et Milieu* 23:81–93.

Médioni, J. (1963) La variabilité des comportements taxiques. Ses principales conditions écologiques et organiques. *Erg. der Biologie* 26:66–82.

Médioni, J. (1967) Les aspects fondamentaus de l'activité motrice. *In*: G. Viaud, Ch. Kayser, M. Klein and J. Médioni (eds): *Traité de Psychophysiologie.* Presses Universitaires de France, Paris, pp 659–737.

Meille, O., Campan, R. and Lambin, M. (1994) Effects of light deprivation on visually guided behavior early in the life of *Gryllus bimaculatus* (Orthoptera: Gryllidae). *Ann. Ent. Soc. Amer.* 87:133–142.

Menzel, R. (1967) Untersuchungen zum Erlernen von Spektralfarben durch die Honigbiene (*Apis mellifica*). *Z. vergl. Physiol.* 56:22–62.

Menzel, R. (1969) Das Gedächtnis der Honigbiene für Spektralfarben. II. Umlernen und Mehrfachlernen. *Z. vergl. Physiol.* 63:290–309.

Menzel, R., Greggers, U. and Hammer, M. (1993) Functional organization of appetitive learning and memory in a generalist pollinator, the honey bee. *In*: D.R. Papaj and A.C. Lewis (eds): *Insect Learning.* Chapman and Hall, New York, London, pp 79–125.

Mezzetti, M.C. and Scapini, F. (1995) Aspects of spectral sentivity in *Talitrus saltator* (Montagu) (Crustacea, Amphipoda). *Mar. Fresh. Behav. Physiol.* 26:35–45.

Mezzetti, M.C., Naylor, E. and Scapini, F. (1994) Rhythmic responsiveness to visual stimuli in different populations of talitrid amphipods from Atlantic and Mediterranean coasts: an ecological interpretation. *J. Exp. Mar. Biol. Ecol.* 181:279–291.

Michieli, S. (1959) Analiza scototakticnih (perigramotakticnih) reakcij pri artropodih. *Acad. Scientiarum et Artium Slovenica, Cl. IV: Historia Naturalis et Medicina* 237–286.

Mieulet, F. (1980) *Etude de la variabilité des modes d'orientation chez le grillon des bois, Nemobius sylvestris, selon des biotopes différents.* PhD Thesis, Toulouse.

Mimura, K. (1986) Development of visual pattern discrimination in the fly depends on light experience. *Science* 232:83–86.

Moericke, V., Prokopy, R.J., Berlocher, S. and Bish, G.L. (1975) Visual stimuli eliciting attraction of *Rhagoletis pomonella* (Diptera: Tephritidae) flies to trees. *Ent. exp. appl.* 18:497–507.

Müller, E. (1931) Experimentelle Untersuchungen an Bienen und Ameisen, über die Funktionsweise der Stirnocellen. *Z. vergl. Physiol.* 14:348–384.

Murlis, J. (1986) The structure of odour plumes. *In*: T.L. Payne, M.C. Birch and C.E.J. Kennedy (eds): *Mechanisms of insect olfaction.* Clarendon, Oxford, pp 27–37.

Murlis, J. and Jones, C.D. (1981) Fine scale structure of odour plumes in relation to insect orientation to distant pheromone and other attractant sources. *Physiol. Entomol.* 6:71–86.

Murlis, J., Elkinson, J.S. and Cardé, R.T. (1992) Odour plume and how insects use them. *Ann. Rev. Entomol.* 37:505–512.

Nayak, M.K., Sehgal, S.S. and Gandhi, J.R. (1990) Aggregating tendency in rice grasshopper *Heiroglyphus banian* Fabricius in response to interaction of some abiotic factors and gravity. *Indian J. Ent.* 52:9–13.

Owens, E.D. and Prokopy, R.J. (1984) Habitat background characteristics influencing *Rhagoletis pomonella* (Walsh) (Diptera, Tephritidae) fly response to foliar and fruit mimic traps. *Z. Ang. Entomol.* 98:98–103.

Papi, F. (1955) Orientamento astronomico in alcuni carabidi. *Atti Soc. Tosc. Sci. Natur.* 62: 83–97.

Papi, F. (1960) Orientation by night: the moon. *Cold Spr. Harb. Symp. Quant. Biol.* 25: 475–480.

Papi, F. and Pardi, L. (1953) On the lunar orientation of sandhoppers (Amphipoda, Talitridae). *Biol. Bull.* 124:97–105.

Papi, F. and Tongiorgi, P. (1963) Innate and learned components in the astronomical orientation of wolf spiders. *Ergeb. der Biol.* 26:259–280.

Pardi, L. (1960) Innate components in the solar orientation of littoral Amphipods. *Cold Spr. Harb. Symp. quant. Biol.* 25:395–401.

Pardi, L. (1963) Orientamento astronomico vero in un isopode marino: *Idotea baltica basteri* (Audouin). *Monit. Zool. Ital.* (NS) 70-71:491–495.

Pardi, L. and Scapini, F. (1983) Inheritanc of solar direction finding in sandhoppers: mass-crossing experiments. *J. Comp. Physiol.* A 151:435–440.

Pardi, L. and Scapini, F. (1987) Die Orientierung der Strandflohkrebse im Grenzbereich Meer/Land. *In*: M. Lindauer (ed.): *Information Processing in Animals 4*. Gustav Fischer Verlag, Stuttgart.

Pardi, L., Ercolini, A., Ferrara, F., Scapini, F. and Ugolini, A. (1984) Orientamento zonale solare e magnetico in crostacei anfipodi di regioni equatoriali. *Atti Accad. naz. Lincei Rc.* 76:312–320.

Pardi, L., Ugolini, A., Faqi, A.S., Scapini, F. and Ercolini, A. (1988) Zonal recovery in equatorial sandhoppers: interaction between magnetic and solar orientation. *In*: G. Chelazzi and M. Vannini (eds): *Behavioral Adaptation to Intertidal Life*. Plenum Press, New York, pp 79–92.

Perttunen, V. (1958) The reversal of positive phototaxis by low temperature in *Blastophagus piniperda* L. (Col., Scolytidae). *Ann. Ent. Fennici* 24:12–18.

Perttunen, V. (1960) Seasonal variations in the light reactions of *Blastophagus piniperda* L. (Col., Scolytidae) at different temperatures. *Ann. Ent. Fennici* 26:86–92.

Popov, A.V. and Shuvalov, V.F. (1977) Phonotactic behavior of crickets. *J. Comp. Physiol.* 119:111–126.

Preiss, R. and Kramer, E. (1984) The interaction of edge fixation and negative phototaxis in the orientation of walking gypsy moths *Lymantria dispar*. *J. Comp. Physiol.* A 154:493–498.

Rabaud, E. (1949) *L'Instinct et le comportement animal. I. Reflexes et tropismes*. Colin, Paris.

Ruiz, P.A. (1991) *Contribution à l'étude de l'utilisation de la lumière polarisée dans l'orientation du grillon Gryllus bimaculatus. Influence de l'expérience individuelle*. D.E.A., Toulouse.

Scapini, F. (1995) Heredity, individual experience, canalization: sandhoppers as a case study. *Polish Archives of Hydrobiology* 42:557–566.

Scapini, F. and Buiatti, M. (1985) Inheritance of solar direction finding in sandhoppers. III. Progeny tests. *J. Comp. Physiol.* A 157:433–440.

Scapini, F. and Fasinella, D. (1990) Genetic determintion and plasticity in the sun orientation of natural populations of *Talitrus saltator*. *Mar. Biol.* 107:141–145.

Scapini, F. and Quochi, G. (1992) Orientation in sandhoppers from Italian populations: have they magnetic orientation ability? *Boll. Zool.* 59:437–442.

Scapini, F., Ugolini, A. and Pardi, L. (1985) Inheritance of solar direction finding in sandhoppers. II. Differences in arcuated coastlines. *J. Comp. Physiol.* A 156:729–735.

Scapini, F. Chelazzi, L., Colombini, I. and Fallaci, M. (1992) Surface activity, zonation and migrations of *Talitrus saltator* on a mediterranean beach. *Mar. Biol.* 112:573–581.

Scapini, F., Lagar, M.C. and Mezzetti, M.C. (1993) The use of slope and visual information in sandhoppers: innateness and plasticity. *Mar. Biol.* 115:545–553.

Scapini, F., Buiatti, M., De Matthaeis and Mattoccia, M. (1995) Orientation behaviour and heterozygosity of sandhopper populations in relation to stability of beach environments. *J. Evol. Biol.* 8:43–52.

Schneider, F. (1952) Untersuchungen über die optische Orientierung der Maikäfer (*Melolontha vulgaris* F. and *M. hippocastani* F.) sowie über die Entstehung von Schwärmbahnen und Befallskonzentrationen. *Bull. Soc. Ent. Suisse*. 25:269–340.

Schneirla, T.C. (1965) Aspects of stimulation and organization in approach/withdrawal processes underlying vertebrate behavioral development. *Adv. Study Anim. Behav.* 1:1–74.

Schöne, H. (1965) Release and orientation of behaviour and the role of learning as demonstrated in Crustacea. *Anim. Behav. Suppl.* 1:135–143.

Schwind, R. (1989) A variety of insects are attracted to water by reflected polarized light. *Naturwiss.* 76:377–378.

Schwind, R. (1995) Spectral regions in which insects see reflected polarized light. *J. Comp. Physiol.* A 177:439–488.

Sergejeva, M.V., Popov, A.V. and Shuvalov, V.F. (1993) Ontogeny of selectivity of positive phonotaxis in female crickets *Gryllus bimaculatus* to temporal parameters of the male calling song. *In:* K. Wiese, F.G. Gribakin, A.V. Popov and G. Renninger (eds): *Sensory systems of Arthropods.* Birkhäuser Verlag, Basel, pp 319–327.

Shuvalov, V.F. (1985) Influence of environmental factors on phonotactic specificity in the cricket *Gryllus bimaculatus* during ontogenesis. *Z. Evolut. Biokhim. Fiziol.* 21:555–560 (in Russian).

Shuvalov, V.F. and Popov, A.V. (1973) Significance of some of the parameters of calling songs of male crickets *Gryllus bimaculatus* for phonotaxis of females. *Z. Evolut. Biokhim. Fiziol.* 9:177–182 (in Russian).

Shuvalov, V.F. and Popov, A.V. (1984) Dependence of phonotactic specificity in crickets of the genus *Gryllus* on the character of preliminary sound stimulation. *Dokladui Akad. Nauk* 274:1273–1276 (in Russian).

Shuvalov, V.F., Rüting, T. and Popov, A.V. (1990) The influence of auditory and visual experience on the phonotactic behavior of the cricket, *Gryllus bimaculatus. J. Insect Behav.* 3:289–302.

Smith, B.H. (1993) Merging mechanism and adaptation: an ethological approach to learning and generalization. *In:* D.R. Papaj and A.C. Lewis (eds): *Insect Learning.* Chapman and Hall, New York, London, pp 126–157.

Suboski, M.D. (1990) Releaser-induced recognition learning. *Psychol. Rev.* 97:271–284.

Thorson, J., Weber, T. and Huber, F. (1982) Auditory behavior of the cricket. II. Simplicity of calling song recognition in *Gryllus* and anomalous phonotaxis at abnormal carrier frequencies. *J. Comp. Physiol.* 146:361–378.

Tinbergen, N. (1951) *The study of instinct.* Clarendon Press, Oxford.

Tongiorgi, P. (1962) Sulle relazioni tra habitat ed orientamento astronomico in alcune specie del gen. *Arctosa.* (Araneae-Lycosidae.) *Boll. Zool.* 28:683–689.

Towne, W.F. and Gould, J.L. (1985) Magnetic field sensitivity in honeybees. *In:* J.L. Kirschwink, D.S. Jones and B.J. MacFadden (eds): *Magnetite biomineralization and magnetoreception in organisms. A new biomagnetism.* Plenum Press, New York, pp 385–406.

Uexküll, Th. v. (1956) *Mondes animaux et mondes humains.* Gonthier, Paris. (French translation of "Streifzüge durch die Umwelten von Tieren und Menschen – Bedeutungslehre." Rowohlt Verlag (1934).)

Ugolini, A. and Chiussi, R. (1996) Astronomical orientation and learning in the earwig *Labidura riparia. Behav. Proc.* 36:151–161.

Ugolini, A. and Macchi, T. (1988) Learned component in the solar orientation of *Talitrus saltator* Montagu (Amphipoda, Talitridae). *J. Exp. Mar. Biol. Ecol.* 121:79–87.

Ugolini, A. and Messana, G. (1988) Sun compass in orientation of *Idotea baltica* (Pallas) (Isopoda, Idoteidae). *Mar. Behav. Physiol.* 13:333–340.

Ugolini, A. and Pardi, L. (1992) Equatorial sandhoppers do not have a good clock. *Naturwiss.* 79:279–281.

Ugolini, A. and Pezzani, A. (1993) Learning of escape direction in *Idotea baltica. Mar. Behav. Physiol.* 22:183–192.

Ugolini, A. and Pezzani, A. (1995) Magnetic compass and learning of the Y-axis (sea-land) direction in the marine isopod *Idotea baltica basteri. Anim. Behav.* 50:295–300.

Ugolini, A. and Scapini, F. (1988) Orientation of the sandhopper *Talitrus saltator* (Amphipoda, Talitridae) living on dynamic sandy shores. *J. Comp. Physiol.* A 163:453–462.

Ugolini, A., Scapini, F. and Pardi, L. (1986) Interaction between solar orientation and landscape visibility in *Talitrus saltator* Montagu (Crustacea, Amphipoda). *Mar. Biol.* 90:449–460.

Ugolini, A., Scapini, F., Beugnon, G. and Pardi, L. (1988) Learning in zonal orientation of sandhoppers. *In:* G. Chelazzi and M. Vannini (eds): *Behavioural Adaptation to Intertidal Life.* Plenum Press, New York, pp 105–118.

Ugolini, A., Talluri, P. and Vannini, M. (1989) Astronomical orientation and learning in *Palaeomonetes antennarius*. *Mar. Biol.* 103:489–493.

Ugolini, A., Felicioni, S. and Macchi, T. (1991) Orientation in the water and learning in *Talitrus saltator* Montagu. *J. Exp. Mar. Biol. Ecol.* 151:113–119.

Ugolini, A., Morabito, F. and Taiti, S. (1995) Innate landward orientation in the littoral isopod *Tylos europaeus*. *Ethol. Ecol. Evol.* 7:387–391.

Varjú, D. (1987) The interaction between visual edge fixation and scototaxis in the mealworm beetle *Tenebrio molitor*. *J. Comp. Physiol.* A 160:543–552.

Viaud, G. (1951) *Les tropismes*. Presses Universitaires de France, Paris.

Vowles, D.M. (1954) The orientation of ants. I. The substitution of stimuli. *J. Exp. Biol.* 31:341–355.

Waddington, C.H. (1975) *The Evolution of an Evolutionist*. Edinburgh University Press, Edinburgh.

Walikonis, R., Schoun, D., Zacharias, D., Henley, J., Coburn, P. and Stout, J. (1991) Attractiveness of the male *Acheta domesticus* calling song to females. III. The relation of age-correlated changes in syllable period recognition and phonotactic threshold to juvenile hormone III biosynthesis. *J. Comp. Physiol.* A 169:751–764.

Watson, J.B. (1913) Psychology as the behaviorist views it. *Psychol. Rev.* 20:158–177.

Weber, T. and Thorson, J. (1989) Phonotactic behaviour of walking crickets. *In:* F. Huber, T.E. Moore and W. Loher (eds): *Cricket Behaviour and Neurobiology*. Cornell Univ. Press, Ithaca, pp 310–339.

Wehner, R. (1981) Spatial vision in arthropods. *In*: H. Autrum (ed.): *Handbook of Sensory Physiology* VII/6a. Springer-Verlag, Berlin, Heidelberg, New York, pp 287–616.

Wehner, R. (1992) Homing in arthropods. *In*: F. Papi (ed.): *Animal Homing*. Chapman and Hall, London, pp 45–144.

Wehner, R., Michel, B. and Antonsen, P. (1996) Visual navigation in insects: coupling of egocentric and geocentric information. *J. Exp. Biol.* 199:129–140.

Wiltschko, W. and Wiltschko, R. (1975a) The interactions of stars and magnetic field in the orientation system of night migrating birds. I. Autumn experiments with European warblers (Gen. *Sylvia*). *Z. Tierpsychol.* 37:337–355.

Wiltschko, W. and Wiltschko, R. (1975b) The interactions of stars and magnetic field in the orientation system of night migrating birds. II. Spring experiments with European robins (*Erithacus rubecula*). *Z. Tierpsychol.* 39:265–282.

Zanforlin, M. (1969) *Inhibition of Responses to Light during Pre-pupation Behaviour in Larvae of the Fly Sarcophaga barbata*. Sansoni, Firenze.

Orientation and Communication in Arthropods
ed. by M. Lehrer
© 1997 Birkhäuser Verlag Basel/Switzerland

The selection and use of landmarks by insects

T. S. Collett[1] and J. Zeil[2]

[1] Sussex Centre for Neuroscience, School of Biological Sciences, University of Sussex, Brighton BN1 9QG, UK
[2] Centre for Visual Sciences, Research School of Biological Sciences, Australian National University, P.O. Box 475, Canberra, ACT 2601, Australia

Summary. Advanced hymenoptera rely heavily on visual landmarks for finding their way between familiar places. We discuss the selection, learning, and use of landmarks, both for pinpointing a location and for route guidance. We emphasise that landmark guidance involves several distinct navigational strategies and that different strategies may best be served by different kinds of landmarks which need to be learnt in different ways.

Introduction

Insects, such as ants, bees or wasps, make extensive use of landmarks when navigating between familiar places. We discuss here what makes a good landmark, how insects select suitable objects as landmarks and the ways in which landmarks are used for navigation. There cannot be a single answer to any of these questions because, as we detail below, landmarks play diverse roles and must operate over a variety of scales. For example, the location of a nest hole on the ground may be defined by a nearby pebble. The significant properties of the pebble for finding a nest are that the pebble should be close to the nest entrance and relatively small, for only localisable and nearby features of the landscape can specify the nest's position accurately. Such small objects can only be detected over a short distance so that other larger objects are needed to help guide the insect to an area where the pebble is visible. At the other end of the scale, large and distant objects like mountains also make useful landmarks but for quite different purposes, such as helping to label a scene or in providing a directional bearing.

Selecting landmarks near a goal: 3-D objects are best

Tinbergen and Kruyt (1938) performed a long series of experiments to work out what objects the digger wasp *Philanthus* prefers for specifying the location of its nest entrance. In a typical experiment, the nest entrance on the ground was surrounded by a ring of flat discs and small hemispheres arranged in an alternating pattern, so decorating what is normally a bare expanse of sand (Fig. 1A). After the wasp had been given ample opportunity to learn this circular array, it was presented with a very dif-

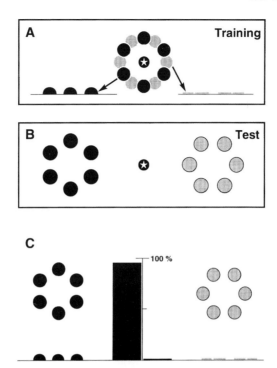

Figure 1. A demonstration that the sand wasp *Philanthus triangulum* prefers to use 3-D objects to pinpoint its nest. (A) A circle of alternating 2 cm diameter black hemispheres and 2 cm diameter flat, black disks surrounded the wasp's nest which is marked by a star. For clarity, disks are shown grey. (B) When the wasp had become accustomed to this array, it was confronted with a ring of disks and a ring of hemispheres placed on either side of the nest entrance. (C) Choices made by nine wasps in these tests. Wasps tend to search for the nest entrance in the ring composed of hemispheres. Modified from Tinbergen and Kruyt (1938).

ferent scene on its return from a foraging trip. The objects were removed from the nest entrance and arranged in two rings on either side of its nest. One was formed by the subset of discs, the other by the hemispheres (Fig. 1 B). In many tests of this kind, wasps searched as though their nest was to be found in the ring formed by objects which stuck up above the ground (Fig. 1 C). Tinbergen and Kruyt stress that, in the absence of anything better, the flat discs are used as landmarks, but wasps take longer to learn to use them. However, given a choice, the protruding objects are consistently preferred.

There are several visual advantages to selecting objects that protrude above the ground. First, when the wasp flies close to the ground, as normally happens during the last stages of its approach, these landmarks will contrast sharply against the sky. Second, the landmarks are mostly in a different depth plane from the background and can be distinguished from it by relative image motion generated by the insect's own movements. The landmarks will thus tend to be highly visible, with boundaries that are well defined by both luminance and motion contrast. Third, cast shadows are to a large extent filtered out. Shadows can be a very prominent part of the visual landscape, but they are usually useless as landmarks (but see Hoefer and Lindauer, 1976), because they move with the sun and alter unpredictably and unexpectedly with changing cloud cover. However, they are often restricted to the ground plane so that the contours of objects above the ground are less liable to be masked by the confusing edges that shadows generate.

Because bees and wasps fly close to the ground both in their final approach to a target on the ground and when first learning about a site, protruding objects will tend to fall within their dorsal visual field. For desert ants, with their head a few millimeters above the ground, all likely landmarks appear above the equator. Dorsal retina might thus have become specialised for landmark learning, following the general tendency of invertebrates to employ different regions of the eyes for different tasks (see review by Land, 1989). Indeed, selective masking of ventral or dorsal retina shows that the dorsal visual field is used in landmark guidance (Antonsen and Wehner, 1995; Wehner et al., 1996). However, landmark navigation does not depend exclusively on dorsal retina. We know that wasps do not altogether ignore objects on the ground (Tinbergen and Kruyt, 1938; Zeil, 1993b). And it is unknown whether ants will navigate using visual features on the ground plane, when protruding landmarks are absent.

The role of learning flights in selecting landmarks near the goal

The precision with which a landmark can specify a place depends upon the proximity of the landmark to that place and on the spatial resolution of the insect's eye. There is thus good reason for insects to pay particular attention to objects that are close to their goal. Experiments similar to those of Tinbergen and Kruyt (1938) showed that honey bees are sensitive to absolute distance and do indeed prefer to be guided by objects near to the goal (Cheng et al., 1987). Bees were trained to feed on the ground among an array of cylinders some of which were small and close to the feeder and others larger and more distant. In tests, the cylinders were arranged so that the sub-arrays of small and large cylinders signalled different places. When the two signalled places were relatively close together, the bees' search was concentrated at the site specified by the small cylinders. When the two sites were well separated, bees searched at both sites, but spent more time at the site associated with the small cylinders. The bees must have paid special attention to the near, small landmarks during training.

Additional evidence that bees and wasps record the absolute distance of landmarks from a goal is that, under some conditions, they will search for their goal at the trained distance from a landmark, unperturbed by changes to its size (Cartwright and Collett, 1979; Zeil, 1993b; Brünnert et al., 1994; Lehrer and Collett, 1994). These two sets of findings mean that distance estimates are not obtained entirely from static image cues, such as the apparent size of the landmark or its retinal elevation. Although conclusive tests are missing, it is very likely that distance information comes from the retinal image motion generated as the insects move through space, as is the case for many other visually guided behaviours (see review by Srinivasan, 1993).

Landmarks are initially selected and their properties learnt during stereotyped flights that are performed when wasps or bees first leave a nest or a foraging site (see review by Zeil et al., 1996). Insects must learn enough about the surrounding landscape in a single flight to ensure a successful return. These learning flights generate a pattern of image motion over the eye that seems ideal for picking out objects that are in the immediate vicinity of the return site. The insect flies away moving in a series of arcs that are roughly centred upon the target (Fig. 2 A). As the insect flies, its body axis (θ in Fig. 2 B) rotates at an angular velocity that matches the velocity at which the arc (β in Fig. 2 B) is described. Consequently, the target is viewed by roughly the same area of retina (ϕ_n in Fig. 2 B) throughout each arc. The target falls on an area about 45° left of the midline during arcs to the left and 45° to the right of the midline during arcs to the right. Essentially, the insect pivots about the target, adjusting its translational velocity so that the pivoting speed is constant and independent of its distance from the target.

One visual consequence of pivoting is that objects close to the pivoting centre will remain relatively stationary on the retina, whereas those further away will move faster, with very distant objects travelling over the retina at the insect's rotational velocity. Thus, objects near to the target can be picked out selectively because they suffer little motion blur and their stationary boundaries stand out against the rapidly moving background. This segregation of close and distant objects by relative motion is illustrated in Figure 3 by two moments from a simulated movie of the scene viewed by a *Cerceris* wasp during a learning flight (Voss, 1995). The upper images in Figure 3 A and B show the 360° panorama around the nest as it is seen by the wasp at moments towards the beginning and the end of the flight. The image sequence was fed through an array of motion detectors sensitive to the horizontal component of image motion. The lower panels of Figure 3 show the output of the array, with black and grey indicating movement of the panorama to the right and to the left, as the wasp moves in the opposite direction. Shearing motion between foreground and background occurs through much of the learning flight so that the retinal location of the shear indicates the position of image features that are close to the goal.

Because shapes learnt through motion contrast alone can be recognised later through colour or luminance channels (Zhang et al., 1995), these motion-enhanced edges should subsequently be recognisable by their static

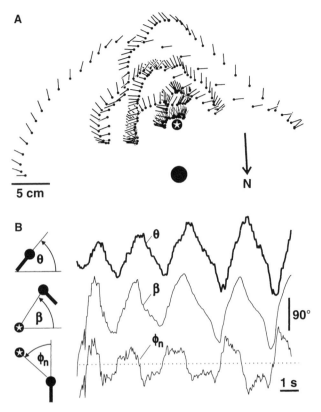

Figure 2. A learning flight of the digger wasp *Cerceris rybyensis* on leaving its underground nest. (A) The flight is shown from above with dots marking the wasp's position and the tails marking the orientation of its body axis every 40 ms. The nest entrance is indicated by a star and the position of a small cylinder that served as a landmark is shown by a black circle. (B) Plots against time of the orientation of the wasp's body axis θ, the angular bearing relative to the nest β, and the retinal azimuth position of the nest entrance ϕ_n. See insets for definition of variables. Replotted from Zeil (1993a).

properties on returns when elaborate scanning movements are absent. Indeed, bees are guided by the absolute distance of landmarks (suggesting the use of motion parallax) only during their first few visits to a newly discovered feeding dish, when their approach flights are relatively complex. Later, when approach flights are relatively simple, bees are guided by the landmark's image size (Lehrer and Collett, 1994).

That learning flights do play a role in acquiring distance information has been shown by training bees to approach a feeder with a cylinder of a constant size placed at a fixed distance behind it (Lehrer and Collett, 1994). Some bees were allowed to see the cylinder only on arrival. It was removed while they fed and replaced after their departure. Other bees were allowed to see the cylinder only on departure during learning flights. It was put in

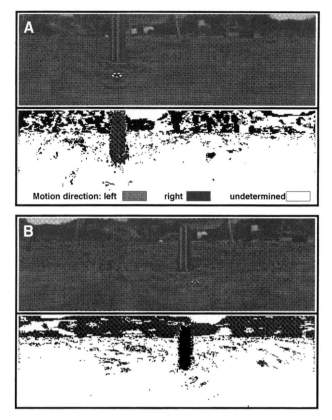

Figure 3. Simulation of the panorama viewed by a wasp (*Cerceris rybyensis*) at two moments in a learning flight. Top panels in (A) and (B) show the 360° horizontal panorama stretching from – 70° in the ventral to 30° in the dorsal visual field of the wasp. The nest entrance is marked by a star. These stills show moments towards the beginning and the end of a flight. The sequence of images obtained during the flight was passed through a network of horizontal motion detectors and the lower panels show the outputs of the network at corresponding times. Grey and black indicate image motion to the left and right, respectively. The two examples show the direction of shearing motion between the cylinder in the foreground and the distant background during arcs to the right and to the left. Modified from Voss (1995).

position while they fed and removed when they had left. After this training, the two groups of bees were guided by different properties of the cylinder. Bees that had viewed the cylinder only on arrival were guided to a position defined by the apparent size of the cylinder as seen from the feeder, even if the cylinder was of a different physical size, so that the absolute distance was wrong. Bees that had viewed the cylinder only on departure behaved in the opposite way. They were guided to a position specified by the absolute distance between feeder and cylinder, even if the apparent size of the cylinder was wrong.

Other features of landmarks, like their colour, or their position relative to the feeder, are learnt equally well on arrival and on departure (Lehrer, 1993). So far, the absolute distance between landmark and goal is the only property that is known to be acquired preferentially during learning flights. It is difficult for an insect to infer this distance on arrival, because a goal, such as an inconspicuous nest entrance or feeder, is often invisible on approach until the insect has almost reached it. Gauging such distances would require specialized scanning movements like those performed on departure.

The use of landmarks in pinpointing a goal

Because it is easy to record on videotape the path of a returning insect while it is flying low on the ground within a meter of its goal, we have the most knowledge of several navigational strategies that are employed during this stage of an approach flight. When many returns are recorded from a single insect, it can be seen that bees and wasps adopt preferred viewing directions when they are close to the goal and that they tend to fly down a standard approach corridor until they reach the goal (Zeil, 1993b; Collett, 1995; Collett and Baron, 1994). The insect, on its final approach, will thus encounter a regular sequence of views of the scene, culminating in a view of the surroundings from the vantage point of the goal. The adoption of a standard viewing direction means that the scene at the goal will be imaged in a relatively fixed position on the retina.

This feature of the approach stresses one of the important principles of landmark guidance in insects: the reliance on retinotopically organised visual memories. Much pattern learning in insects is turning out to be retinotopically coded, such that a pattern is recognised only when viewed through the same region of retina that was exposed to it during learning (Wehner, 1981; Dill et al., 1993; Dill and Heisenberg, 1995). This view dependence of pattern recognition has the benefit of giving insects a relatively simple method of pinpointing a place: the insect stores a view from the goal and then governs its later approaches by moving so as to reacquire that view. Such a mechanism of landmark guidance is consistent with the results of experiments in which bees and ants learn to locate a foraging site or their nest by means of an array of landmarks. When the array is altered and the goal removed, an insect will often search for the absent target where it finds the best 2-D match between its current image and that which it is accustomed to view from the goal (see reviews by Collett, 1992; Wehner, 1992).

A major problem with image matching as a navigational tool is that when the insect is at all far from its goal there will be a large discrepancy between the insect's current retinal image and that defining the goal. It may then be difficult for the insect to know how it should move in order to improve the match between its current and stored images. The nature of the problem

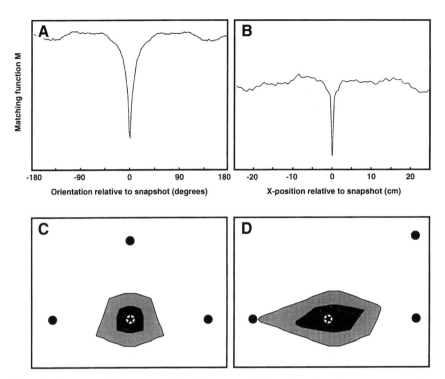

Figure 4. Image matching at increasing distances and orientations from the vantage point of a reference image. (A) and (B) illustrate the limitations of pixel matching. An image was record- ed with a video camera 5 cm off the ground in the centre of a triangular array of landmarks. The camera was subsequently turned on the spot (A) or shifted laterally on a guide rail (B). Images were subtracted from the reference image and the matching function M was calculated as the sum of absolute values of differences between pixel values P_{ij} of a given image and the pixel values R_{ij} of the reference image by $M = 1/_{ij} \sum_{ij} |P_{ij} - R_{ij}|$. (A) The matching function M as the viewing direction moves away from the reference direction at $0°$. (B) The matching function M for images taken from different positions along a straight line either side of the position of the reference image at 0 cm. The camera orientation was kept constant at $0°$. See text for further details. Modified from Voss (1995). (C) and (D) Matching using identified edges. Reference image is taken at the star. The light grey and black regions represent the areas within which all three landmarks (indicated by the black circles) fall within $\pm 20°$ (grey) or within $\pm 10°$ (black) of their angular position in the reference image.

that the insect visual system must solve is illustrated in Figure 4 A and B which has been taken from Voss (1995). A reference image was recorded outdoors at the centre of a triangular array of three landmarks with a video camera. The camera was then either turned through $360°$ on the same spot or shifted laterally on a guide rail. A matching function was calculated as the sum of the differences of pixel values between the reference image and the images taken in different directions or at different points along the path of the camera. This function rises steeply from a minimum when the camera is in the reference position to reach a plateau with only a few centimetres of translation or a few degrees of rotation. The residual match with translational shift comes from the distant panorama which has negligible parallax over the range of movement of the camera. Once on such a plateau, there is no possibility of using matching algorithms which involve gradient descent. The extreme narrownes of these troughs thus tells us that image matching is only likely to work after the scene has undergone considerable processing and filtering so that, for example particular edges in the image can be identified and matched.

If particular objects or features in the image can be labelled, then their arrangement on the retina distorts more gradually with distance from the goal. Figures 4 C and D illustrate this point by plotting areas within which all the landmarks are less than 10° or 20° of their retinal positions at the goal. These areas scale roughly with the distance of landmarks from the goal and their shapes depend upon the layout of objects in the scene. Consequently, one possible way to extend the range of image matching is to use snapshots which cover different spatial scales. A stored image containing only relatively distant landmarks might bring an insect to the approximate region of the goal, from where landmarks near to the goal can be successfully exploited (Cartwright and Collett, 1987).

One of the functions of learning flights might be to acquire a stack of differently filtered snapshots. Early in the flight, when the insect is close to the ground and to the goal, objects near to the goal are prominent and are likely to be recorded. As the radius of the arcs increases and the insects flies higher (Zeil, 1993a), small objects close to the nest are less noticeable. Instead, larger objects that are a little more distant from the goal become more salient as their images travel progressively more slowly over the retina and so come to stand out from the moving background. By recording such filtered views, an insect might assemble a sequence of snapshots of steadily increasing catchment area which together can guide the return over an appreciable distance (Cartwright and Collett, 1983; Zeil, 1993a).

There is little experimental evidence for a against such a scheme. However, it has been established that a location is defined both by large and distant objects and by small and close ones. Foraging stingless bees, for example, had difficulty in locating their nest when it was moved a short distance from its usual site. The bees found it readily, however, when the large objects that made up the more distant scene were shifted by the same amount so that the information provided by close and distant landmarks was congruent (Zeil and Wittmann, 1993). As yet, it is unclear whether the different landmarks exert their influence sequentially or concurrently.

One way in which insects do cope with the limitations of image matching is by employing a sequence of different navigational strategies. On first approaching the goal area, a wasp or bee tends to face and to aim for a prominent object close to the goal (Collett and Baron, 1994; Collett, 1995). Such beacon aiming allows the insect to ignore much of the swiftly changing scene on its retina. The insect need only be familiar with the beacon.

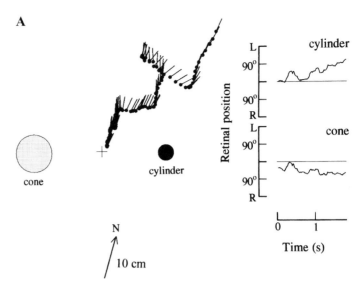

Figure 5. From aiming at beacons to image matching. A single bee is trained to forage at a feeder (+) with a cone 15 cm to the east and a cylinder 15 cm to the west. (A) Approach flight videotaped from above. Circles and tails show the position of the bee's head and the orientation of its body axis every 20 ms. Graphs show the retinal position of the cylinder and cone during the course of the flight. Bee first fixates cylinder and then the cone. Finally, both objects travel towards the lateral visual fields of the left and right eyes, as the bee moves towards the feeder. Arrows point north and their length represents 10 cm on the ground. (B) Tests with one object and no feeder. Bee fixates object and then flies so that object moves laterally over the retina. In tests with the cylinder, the bee moves so that the cylinder usually travels across the left eye, whereas in tests with the cone the bee moves so that the cone travels across the right eye. Bees thus link a different movement to each object. Bottom panels show the retinal trajectories of cylinder and cone obtained from several tests together with the accompanying change in body orientation. From Collett and Rees (1997).

Once the insect is close to the beacon it moves towards the goal following a more or less standard trajectory. This switch in strategy can be seen in a bee that has been trained to forage at a site midway between two objects: a black cone 15 cm west of the feeder, and a blue cylinder 15 cm to the east (Collett and Rees, 1997). Over some hours, the bee approaches the array from a constant direction, in this case roughly from the north. It aims sometimes at the cone, sometimes at the cylinder and sometimes at both (Fig. 5A). Occasional tests are introduced in which the feeder and one of the flanking objects are absent. When only the cone is present, the bee approaches it and moves so as to place the cone to its west, and when only the cylinder is present the bee moves to place the cylinder to its east. Superimposed trajectories of the object's path across the retina accumulated over several tests show that the object travels from the front towards the periphery of the retina. Movement is across the right eye in tests with the cone, and across the left eye in tests with the cylinder (Fig. 5B). Thus, bees seem

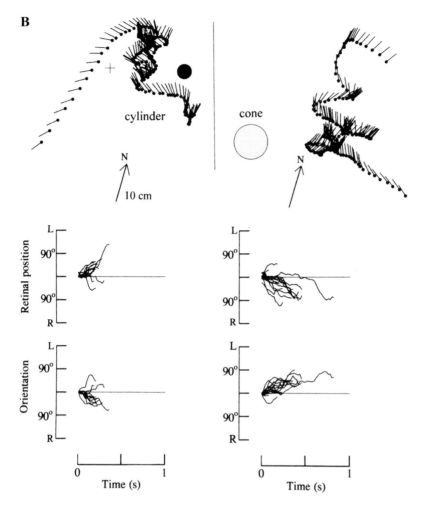

Figure 5. (continued).

to have learnt to distinguish the two objects and to have associated a different trajectory with each of them. This trajectory will take a bee close enough to the goal for image matching to operate successfully. It provides for an orderly transition between beacon aiming and image matching and means that image matching can be delayed until the bee is close to the goal.

Landmarks for specifying routes

Because insects often forage repeatedly within the same area, they have the opportunity to guide themselves by familiar landmarks viewed on their route to and from the nest. Although dead reckoning (see Wehner, 1992) is their primary means of guidance, route landmarks are used in addition to correct inevitable errors in dead reckoning, to help when cloud-cover or dense foliage deprives animals of celestial compass information, and to mark a route through cluttered terrain.

One difficulty in studying the use of route landmarks in flying insects is that they fly too high and too fast for it to be easy to follow their path. Baerends (1941) overcame this difficulty by working with a hunting wasp, *Ammophila*, which often catches prey that is so heavy that the wasp has to drag it to the nest over the ground. A wasp so encumbered can easily be followed and its path recorded. Once the wasp had reached the nest, Baerends caught it and carried it, still grasping its prey, to a new location where it was released. Individual wasps were found to have a restricted knowledge of the surroundings of their nest and to keep to idiosyncratic routes. When released outside their familiar corridor, they searched extensively until they happened upon a familiar route. By accustoming a wasp to artificial landmarks and then displacing them, Baerends showed that their route was guided both by individual objects and by rows of landmarks.

Ants are perhaps a more convenient animal for investigating landmark guidance. Initial evidence that ants use route landmarks came from the observation that individuals follow the same complex path over many journeys, even after a winter's confinement to the nest (Rosengren, 1971). These stereotyped paths were disrupted when trees were felled during the previous winter, suggesting that they are shaped in part by the guiding influence of familiar landmarks (Rosengren and Pamilo, 1978).

Figure 6 shows examples of the homeward paths of *Cataglyphis bicolor* on an empty desert terrain on which some artificial landmarks had been placed for several days (Collett et al., 1992). A single individual will typically stick to a single route over many journeys, but its route is idiosyncratic and may well differ from that of a nestmate foraging in the same place. For example, one animal goes consistently to the left (Fig. 6A) and

Figure 6. Homeward trajectories of ants after several days of foraging at the same site. Four visually distinct landmarks were positioned near to the direct path between food and nest. (A, B) Superimposed individual trajectories from two ants, one that habitually went to the left, the other to the right of the first landmark. (C) Mean trajectories of five ants that passed to the left of the first landmark. (D, E) Simulated ant trajectories. At each step, the simulated path was controlled by the landmark that had the largest apparent size and was less than 70° from the centre of the visual field. The direction of each step was specified by a linear combination of (i) dead reckoning (turning tendency was proportional to the angular difference between the ant's long axis and the home vector), (ii) aiming at beacons (turning tendency was proportional to the angular difference between the ant's long axis and the retinal bearing of the relevant landmark), (iii) biased detours (turning tendency to the left or right of a landmark was proportional to the retinal width of the relevant landmark). In (E), dead reckoning made no contribution to the ants's behaviour and the path became more sinous. (A–C) adapted from Collett et al. (1992).

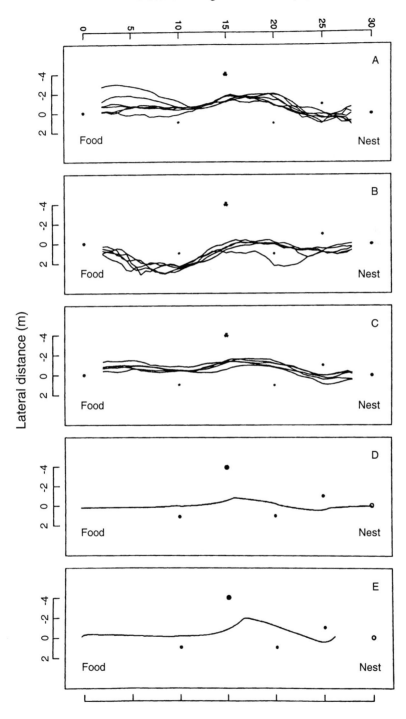

another to the right of the landmark closest to the food (Fig. 6B). This individual variation stresses that within broad limits the exact route is not important so long as the ant travels efficiently between the two places. However, individual stereotypy has the advantage that the same sequence of landmarks is always seen from the same succession of vantage points, making it possible for insects to be guided by a sequence of views. The second notable point about these routes is that different individuals passing to the same side of the first landmark, follow very much the same route (Fig. 6C), suggesting that there are likely to exist relatively simple rules controlling the ant's path through these landmarks.

To obtain direct evidence that desert ants learn the appearance of familiar landmarks on their route, animals were trained to forage at a fixed location (Collett et al., 1992). Two artificial landmarks were placed on the terrain, one close to the nest, the other half-way along the 30 m route. These landmarks were of very different shape and were positioned either side of the direct path between feeding site and nest. The ants, on their homeward route, usually passed to the left of the first landmark and to the right of the second. After several days of experience with this arrangement, individual ants that were about to set off home were carried to an empty testing area. On their release they typically walked in the usual homeward direction as dictated by their 'home vector'. But when a copy of one or the other of the route landmarks was placed in the direct path of the ant's expected trajectory, ants consistently detoured to the left of the landmark they expected to see on their right, and to the right of the landmark they expected to see on their left. They thus recognise the appearance of these objects and use them to correct their path.

The requirements of a good route landmark are somewhat different from those for specifying places. Whereas small, nearby objects may be significant for defining a destination, there is little virtue in learning and exploiting objects like small stones when travelling over routes, that in the case of desert ants, may be several hundred metres long. Such objects could only be useful signposts over a very small distance, and, unless the route is specified on an unnecessarily fine scale, the insect may not even come close enough for the same object to be visible on each trip. Indeed, it may often be sensible to limit the precision with which a route is defined by selecting as landmarks large objects which can act as guideposts over a long stretch of the route. Large objects, when seen at a distance, have the additional advantage that their appearance will change relatively slowly per unit distance travelled.

Although large, distant landmarks can help control the direction of travel, to ensure an insect's adherence to a particular route, a landmark needs to be sufficiently close to the path that, should the insect stray from it, an error signal is generated that can drive corrective manoeuvres. Consequently, objects should also be selected according to their proximity to the route.

Relations between dead reckoning and route landmarks

Under natural conditions, route landmarks are learnt and employed as an adjunct to dead reckoning (see review by Wehner et al., 1996). With large discrepancies between the information from landmark cues and dead reckoning commands, dead reckoning is often dominant. For example, Chittka and Geiger (1995) trained bees to follow a line of tents laid out between the hive and a feeding station. In tests, the direction of the line was rotated about the hive by varying amounts. One feeder was placed at the end of the displaced line of tents and another at the usual feeding station. With a discrepancy of no more than a few degrees between the directions signalled by the tents and that commanded by dead reckoning, bees landed on the feeder specified by dead reckoning and mostly ignored the feeder at the end of the row of tents. On cloudy days, when directional cues are less reliable, the line of tents influenced the insects' choice of feeder over larger discrepancies (Fig. 7).

There is an analogous phenomenon in the interaction between landmarks marking the nest and the distance that ants travel along a homeward trajectory. Ants accustomed to seeing an array of landmarks close to the nest were caught at their feeding site and released at a test site. Copies of the

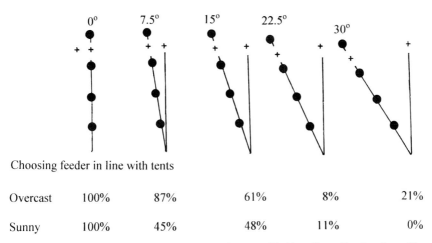

Figure 7. Resolving discrepancies between directions specified by a line of landmarks and by sun compass cues. Bees were trained to a feeder 262.5 m SE of their hive. The direct path between hive and feeder (+) was marked by a line of yellow tetrahedral tents 3.46 m high (shown by circles). In tests, the line of tents was rotated about the hive through 7.5°, 15°, 22.5° or 30°. The vertical line represents the usual flight path from hive at bottom to feeder at the cross. One observer counted the number of bees landing at the standard feeder, a second observer counted bees landing at a feeder situated at the accustomed distance along the line of tents. In control tests (0°), the direction of the line of tents was as in training and bees chose between two feeders, one in the training location and a second 7.5° away. Tests were conducted both on sunny days and days with complete cloudcover. Adapted from Chittka and Geiger (1995).

nest landmarks were positioned either at the expected distance along the home vector or at some point nearer to the start. The landmarks only influenced the ants' trajectory when they were placed near to the end of the home vector (Burkhalter, 1972; Wehner et al., 1996). This means that ants will ignore distracting objects on their path which may bear some fortuitous similarity to landmarks at the goal. A frequent suggestion is that home vectors provide contextual cues, in the sense that particular land- marks are expected at particular distances along the home vector. A more likely explanation in this case is that an ant follows its home vector, ignoring other signals that tell it to stop, until it approaches the end of its preordained trajectory.

How are landmarks used in route guidance?

With a full understanding of the way in which landmarks contribute to specifying routes, it would be possible to simulate the stereotyped paths taken by a returning ant through a complex environment like that illustrat- ed in Figure 8. This goal is still someway off. But we can list some of the navigational strategies that may need to be included in such a simulation.

Use of beacons
Prominent objects along a path are used as guideposts at which an insect aims (see for example, von Frisch, 1967; Chittka et al., 1995a, b). Insects will also aim at particular features of the skyline. For example, if a beehive is on the side of a mountain and bees from it forage on its lower slopes, or

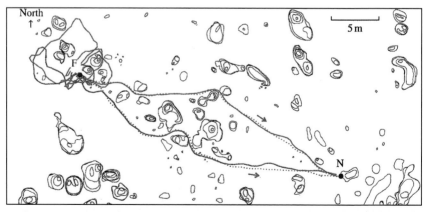

Figure 8. Routes followed by two desert ants returning from a feeding site (F) through desert scrub to their nest (N). Shrubs are marked by contour lines at 15 cm height intervals. Dotted green and red lines show the normal idiosyncratic path of each ant. Solid lines show the paths followed by the same ants when they were carried back to the feeding site after reaching the nest. The ant's path after displacement and release is the same as it is on normal return journeys. From Wehner et al. (1996).

on the plain below, they can use the mountain to find their way home even when they are displaced by as much as 10 km. In contrast, bees nesting in an environment with a flat skyline do not home successfully after such large displacements (Southwick and Buchman, 1995).

Following edges or lines of landmarks

There are many examples of bees or wasps following lines of trees or the edges of lakes or forests (Baerends, 1941; von Frisch and Lindauer, 1954; Dyer and Gould, 1981).

Biased detours

When a desert ant approaches close to a beacon, it turns so that it passes consistently to one side of the beacon, as already discussed. This fixed manoeuvre will ensure a standard view of the next stage of the route (Collett et al., 1992).

Linking vectors to landmarks

Somewhat similarly, bees flying through mazes will learn to associate a trajectory of a given distance and direction with the view of a visual stimulus, such as a patch of colour, or a pattern of stripes, so that when they view the stimulus they generate the linked trajectory (Collett et al., 1993, 1996). Although this linkage has so far only been studied in small, artificial environments, it is likely to apply also to larger scale conditions such as those of Figure 8.

Despite the difficulty of simulating routes through rich environments, it is possible to mimic an ant's homeward path through less complex surroundings like those sketched in Figure 6. The trajectories of Figure 6C can be simulated by combining dead-reckoning, aiming at beacons and biased detours in an additive manner (Fig. 6D). If the contribution of dead-reckoning is set to zero, as might conceivably happen when ants are returned to the feeding site after they have reached their nest, homeward paths are still generated. However, the curves are no longer damped by the homevector and become more sinuous (Fig. 6E). This contrasts with the behaviour of real ants in a cluttered environment. Real ants follow exactly the same route when they are returned to the feeding site with zero home vector as they do when they home normally (Fig. 8; Wehner et al., 1996). Does landmark information in a richer environment specify the ant's route more precisely or is the home vector retrieved from long-term memory once the ant has recognised its route? One experimental approach to this question is to take an ant that has reached its nest so that its homing vector is zero, return it to the feeding site and allow it to start on its homeward route. After the ant has followed the route sufficiently far for an observer to be sure that it is heading homewards, displace the ant to an empty test ground. If it then continues to walk in the homeward direction, its experience with the landmarks when near to the feeder must have evoked its home vector.

Landmarks for learning scenes

In addition to using landmarks to pinpoint a place, it is also helpful to learn scenes on a coarser scale by recording the more distant panorama of hills and trees that does not change its appearance with small-scale shifts of the insect's position. In this case the insect is not guided by particular features of the scene. Instead, the panoramic view labels or identifies a place, thus facilitating the performance of the particular behaviour that is appropriate to that place. Consider bees which forage regularly at several sites along a complex route. When they leave each foraging site, they must recall the trajectory that will take them to their next stop. It would clearly be advantageous if a view of the surrounding scene when leaving one foraging site were to help trigger the trajectory that leads to the next feeding site or guides the animal back home.

If bees and wasps do learn more distant panoramas of this kind during the learning flights which we have described earlier, one might expect that they do so when they are flying high above the ground before their final departure. The design requirements of this 'circling' phase of the learning flight differ from the pivoting observed in its early stages. To pick out distant landmarks and to keep them relatively stationary on the retina, it would be useful to have segments of flight in which the insect flies facing in a constant direction. Unfortunately, it is difficult to ascertain whether the final 'circling' phase of learning flights is in fact constructed from segments during which the insect's rotational velocity is clamped at zero.

Nonetheless, displacement and release experiments suggest that bees to have this broader knowledge of a scene. In such tests, bees are taken from their hive to one of several places with which they are known to be familiar and are then set free. Under certain circumstances, their vanishing bearings when they fly away are in the direction of their hive, even though in some cases neither the hive nor landmarks lying close to the hive are visible from the release site.

One plausible interpretation of this behaviour is that bees have learnt the surrounding scene and have associated with that scene the direction in which they normally fly to reach home. However, it is often hard to be sure that the bees are not attracted towards landmarks situated between the hive and the release site or towards appropriately positioned features on the skyline. Bees clearly do respond to such features, which are acquired rapidly on departure flights (Becker, 1958; Capaldi and Dyer, 1995). Menzel et al. (1996) have endeavoured to avoid such complications by a careful choice of terrain. The landscape between hive and release sites was bare and the release site was positioned far from the hive. Bees were trained to forage at one site in the morning and at another in the afternoon, so that the two sites were equally familiar. But when bees caught at the hive were released at the morning site in the afternoon, or at the afternoon site in the morning, their vanishing bearings were appropriate to the release site rather than to the time of day. Release at an intermediate site resulted in trajectories of intermediate directions. This behaviour suggests that bees at the intermediate site recalled both trajectories and elaborated a mean trajectory. On the assumption that bees were not familiar with the intermediate site and that there were no beacons to guide them, such a trajectory requires bees to have recognised features of the two training sites from the intermediate site. Furthermore, these recognised features must be large objects that are visible at a distance.

The behaviour of bees in a small-scale environment provides support for the supposition that intermediate trajectories are generated by averaging two stored trajectories (Collett and Baron, 1995; Collett et al., 1996). Bees were trained to fly through two compartments of a maze to collect sucrose. In one compartment, bees viewed an almost panoramic pattern of diagonal stripes and had to fly to the right to reach the exit hole of that compartment. In the second compartment, they viewed a similar pattern of stripes oriented along the opposite diagonal and flew to the left to exit from the second compartment. When they were presented in one compartment with patterns of vertical or slightly oblique stripes which they had never seen in training, their direction of flight was intermediate (Fig. 9). As in the experiments of Menzel et al. (1996), intermediate scenes evoked intermediate trajectories.

Landmarks in context

Bees learn readily that one pattern or colour is linked to food in one place and a different pattern or colour is associated with food in a second place, so that when they are given a choice they will select the stimulus appropriate to the place that they are in (e.g., Menzel et al., 1996). Results of this kind suggest that particular local memories are primed by the broader spatial context provided by the surrounding panorama. Local landmarks are similarly embedded in a broader spatial context. For example, individual bees were trained outdoors to search for sugar water in two different places separated by 40 m (Collett and Kelber, 1988). In each place, the location of the sugar water on a 2 m square white platform on the ground was defined by one or two landmarks which differed in colour and shape between platforms. On one platform, food was to the east of the landmarks and on the other it was to the west. In occasional tests, landmarks were swapped between platforms and the feeder was absent. Despite the landmarks' unusual appearance, bees searched on the side of the landmark that was appropriate to the platform. Their search was thus governed by the spatial context of the platform, rather than by the appearance of the landmark array. In the same way, bees that are trained to forage on just one platform will search in the appropriate direction from the landmarks when they

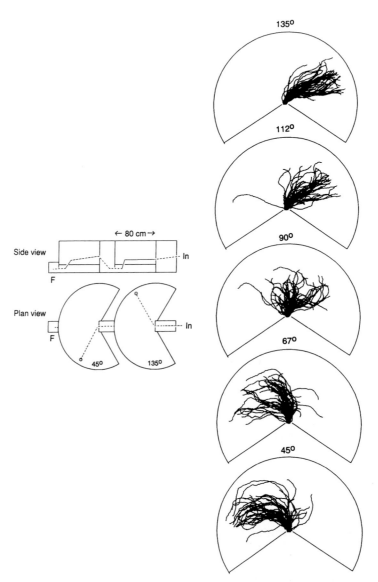

Figure 9. Averaging vectors. Bees were trained to fly through a two-compartment maze to collect sucrose. Each compartment was a 240° segment of a cylinder and its inner wall was lined with a pattern of diagonal stripes. These were oriented at 135° relative to the horizontal in the first compartment and at 45° in the second compartment. Bees entered the first compartment at the centre and flew to their right towards an exit hole in the floor of the compartment from which they could reach the entrance to the second compartment. In the second compartment, the exit hole in the floor was to their left and from this they could reach the food. Side and plan views of the maze and the bees' path through it (dotted line) are shown to the left. In: entrance to the maze. F: feeding dish. Panels in the right column show superimposed flight paths during tests in compartment 1 with different stripe orientations on its inner wall. Trajectories evoked by 135° stripes were to the left and those evoked by 45° stripes were to the right. Intermediate orientations evoked trajectories of intermediate directions. Adapted from Collett et al. (1996).

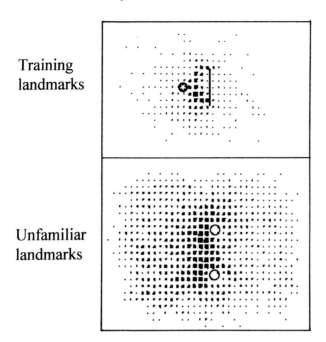

Figure 10. The influence of spatial context on landmark guidance. Bees were trained to find a bottle cap filled with sucrose solution (denoted by a star) to the west of an upright blue triangle (29 cm base, 25 cm high, indicated by the line in the black and white figure). Both triangle and bottle cap were placed on a 2 m² platform on the ground. The positions of the triangle and the bottle cap on the board were changed between foraging visits to ensure that the triangle was used to define the feeding site. When bees were tested with the feeder missing, they searched to the west of the blue triangle (top panel). In occasional tests, the triangle was replaced by two yellow cylinders (7 cm in diameter, 25 cm in height, shown by circles in the black and white figure). Despite the substitution of novel landmarks for the usual ones, bees still searched to the west of the landmarks (bottom panel). Relative amount of time spent in each 4.5 cm by 5 cm cell of the imaginary grid is shown by the area of black in the cell. Adapted from Collett and Kelber (1988).

are replaced by unfamiliar ones differing in shape and colour (Fig. 10). These experiments show that the context of a distant panorama primes the recognition of local landmarks to such an extent that objects which bear little resemblance to the original ones can substitute for the latter, and bees will nonetheless search in roughly the correct location relative to the substitutes. This latitude means that under normal circumstances insects can respond to local landmarks when their appearance differs markedly from what has been memorized.

Setting landmarks in context may thus be one way of overcoming some of the limitations arising from the difficulty of generalising across views and of recognising landmarks from several vantage points. The success of this strategy relies on the bees' ability to pick out the most conspicuous objects during the learning phase so that the same conspicuous objects can guide the beginning of the bees' local search even if the objects viewed from the current vantage point bear a poor resemblance to their appearance as memorized at the goal. This does not of course mean that the final matching with the correct landmark is sloppy. The detail with which individual local landmarks in an array are learnt and guide a bee's search is evident, for example in the behaviour illustrated in Figure 5.

The benefits of such contextual priming become clear when we consider bees foraging at several sites. The spatial relationships between target and local landmarks are likely to differ between sites. If bees in one place were to recall local memories that are appropriate to somewhere else, they would attempt to match the scene to the wrong template. Stored panoramic views thus provide a zooming mechanism for focusing down from a coarse scale to retrieve the correct information for navigating on a fine spatial scale. Content addressable recall unsupported by context is chancy and would probably lead to confusion, partly because there is likely to be a resemblance between different local arrays, and partly because of the difficulty of recognising arrays from an unfamiliar vantage point. The same argument emphasises the benefits of learning a sequence of views along a route and of not relying simply upon content-addressable recall to retrieve the next view (Collett et al., 1993; Pastergue-Ruiz and Beugnon, 1994).

Conclusion

We are thus led to the following somewhat impressionistic understanding of an insect's 'mappa mundi'. The association of trajectories with panoramic views provides a spotty and partial content addressable map that consists of a web of interlinked places. An insect that is placed or finds itself in a particular site may recognise the visual scene that it encounters and so recall movement vectors that are linked to that scene and that will direct it towards other significant places. Temporal or motivational contexts may operate switches that select among different vectors which are linked to the same scene (Menzel et al., 1996). A single node of the insect's graph can thus be connected to several others.

Although an insect's topographical knowledge of a landscape may be punctate, several of the navigational strategies that we have discussed allow insects to navigate as though they have a continuous and graded representation of some parts of their surroundings. All the ones that we now list depend upon the environment containing large and prominent landmarks which can be recognised over a reasonably large area. 1) Because views from different familiar places may have distant landmarks in common, insects at a new intermediate site can average movement vectors linked to the familiar places and so can generate an appropriate movement vector from the intermediate place (Collett and Baron, 1995; Menzel et al., 1996). 2) Large landmarks close to a route or a goal can be used as beacons to attract insects over a wide area (e.g., Dyer, 1991). 3) Image matching using distant landmarks allows insects to reach a site from a wide area around it, even in the absence of beacons to mark the goal. 4) Finally, and rather differently, path integration allows an animal to explore an unfamiliar part of its environment and nonetheless find its way directly home, detouring around large obstacles when necessary (review Wehner, 1992). By these simple stratagems, animals with limited spatial knowledge can operate effectively within a surprisingly large environment.

Acknowledgment
We thank Miriam Lehrer for her very detailed and helpful editiorial commentary.

References

Antonsen, P. and Wehner, R. (1995) Visual field topology of the desert ant's snapshot. *Proc. Neurobiol. Conf. Göttingen* 23:42.

Baerends, G.P. (1941) Fortpflanzungsverhalten und Orientierung der Grabwespe *Ammophilia campestris* Jur. *Tijdschr. Entomol.* 84:68–275.

Becker, L. (1958) Untersuchungen über das Heimfindevermögen der Bienen. *Z. vergl. Physiol.* 41:1–25.

Brünnert, U., Kelber, A. and Zeil, J. (1994) Ground-nesting bees determine the location of their nest relative to a landmark by other than angular size cues. *J. Comp. Physiol.* A 175:363–369.

Burkhalter, A. (1972) Distance measuring as influenced by terrestrial cues in *Cataglyphis bicolor*. *In*: R. Wehner (ed.): *Information processing in the visual system of arthropods.* Springer-Verlag, Berlin, pp 303–308.

Capaldi, E.A. and Dyer, F.C. (1995) Contents and acquisition of large-scale spatial memory by honey bees. *In*: M. Burrows, T. Matheson, P.L. Newland and H. Schuppe (eds): *Nervous systems and behaviour.* Thieme Verlag, Stuttgart, New York, p 124.

Cartwright, B.A. and Collett, T.S. (1979) How honey-bees know their distance from a nearby visual landmark. *J. Exp. Biol.* 82:367–372.

Cartwright, B.A. and Collett, T.S. (1983) Landmark learning in bees: experiments and models. *J. Comp. Physiol.* 151:521–543.

Cartwright, B.A. and Collett, T.S. (1987) Landmark maps for honeybees. *Biol. Cybernetics* 57:85–93.

Cheng, K., Collett, T.S., Pickhard, A. and Wehner, R. (1987) The use of visual landmarks by honeybees: Bees weight landmarks according to their distance to the goal. *J. Comp. Physiol.* A 161:469–475.

Chittka, L. and Geiger, K. (1995) Honeybee long-distance orientation in a controlled environment. *Ethology* 99:117–126.

Chittka, L., Geiger, K. and Kunze, J. (1995a) The influence of landmarks on distance estimation of honey bees. *Anim. Behav.* 50:23–31.

Chittka, L., Kunze, J., Shipman, C. and Buchmann, S.L. (1995b) The significance of landmarks for path integration in homing honeybee foragers. *Naturwiss.* 82:341–343.

Collett, T.S. (1992) Landmark learning and guidance in insects. *Phil. Trans. R. Soc. Lond.* B 337:295–303.

Collett, T.S. (1995) Making learning easy: the acquisition of visual information during orientation flights of social wasps. *J. Comp. Physiol.* A 177:737–747.

Collett, T.S. and Baron, J. (1994) Biological compasses and the coordinate frame of landmark memories in honeybees. *Nature* 368:137–140.

Collett, T.S. and Baron, J. (1995) Learnt sensori-motor mappings in honeybees: interpolation and its possible relevance to navigation. *J. Comp. Physiol.* A 177:287–298.

Collett, T.S. and Kelber, A. (1988) The retrieval of visuo-spatial memories by honeybees. *J. Comp. Physiol.* A 163:145–150.

Collett, T.S. and Rees, J.A. (1997) View-based navigation in Hymenoptera: multiple strategies of landmark guidance in the approach to a feeder. *J. Comp. Physiol.* A 181:47–58.

Collett, T.S., Dillmann, E., Giger, A. and Wehner, R. (1992) Visual landmarks and route following in desert ants. *J. Comp. Physiol* A 170:435–442.

Collett, T.S., Fry, S.N. and Wehner, R. (1993) Sequence learning by honeybees, *J. Comp. Physiol.* A 172:693–706.

Collett, T.S., Baron, J. and Sellen, K. (1996). On the encoding of movement vectors by honeybees. Are distance and direction represented independently? *J. Comp. Physiol.* A 179: 395–406.

Dill, M. and Heisenberg, M. (1995) Visual pattern memory without shape recognition. *Phil. Trans. R. Soc. Lond.* B 349:143–152.

Dill, M., Wolf, R. and Heisenberg, M. (1993) Visual pattern recognition in *Drosophila* involves retinotopic matching. *Nature* 365:751–753.

Dyer, F.C. (1991) Bees acquire route-based memories but not cognitive maps in a familiar landscape. *Anim. Behav.* 41:239–246.

Dyer, F.C. and Gould, J.L. (1981) Honey bee orientation: A backup system for cloudy days. *Science* 214:1041–1042.

Frisch, K. von (1967) *The dance language and orientation of bees.* Oxford University Press, London.

Frisch, K. von and Lindauer, M. (1954) Himmel und Erde in Konkurrenz bei der Orientierung der Bienen. *Naturwiss.* 41:245–253.

Hoefer, I. and Lindauer, M. (1976) Der Schatten als Hilfsmarke bei der Orientierung der Honigbiene. *J. Comp. Physiol.* 112:5–18.

Land, M.F. (1989) Variations in the structure and design of compound eyes. *In*: D.G. Stavenga and R.C. Hardie (eds): *Facets of vision.* Springer-Verlag, Berlin, Heidelberg, New York, pp 90–111.

Lehrer, M. (1993) Why do bees turn back and look? *J. Comp. Physiol.* A 172:549–563.

Lehrer, M. and Collett, T.S. (1994) Approaching and departing bees learn different cues to the distance of a landmark. *J. Comp. Physiol.* A 175:171–177.

Menzel, R., Geiger, K., Chittka, L., Joerges, J., Kunze, J. and Müller, U. (1996) The knowledge base of bee navigation. *J. Exp. Biol.* 199:141–146.

Pastergue-Ruiz, I. and Beugnon, G. (1994) Spatial sequential memory in the ant *Cataglyphis cursor. In*: A. Lenoir, G. Arnold and M. Lepage (eds): Proceedings of 12th Cong. Int. Union. Study Social Insects. Univ. Paris Nord, Paris, p 490.

Rosengren, R. (1971) Route fidelity, visual memory and recruitment behaviour in foraging wood ants of the genus *Formica* (Hymenoptera, Formicidae). *Acta Zool. Fenn.* 133: 1–106.

Rosengren, R. and Pamilo, P. (1978) Effect of winter timber felling on behaviour of foraging wood ants (*Formica rufa*) in early spring. *Memorabilia Zool.* 29:143–155.

Southwick, E.E. and Buchmann, S.L. (1995) Effects of horizon landmarks on homing success in honey bees. *Am. Nat.* 146:748–764.

Srinivasan, M.V. (1993) How insects infer range from visual motion. *In*. F.A. Miles and J. Wallman (eds): *Visual motion and its role in the stabilization of gaze.* Elsevier Science Publishers, Amsterdam, pp 139–156.

Tinbergen, N. and Kruyt, W. (1938) Über die Orientierung des Bienenwolfes (*Philanthus triangulum* Fabr.). III. Die Bevorzugung bestimmter Wegmarken. *Z. vergl. Physiol.* 25: 292–334.

Voss, R. (1995) *Information durch Eigenbewegung: Rekonstruktion und Analyse des Bildflusses am Auge fliegender Insekten.* Doctoral Thesis, University of Tübingen.

Wehner, R. (1981) Spatial vision in arthropods. *In*: H. Autrum (ed.): *Handbook of sensory physiology*, Vol. VII/6C, Springer-Verlag, Berlin, Heidelberg, New York, pp 287–616.

Wehner, R. (1992) Arthropods. *In*: F. Papi (ed.): *Animal homing.* Chapman and Hall, London, New York, pp 45–144.

Wehner, R., Michel, B. and Antonsen, P. (1996) Visual navigation in insects: coupling of egocentric and geocentric information. *J. Exp. Biol.* 199:129–140.

Zeil, J. (1993a) Orientation flights of solitary wasps (*Cerceris*; Sphecidae; Hymenoptera): I. Description of flight. *J. Comp. Physiol.* A 172:189–205.

Zeil, J. (1993b) Orientation flights of solitary wasps (*Cerceris*; Sphecidae; Hymenoptera): II. Similarities between orientation and return flights and the use of motion parallax. *J. Comp. Physiol.* A 172:207–222.

Zeil, J. and Wittmann, D. (1993) Landmark orientation during the approach to the nest in the stingless bee *Trigona (Tetragonisca) angustula* (Apidae, Meliponinae). *Ins. Soc.* 40:381–389.

Zeil, J., Kelber, A. and Voss, R. (1996) Structure and function of learning flights in bees and wasps. *J. Exp. Biol.* 199:245–252.

Zhang, S.W., Srinivasan, M.V. and Collett, T.S. (1995) Convergent processing in honeybee vision: Multiple channels for the recognition of shape. *Proc. Natl. Acad. Sci. USA* 92: 3029–3031.

Orientation and Communication in Arthropods
ed. by M. Lehrer
© 1997 Birkhäuser Verlag Basel/Switzerland

Course control and tracking: Orientation through image stabilization*

K. Kirschfeld

Max-Planck-Institut für biologische Kybernetik, Spemannstraße 38, D-72076 Tübingen, Germany

* Dedicated to Bernhard Hassenstein on the occasion of his 75th birthday.

Summary. Course control and tracking are based on visual detection of the position and movement of objects. A disadvantage of biological movement detectors is that they cannot provide a signal proportional to the speed at which the image of an object moves over the retina. Other image parameters, such as brightness, contrast, and texture, strongly affect the magnitude of the detectors' output signals. To function well, the optomotor control circuit must solve these problems. One possible solution, realized in Diptera, is the principle of "gain control by feedback oscillations" described in this chapter.

The optomotor system serves for course control by stabilizing the image of the visual panorama on the eye, and for tracking a moving object by stabilizing the object's image on the eye. When an object moves in front of a structured background, it is impossible for the images of both object and background to be stabilized simultaneously. Arthropods and vertebrates usually employ the same strategy to cope with this problem: saccadic tracking. In Diptera, the neural substrate for saccadic tracking is partially understood.

Introduction

An animal moving around does so in order to arrive at some goal, be it a particular, ecologically relevant site in space, or a particular target (food source, prey, or a potential mate) relevant to the animal's survival and reproduction. As the animal moves, some neural mechanisms are expected to be active that will enable it to maintain its intended course of locomotion. Since the beginning of our century (e.g., Radl, 1903), a large body of experimental evidence has accumulated to show that one of these mechanisms involves stabilization of the retinal image perceived by the animal from its visual environment in the course of locomotion.

When an animal rotates passively (for example, due to air current) with respect to its surroundings – but also, as will be shown below, when it moves actively – it tends to stabilize the retinal image of the surroundings by turning its eyes, head or body in the direction of the image motion (the *large-field optomotor response*). If, at the same time, the animal performs a translational movement (for example, forwards), the retinal image cannot be stabilized in all parts of the retina. However, the visual flow field produced by the animal's own movement can provide important information about that movement, from which signals useful for course stabilization can be extracted.

If only a single small object is moving in an animal's visual field, the animal, again, will often turn its eyes, head or body in an attempt (i) to bring the image of the object onto a particular part of the retina (for example, the region of a fovea), and (ii) to keep it there. This *tracking* (a *small-field optomotor response*) is a relatively simple task when the object is moving against a homogeneous background. In the natural environment, however, the background is usually structured, which renders the problem considerably more difficult.

This chapter is not intended to provide a summary of all that is known about course control and tracking in arthropods; excellent reviews have been published elsewhere (Buchner, 1984; Wehrhahn, 1985; Land, 1992; Collett et al., 1993). Here, instead, two problem areas are addressed, one involving course control, the other involving tracking. We will see that the two are closely related to each other.

The control system for the large-field optomotor response, for which the input variable is the angular velocity of the surroundings, would ideally require that the angular velocity with which the retinal image is displaced by movement of the surroundings be measured accurately (Fig. 1A). Biological movement detectors, however, are not accurate speedometers: their signals depend not only on angular velocity, but also on other parameters of the image, such as its structure. It follows that the accuracy of stabilization of the retinal image should also depend on image parameters, but it has been demonstrated, at least in certain examples, that this is not the case. The question is: how can the control system stabilize the retinal image satisfactorily despite the non-ideal properties of the movement detectors?

The problem involved in visual tracking of a moving object arises from the fact that the animal must, at the same time, control its course of loco-

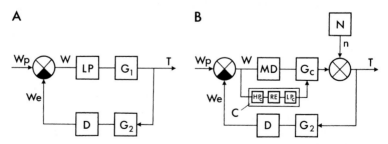

Figure 1. Control systems for the *large-field optomotor response*. (A) Block diagram of the control system used for the model calculations of Figure 2. w_p, angular velocity of the pattern; T, turning tendency of the animal; w_e, angular velocity of the eye; w, retinal slip speed (w = $w_p - w_e$); LP, low-pass filter; D, time delay; G_1, G_2, elements with gain k_1, k_2 (see text). (B) Modified control system with automatic gain control. MD, movement detectors; G_c, element with gain adjustable by the filter C; N, noise generator; n noise; HP_c, high-pass filter; RE, rectifier; LP_c, low-pass filter. From Kirschfeld (1991).

motion. When an object moves in front of a structured background, the two mechanisms, namely, the one that is expected to stabilize the large-area background on the retina, and the one that is expected to stabilize the small object, compete with each other. Here the queston is: which is stabilized in this complex situation, the object or the background?

Elementary movement detectors: Theory

When an image moves across the retina, in order for the eye to move in such a way that this image is stabilized as accurately as possible, the animal's visual system must accomplish two different tasks. It must determine, with sufficient accuracy, (i) the *direction* of the movement vector, characterized by an angle relative to local coordinates, and (ii) the *angular velocity* of the movement of the image across the retina (i.e., the length of the movement vector).

Let us consider the simple case in which the image is moving horizontally over the retina at a constant angular velocity, a type of displacement that occurs, for example, when the animal rotates about a vertical axis. To determine the direction of image movement, the system must possess movement detectors with at least two input elements that sample two different places in the image. A single input element, although it could signal that something is moving (in that its output fluctuates according to the intensity distribution in the moving pattern), cannot extract the direction of movement. Furthermore, a single input element is not expected to distinguish between actual movement and fluctuations in the intensity (flicker) of a motionless pattern (which is admittedly, unlikely to occur under biological conditions). If an image moves across two input elements (Fig. 2), however, both input elements generate identical signals, offset from each other in time. Depending on the direction of the movement, one signal or the other will be delayed. To infer the direction of movement from these non-simultaneous signals, the following arrangement can be used. A delay, for example a low-pass filter or the like, is inserted into one of the channels. The signals from this and the second input element are then non-linearly combined, e.g., by multiplication, with the result that the unit consisting of the two elements becomes directionally sensitive: movement in one ("preferred") direction gives rise to a signal (Fig. 2 A), whereas movement in the opposite ("null") direction does not (Fig. 2 B). Figures 2 C and D show how the combination of two such antisymmetrical units would produce an elementary movement detector (EMD) that generates a positive signal for movement in one direction, and a negative signal for such in the opposite direction. The movement detector shown in Figure 2 is a simplified model of the movement detector based on the correlation principle developed by Hassenstein and Reichardt (1956) from their analysis of the

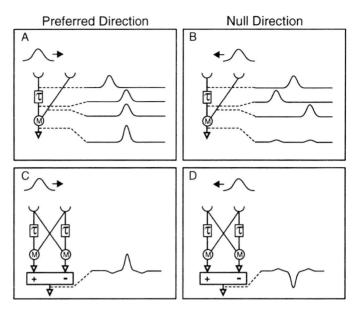

Figure 2. Motion detection by a correlation-type movement detector. Arrows indicate direction of motion of an intensity-modulated pattern. (A, B) Subunit of the elementary movement detector (EMD) shown in (C) and (D). τ, time constant; M, multiplication unit. The time-course of the signal at the different processing stages is shown at the right of the unit. When a stimulus pattern passes the two detector input channels, they are activated one after the other with a certain time shift. When the pattern moves in the preferred direction, as in (A), the temporal separation of the signals in both input channels is compensated for by a time delay (or a low-pass filter, τ). In this case, the two signals coincide at the multiplication stage and lead to a strong response. When the pattern moves in the null direction (B), the signals do not coincide at the multiplication unit and the response, if any, is only weak. (C, D) Two subunits are combined, one as shown in (A), the other with mirror image symmetry, and their output signals are subtracted from each other. Such a combined EMD generates a strong response to pattern motion in both directions, but the signals have opposite signs, depending on whether motion is in the one or the opposite direction. Modified from Egelhaaf and Borst (1993).

optomotor behaviour of insects[1]. It should be noted, however, that additional features, such as thresholds or nonlinear characteristics, must be introduced for a more quantitative modelling of movement-sensitive cells in the insect brain (Egelhaaf et al., 1989).

When a striped pattern with a sinusoidal intensity distribution is moved with a constant angular velocity w past an array of movement detectors of

---

[1] For historical accuracy (J. Thorson, personal communication) it might be noted that parallel formulations were developed in radiophysics at least as early as in the work of Briggs et al. (1950). These studies involved the measurement of the velocity of drift of the ionosphere from correlations in time among signals from multiple, spaced radio receivers.

the correlation type shown in Figure 2, after the initial transients the mean response R is given by

$$R = C \cdot \underbrace{\frac{2\pi\,\tau\,w/\lambda}{1 + (2\pi\,\tau\,w/\lambda)^2}}_{A} \cdot \underbrace{\sin \frac{2\pi\,\Delta\varphi}{\lambda}}_{B} \tag{1}$$

where C is a factor accounting for the dependence of the response on variables unrelated to the movement, such as the contrast and mean brightness of the pattern, and also the optical properties of the ommatidia; τ is the time constant of the first-order low-pass filter, here assumed to be linear (Fig. 2), w is the angular velocity of the pattern, λ is its spatial wavelength (period), and $\Delta\varphi$ is the angular separation of the two input elements. This equation, representing the steady state response R, has been already derived by Reichardt and Varjú (1959) (see also Varjú, 1959; Varjú and Reichardt, 1967; Buchner, 1984). A more recent treatment of the topic by Egelhaaf and Borst (1989) considers, in addition, transient responses, and is particularly recommended as a reference for the mathematical formalism.

Note that, in Equation (1), the angular velocity w always appears in an expression having the wavelength λ as its denominator. w/λ represents the frequency with which the signals oscillate in time in the input channels. It is called the temporal frequency, or contrast frequency, of the moving pattern.

The factor B in Equation (1) is the so-called "interference function". It contains no time-dependent parameters and accounts for the geometric interference between the wavelength λ of the pattern and the sampling distance $\Delta\varphi$ between the two input elements of the movement detector. For example, if $\lambda = \Delta\varphi$, it is clear that movement direction cannot be defined, because the term B is then $\sin 2\pi = 0$, and hence the mean response R will also equal zero. In this case, the signals in the two input channels oscillate in synchrony.

If a moving striped pattern is described one-dimensionally as a function more complicated than a sinusoid, then the response can be represented as the sum of the responses to that function's Fourier components. According to the theory, the response is independent of the phase of the Fourier components (Varjú, 1959; Reichardt and Varjú, 1959).

Up to this point we have been considering the mean response after the transients have died out. The theory of movement detection according to the correlation model, of course, also allows prediction of the response under dynamic conditions, for example, at the very onset of uniform movement of a sinusoidal pattern (Egelhaaf and Borst, 1989).

Large-field optomotor responses: Experiments under open- and closed-loop conditions

Angular velocity and contrast frequency

Very soon after the development of the theory outlined above, one of its most striking predictions was verified experimentally, namely that the maximum of the mean response to uniformly moving sinusoidal patterns must always occur at the same contrast frequency, regardless of the spatial wavelength λ of the moving pattern. This was first shown in behavioural experiments (Kunze, 1961; Götz, 1964) (Fig. 3), and was later also demonstrated electrophysiologically for movement-sensitive large-field neurons which, according to everything we know (Hausen and Wehrhahn, 1990), contribute to the large-field optomotor response (Eckert and Hamdorf, 1981; Buchner, 1984; Hausen, 1984).

The dependence of the optomotor response on the temporal frequency of the moving pattern is illustrated in Figure 3B. It is evident that, as temporal frequency increases, the response rises to a maximum and then declines. As can be inferred from Equation (1) (Factor A), the maximum of this response, regardless of the spatial wavelength λ of the pattern, occurs at a particular temporal frequency (in this case $w/\lambda = 1$), and not at a particular velocity w of the image movement (Fig. 3A). Very similar results have

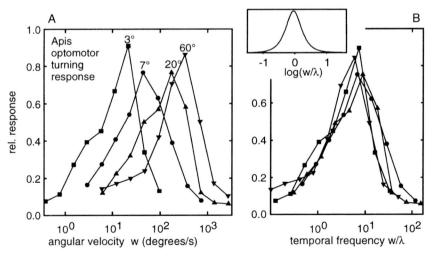

Figure 3. Dependence of relative response (torque) on the angular velocity w (A) and the contrast frequency (w/λ) (B) of the moving pattern. Abscissa in logarithmic units. A sinewave-pattern of various spatial wavelengths λ, indicated for each curve in (A), is moved around a tethered flying bee at different angular velocities w under open-loop conditions. Modified from Kunze (1961). The inset in (B) shows the steady-state response as a function of contrast frequency of an array of EMDs of the type shown in Fig. 2C, D (calculation according to Eq. (1)).

been obtained in vertebrates, e.g., for movement-sensitive cells of the so-called *accessory visual system* of pigeons. The relation between response and contrast frequency was observed not only in large-field neurons in a nucleus of the brain involved in controlling eye and head movements (the nucleus of the basal optic root), but even in the retinal ganglion cells that represent the input elements to this nucleus (Wolf-Oberhollenzer and Kirschfeld, 1994; Türke, 1996).

Even in humans, contrast sensitivity to moving sinusoidal patterns under "stabilized image" conditions (i.e., open-loop), measured using psychophysical methods, is dominated by the contrast frequency, and not by the angular velocity of the image (Kelly, 1979). Similarly, the latency of eye movements elicited by the onset of a uniform movement of a sinusoidal pattern depends on contrast frequency, rather than on velocity (Miles and Kawano, 1987). Within the latency period, this reaction, too, can be interpreted as being measured under openloop conditions.

Finally, the theory also predicts the time-course of the response during the initial transients. When uniform movement of a sinusoidal pattern begins, typical oscillations occur at a frequency corresponding to the contrast frequency of the moved pattern. Such rapid response components cannot readily be measured in the behaviour, but they do appear in the responses of neurons that trigger the large-field optomotor response. In both flies (Egelhaaf and Borst, 1989) and pigeons (Wolf-Oberhollenzer and Kirschfeld, 1994), these transient processes correspond basically to the reactions calculated from the model. However, the simple model shown in Fig. 2 must be expanded in order to render results that match the experimental data quantitatively. For example, according to the experimental results, the time constants τ of the filters are variable (Ruyter van Steveninck et al., 1986; Borst and Egelhaaf, 1987).

The gain in the optomotor control system

As has been shown above, the biological movement detectors cannot deliver a signal proportional to the angular velocity of the pattern movement. For any given pattern velocity, the signal can vary depending, for example, on the amount of contrast contained in the pattern, or, in the case of periodical patterns, on the spatial wavelength λ. This means that, in a control system like that illustrated in Figure 1, the accuracy of pattern stabilization ought to depend on the pattern parameters. In other words, if the response of the movement detectors to patterns moving with the same slip speed w (w = $w_p - w_e$, see Fig. 1A) differs because the patterns differ in, say, contrast, then this corresponds to differences in the gain k_1 of element G_1 of the control system.

To clarify this point, let us suppose that a pattern is moved at a constant speed past a visual system represented by the control system shown in

Figure 1. After the initial transients have died out, the angular velocity of the eye (w_e) is related to that of the pattern (w_p) by

$$w_e = \frac{k_1 \cdot k_2}{1 + k_1 \cdot k_2} \, w_p \tag{2}$$

Assuming that k_2, the proportionality constant relating the torque (T in Fig. 1) to the angular velocity of the eye (w_e in Fig. 1) is constant, it is evident that w_e depends on k_1 and that it approaches w_p the more closely the larger k_1 is. However, very large gains are not practicable in biological control systems, because the delays and phase-shifts that are always present must necessarily lead to instabilities (Kirschfeld, 1991). Assuming a realistic gain, w_e should therefore depend on pattern parameters, such as the spatial wavelength of the moving pattern.

Surprisingly, experiments on the optokinetic movements of human eyes in closed-loop conditions showed that their angular velocity depends only on the *angular velocity* of a sine wave pattern, and *not* on its *spatial wavelength* (Fig. 4). A similar result under closed-loop conditions was obtained in experiments on the optokinetic reaction of goldfish eyes (Schaerer et al., 1996). Also relevant in this context are experiments performed by Srinivasan et al. (1991), who analysed the flight path of bees flying freely through channels with various patterns displayed on the side walls. Their results suggest that the parameter determining the flight path is not the *contrast*

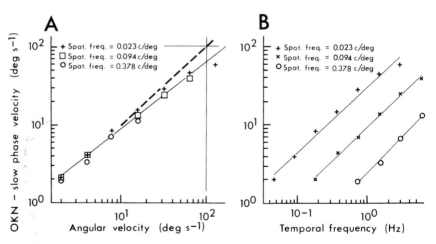

Figure 4. Optokinetic nystagmus (OKN) of the human eye (closed-loop). Slow-phase eye movement velocity is plotted as a function of pattern angular velocity (A) and temporal frequency (B). Parameter for the different curves is the spatial frequency (inset) of the sinewave stripe pattern that was rotated around the human observers. The dashed line in (A) has the slope 1 (perfect correlation between eye velocity and pattern velocity). Modified from de Graaf et al. (1990), and from Kirschfeld (1991).

frequency of the pattern, but rather the *velocity* of the animal relative to the laterally presented patterns (see also M.V. Srinivasan and S. Zhang, this volume). And, finally, if a horizontal cylinder, patterned with a helical stripe, is rotated around a free-flying *Drosophila*, producing apparent movement, the fly's response depends on the *velocity* of the pattern, and not on its *contrast frequency* (David, 1982)

Thus, it appears that, under *open-loop* conditions, the crucial parameter is the *contrast frequency* of the moving pattern, but that when the *loop* is *closed*, pattern *velocity* is most effective. The accuracy of image stabilization in the latter case is largely independent of the structure of the moved pattern.

This finding, of course, raises questions. What neural mechanisms account for the difference between open- and closed-loop responses, and how can the optomotor control system compensate for the obvious inability of the elementary movement detectors to measure velocity? One way of approaching these questions is to look at the gain of the optomotor response.

Variable gain in the optomotor response

Given a constant pattern velocity, the EMD response decreases as contrast decreases, both in theory (the dependence of the response on contrast is included in the factor C in Eq. (1)), and – at least in a certain range of pattern contrast – also in experiments. If, for a particular angular velocity, the EMD response is reduced due to reduced contrast, then this is equivalent to a reduction of gain in the feedback loop (Fig. 1). If the gain in the control system in element G_c (Fig. 1B) were to be suitably increased, then, at least theoretically, the accuracy of pattern stabilization could become independent of contrast and other pattern parameters, which could explain results like those shown in Figure 4.

Implementation of such a mechanism in real nervous systems would require that the gain in the optomotor control system be variable. A simple experiment, illustrated in Figure 5, shows that the gain of the optomotor response is, indeed, not at all constant. When a fly views a pattern of horizontal stripes moved upwards or downwards, the velocity with which it turns its head upwards or downwards varies depending on the stimulation conditions. When the pattern repeatedly moves upwards for 4 s, and then downwards for 4 s (Fig. 5B, stimulus *a*), the head follows the changes in direction relatively slowly; i.e., the gain is small. When, however, the pattern moves, at the same speed, only for 1.5 s in each direction, pausing for about 2.5 s before each change in direction (Fig. 5B, stimulus *b*), the induced head movement is considerably more rapid; the gain is about four times larger than before. This experiment corresponds to an open-loop situation, because the pattern velocity is much higher (ca. 80°/s) than is the velocity of head turning (ca. 10°/s).

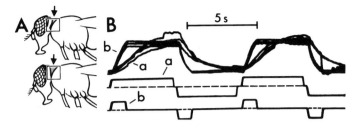

Figure 5. Optomotor roll response of *Musca*. The fly's head movements in response to a striped pattern ($\lambda = 10°$; $w = 80°/s$) moving up and down in front of the fly's head are shown. (A) Alignment of the aperture (square, marked by arrow) for monitoring head movements. The light flux through the rectangular aperture, indicating changes of head position, was measured using a photomultiplier. Red light, invisible to the fly, was used. (B) Time-course of stimuli *a* and *b* (*lower tracks*) and of the fly's corresponding responses (*upper tracks*). Three superimposed episodes of 4.5 s duration are shown, each from long lasting stimulus presentations of type *a* and *b*, respectively. In stimulus *b*, movement from ventral to dorsal (upwards in the track) and from dorsal to ventral (downwards in the track) was separated by a time interval in which no motion occurred. Modified from Kirschfeld (1989).

Two questions arise here: (i) does the gain change not only in the case of this type of head movement, but also in the case of rotation of the whole body about the vertical axis (yaw response), which is particularly important for course control, and, mainly, (ii) which parameters control the gain, i.e., how is the gain value set in closed-loop conditions?

Automatic gain control by feedback oscillations

Analyses of the results of optomotor studies on various arthropods have led to the concept of *Automatic gain control by feedback oscillations*. According to this hypothesis, within the optomotor control system, the gain increases spontaneously until the system, under the influence of noise signals, begins to oscillate. The oscillation, a consequence of the closed loop, is used to prevent a further increase in gain. Thus, the larger the oscillation amplitude, the more the gain is reduced. With the parameters adjusted suitably, the system is maintained in the state shown in Figure 6A for gain $k_1 = 5$. In other words, the system, although already tending to oscillate, remains stable because the oscillations induced by a sudden disturbance die out spontaneously.

The model calculation of Figure 6A was done using the model shown in Figure 1A, i.e., without the implementation of an automatic gain control mechanism. Therefore, increasing the gain k_1 to a value of 10 leads to instability. The model calculation in Figure 6B, on the other hand, shows the improvement introduced by automatic gain control. Here, when a gain factor is suddenly increased so much that instability ought to result, the

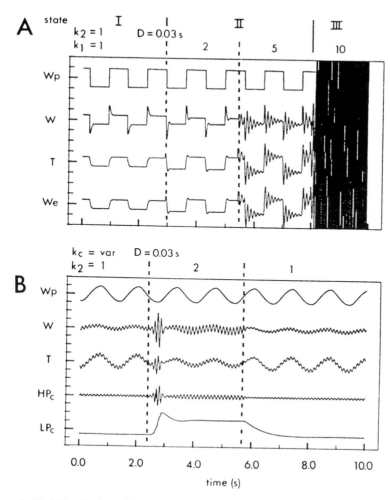

Figure 6. Modelling the large-field optomotor response. (A) Model calculations based on the linear control system shown in Figure 1 A. Pattern velocity w_p, slip speed w, torque T and eye velocity w_e are plotted as a function of time. The time delay D (see Fig. 1) was kept constant at 0.03 s; k_2 was kept constant at a value 1, whereas k_1 was varied. During Episode I, k_1 was 1, during Episode II, k_1 was first 2, then 5; during Episode III, k_1 was 10. For $k_1 > 6$, the system becomes unstable. (B) Model calculations based on the nonlinear control system with gain control shown in Figure 1 B. In addition to w_p, w and T, the outputs of HP_c and LP_c (see Fig. 1B) are shown. Here, k_2 was varied, in order to simulate the response of the system to changes in parameters of the moving pattern. k_c is the gain which is controlled by the filter C, Figure 1 B. When the model calculations are done without considering gain control and with k_c constant, the system becomes unstable for $k_2 = 2$. From Kirschfeld (1991).

gain control mechanism comes into play. As shown, in this case the gain k_c of element G_c (Fig. 1B) is reduced automatically, so that stability is preserved and the slip signal, i.e., the accuracy of stabilization, remains approximately constant. There are oscillations superimposed on both the slip signal and the torque exerted by the animal, but these are functionally negligible if they are of small enough amplitude, or of such high frequency that they are beyond the range relevant to the natural function of the system (Kirschfeld, 1991).

Experimental evidence for gain control in the optomotor system

The idea that gain control is, indeed, implemented in the optomotor system of at least some arthropods is supported by the following findings.

(i) Under closed-loop conditions, the crustacean *Leptograpsus* (Sandeman, 1978), as well as the fly *Drosophila* (Fig. 7A, open circles; Fig. 7B), exhibit high-frequency oscillations in the slip signal.

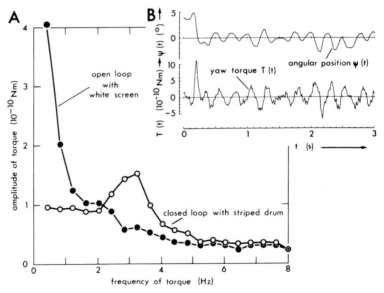

Figure 7. (A) Fourier spectra of "noise" generated by flying *Drosophila* suspended from a torque meter. The spectra indicate the amplitudes of yaw-torque fluctuations at various frequencies. Open-loop experiment (filled circles) was conducted using a white screen, and closed-loop experiment (open circles) using a striped drum. Under closed-loop conditions, low frequencies in the torque are suppressed, whereas frequencies close to 3 Hz are enhanced. This behaviour is characteristic of a feedback loop of the type shown in Figure 1, in which the gain is so high that the loop is on the verge of instability (see Fig. 6A at $k_1 = 5$). (B) High-resolution traces of pattern position ψ and yaw torque T in a *Drosophila* flying in a flight simulator under closed-loop conditions (see Fig. 11A). From this type of data, the Fourier spectra in A (circles) have been determined. Redrawn from Wolf and Heisenberg (1990) and Kirschfeld (1991).

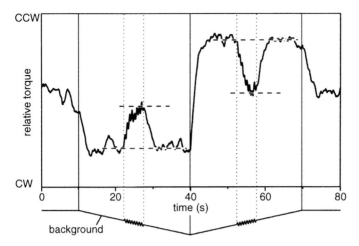

Figure 8. Attenuation of the large-field optomotor response under open-loop conditions by superimposing high-frequency oscillations on the pattern movement. Torque produced of tethered flying female *Musca* (upper trace). A cylinder with a noise pattern (pixels black and white, size 2.5°) was first stationary for 10 s, then rotated about the fly for 30 s (at 40°/s) to the right (clockwise direction, cw), then for 30 s (at the same speed) to the left (counterclockwise, ccw), and was then stationary again (lower trace). For 5 s during rotation in each direction, a sinusoidally oscillating movement was superimposed on the uniform movement of the cylinder (frequency 3.1 Hz, amplitude ±8°). The oscillation weakens the response (dashed lines), i.e., the gain is reduced. The data are the average of 107 experiments performed with the paradigm described. 10 flies have been used for the experiments. Torque strength (ordinate) is given in relative units. Torque zero (not indicated) roughly corresponds to the response to the stationary background pattern (time otolos; 70 to 80 s). From Kirschfeld et al., in preparation.

(ii) Even under open-loop conditions, broad-band noise appears in the torque signal (Fig. 7A, filled circles), and with the loop closed it changes in the manner to be expected from a feedback system on the verge of instability (Fig. 7A, empty circles).

(iii) In experiments in a flight simulator (see experimental set-up in Fig. 11A), when the gain of the link between a fly's torque and the servo-motor that moves the pattern is reduced, the fly's response increases, so that the *overall gain* stays approximately constant (Kirschfeld, 1991). At least the data obtained for *Drosophila* (Wolf and Heisenberg, 1990) can be thus interpreted.

One prediction of the hypothesis of feed-back-based gain control pro-posed above is that, under open-loop conditions, the large-field optomotor response ought to be attenuated if relatively high-frequency oscillations (~3 Hz) are superimposed on the pattern movement. Figure 8 illustrates a case in which this prediction is, indeed, fulfilled. Movement detectors of the type shown in Figure 2, on the other hand, do not exhibit such a behaviour (see Fig. 4 in Kirschfeld, 1991). Thus, there is a large body of evidence to support the hypothesis.

These results mainly show that optomotor behaviour in closed loop conditions cannot be predicted from results measured under open loop. Only under closed loop, eye oscillations in the range of 3 Hz are generated by the animal, and these oscillations dramatically modify the strength of the response (Figs. 7 and 8). The following considerations illustrate the function of the postulated feedback loop with gain control in more detail.

Theoretical considerations and model calculations

Let us assume the optical panorama to be of a relatively high contrast, so that the angular velocity w_e of the eye (in steady state) is 0.9 of the angular velocity w_p of the pattern. This corresponds to a slip speed $w = 0.1$ (i.e., 10%) of w_p. These values are calculated according to Equation (2) if k_1 and k_2, the gains of the elements G_1 and G_2, respectively, are both arbitrarily set to 3. When the contrast of the pattern is reduced, so that the signal of the movement detectors is also reduced say, to 1/3, than w_e, according to Equation (2), will be reduced to 0.75 of w_p (75%), and the slip signal w will increase to 0.25 of w_p (25%). Under these conditions, the overall gain in the loop $(k_1 \cdot k_2)$ will be reduced to 1/3, and therefore the amplitude of the eye oscillations (Fig. 7B) will also be reduced. If, in the feedback loop, a gain control element G_c is included, as in Figure 1B, then, under the same conditions, the gain of this element will be increased. As a consequence, retinal stabilization will again improve, i.e., the slip signal w will become smaller. These considerations show how the detrimental effect for visual stabilisation due to reduced pattern contrast could be compensated for by an automatic gain control mechanism in the optomotor loop. How this compensation for the reduced contrast looks like quantitatively depends upon the relationship between the gain G_c and the amplitude of the oscillations. This relationship still remains to be determined in detail in experiments of the type illustrated in Figure 8. The model calculation of Figure 6B illustrates a quantitative example, in which the parameters, however, have been set arbitrarily, just to illustrate the principle.

The fact that high-frequency oscillations reduce the gain (Fig. 8) was not known when Equation (1) was derived. Hence, Equation (1) needs to be modified to take this finding into account. The exact anatomical location of the site at which the value of the gain is adjusted also remains to be determined. It seems possible that the gain control mechanism, causing the decrease in torque response to oscillating stimuli (Fig. 8), is located in giant neurons of the lobula plate. These neurons are target cells of EMDs. Ca^{2+}-ions accumulate in the dendrites of these giant neurons during application of motion stimuli to the eyes (Borst and Egelhaaf, 1992). Ca^{2+}-ions are capable of opening particular potassium channels with the consequence that neurons are hyperpolarized, i.e., inhibited. This effect corresponds to a reduction in gain, and, if realized, reflects not a property of the EMDs, but rather one of a later stage of neural processing.

Image stabilization during voluntary flight

When flies of the genus *Musca* are flying freely along a curved path, they execute brief, rapid turns (saccades) separated by relatively long periods of flight with little or no rotation of the body (Fig. 9A). The stabilization of body angle between the saccades could be brought about by visual signals (large-field optomotor response), as well as by mechanical signals (coming, for example, from the halteres). Under physiological conditions, both of these inputs presumably make a contribution. Saccades very similar to those observed in free-flying flies are also discernible in the head movements of pigeons flying on a curved path (Fig. 9B).

These findings suggest that the large-field optomotor response, usually considered to serve for stabilizing the retinal image (and hence the course of locomotion) during involuntary displacements of the animal (caused, for example, by air turbulences), is active even during voluntary locomotion. When a fly tries to follow a curved path, the image motion perceived from the panorama is compensated by the optomotor response. Therefore, the animal is only capable of flying straight on. After a short time, however, it seems to "realize" that, by having done so, it has deviated from its intended course of locomotion, and therefore it makes a saccadic turn (Fig. 9A). During this turn the optomotor response is suppressed (Heisenberg and

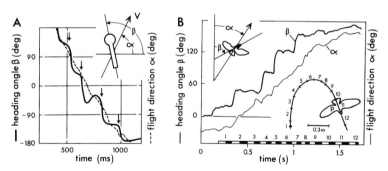

Figure 9. Saccadic compensation for the rotational component of image motion. Insets show definition of angles α and β. (A) Cruising flight of a female fly (*Musca*). Time course of horizontal flight direction (broken line), orientation of the animal's longitudinal body axis (continous line). Although the flight path is circular, the changes of long-axis orientation are step-like (arrows). The head of the fly is represented by a circle, the longitudinal body axis by a bar. Modified from Wagner (1986). (B) Angular movement of a pigeon's head and trunk in the horizontal plane during slow flight on a curve to the right. As in (A), motion of the head is saccadic (thick line), although the flight path is circular (thin line). Redrawn from Bilo (1992), both figures from Kirschfeld (1994).

Wolf, 1984; Egelhaaf, 1985, 1990; Land, 1992). In animals capable of rapid eye or head movements, the situation is simpler: the body can follow a curved path and just the eyes or head exhibit saccades (Fig. 9B). We shall return to this point further below.

Tracking of small objects in front of various backgrounds

Saccadic turns similar to those described above for the large-field opto-motor response are also involved in the task of tracking small moving objects (small-field optomotor response). We here propose that similar mechanisms are involved in these two types of behaviour.

Male flies, as well as females, pursue targets in the air. They do so regardless of whether the background against which the target moves is homogenous or structured (Wagner, 1986). In the latter case, the animal is faced with the problem of stabilizing either the image of the background or that of the target on the eye.

The effect of a structured background on the tracking behaviour has been most directly demonstrated in another insect, the praying mantis (Rossel, 1980), and recently also in a primate species, *Macaca mulatta* (Ilg et al., 1992), (Fig. 10). In the case of the mantis, the measured response was turning of the head. When a small object is moving in front of a homogenous

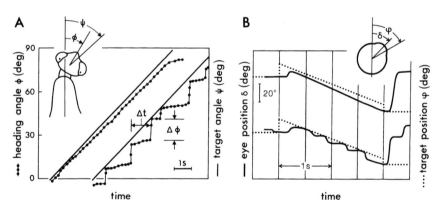

Figure 10. Visual tracking of small moving objects. (A) Head movements of praying mantis responding to a target 10° in diameter, presented at a distance of 12 cm from its head. The target was moved in front of either an unstructured (left-hand pair of curves) or a structured background (right-hand pair of curves). Target angle ψ (continuous line); heading angle ϕ (line through dots). From Rossel (1980). (B) Eye movements of a Rhesus monkey (*Macaca mulatta*) responding to a small white disk that was moved in front of either a dark, homogeneous background (upper pair of curves) or a randomly dotted pattern (bottom pair of curves). In each case, the target was first moved rapidly 10° to the left (upward deflection) and then to the right at 15°/s (dotted line). The continuous line shows the change in angular position of the eye. After Ilg et al. (1992), both figures from Kirschfeld (1994).

(unstructured) background, it is tracked with a smooth head movement (Fig. 10A, left). If the background is textured, however, the system switches to saccadic tracking (Fig. 10A, right). The Rhesus monkey, *Macaca mulatta*, behaves similarly (Fig. 10B). In this case, the measured response was movement of the eyes. During visual fixation, the monkey's eyes follow the object smoothly when the background is unstructured, but by saccadic tracking, as in the mantis, when the background is structured.

Why does a homogenous background elicit smooth tracking, whereas a structured background gives rise to saccadic turns? The following sections provide an answer that might contribute to our understanding of the mechanisms underlying both course control and tracking.

Object detection in flies is not due to relative movement between object and background

Many of the experiments on flies involving tracking or fixation responses have employed the following paradigm. The flying animal was attached to a torque-measuring system (Fig. 11A, open-loop). When a narrow, vertical

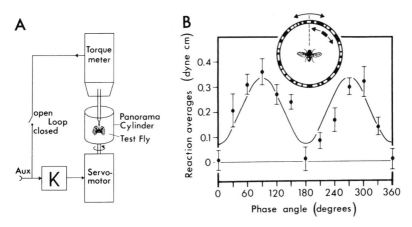

Figure 11. Dependence of object detection on the phase angle between oscillation of the object and that of the background. (A) Schematic view of a flight simulator. The insect is flying tethered in the centre of a rotating, patterned drum. The strength of the torque generated by the animal as a response to the moving pattern is measured by a torque-meter and can be used to rotate a cylinder bearing the optical panorama, i.e., stimulation is under closed-loop conditions. K indicates the gain with which the torque is converted to angular velocity. When the switch connecting the torque-meter with the drum-motion control is open, the torque generated by the fly can be measured under open-loop conditions. (B) Mean torque generated by a fly at the torque-meter under open-loop conditions when a 3°-wide vertical stripe positioned 30° to the right of the long axis of the fly's body is moved sinusoidally over ±1° at a frequency of 2.5 Hz (inset). The background (random dot pattern) is moved simultaneously with the stripe, also at 2.5 Hz over ±1°, but with various phase discrepancies between pattern motion and figure motion (abscissa). From Kirschfeld (1994), modified from Reichardt and Poggio (1979).

stripe (the object) positioned 30° to the right or left of the animals long body axis is oscillated sinusoidally in front of a stationary background, the animal continually generates torque in the direction in which the stripe is moving. When the background is also moved, with the same amplitude and frequency, detection of the bar, as expressed by the fly's following it, depends on the phase relation between object and background (Fig. 11B). Following is most conspicuous when the oscillation of the object is 90° or 270° out of phase with respect to the motion of the background. The finding that the object is not detected when its motion is in exact counterphase (180°) to that of the background suggests that, under these experimental conditions, object detection is not due to the relative motion between object and background.

The possible mechanism underlying object-ground discrimination

Two mechanisms are conceivable to explain the detection and tracking of an object moving against a patterned background. One is based on findings in human vision showing that the visual system can recognize an object as being distinct from a background not only if it differs from the background in some property such as contrast or colour, but also if it moves relatively to the background (Metzger, 1975). The fly's performance in the task of figure-ground discrimination described above has been interpreted analogously, i.e., by the use of relative movement between object and background (Reichardt and Poggio, 1979; Reichardt et al., 1983; Egelhaaf, 1985; Egelhaaf et al., 1988). Similar conclusions were drawn from studies on object-ground discrimination in freely flying honeybees (see M. Lehrer, this volume).

Flies, however, as we have seen, do not respond to a moving object when its direction of motion is exactly opposed to that of the background (180° out of phase; see Fig. 11B), although, in this case, the amount of relative motion is maximal. This and further experimental results gave rise to an alternative hypothesis (Kirschfeld, 1994), namely, that the crucial parameter for detecting and tracking a moving object is not the relative motion between the moving object and the moving background, but rather the relative motion between the moving object and the eye (see also Lehrer and Srinivasan, 1992). In other words, the object can only be detected when there is no motion of the background on the eye, i.e., when the image of the background is stabilized.

The basic considerations underlying this hypothesis are as follows. When the fly's eye moves relative to a patterned background on a circular course, for example in a curve, all of the movement detectors are activated, and thus very complicated calculations on the part of the nervous system would be needed in order to detect an object that moves at the same time against the same background. A much simpler strategy to detect the moving object would be to compensate for the rotational component of the movement to a great extent by performing saccades, so as to stabilize the image of the background on the eye (see Fig. 9A). As a result, the overall activity of the visual system is dramatically reduced at the most peripheral level, the retina. Only objects that move with respect to the *stabilized background* induce a strong response of the motion detection system, and this response can be isolated from the response to the remaining background motion by, for example, a simple threshold mechanism.

A neural model for the interactions between the large-field and the small-field optomotor systems

Figure 12 is a diagram of the possible neural substrate underlying the mechanism outlined above. It is based on the following findings. Analysis of the fly visual system has revealed, as already mentioned in the Introduction, that the optomotor system consists of two subsystems: a *large-field compensatory optomotor system* (LF) and a *small-field tracking system* (SF). It is generally agreed that the LF system enables the fly to execute flight manoeuvres that compensate for passive displacement (drift), whereas the SF system is employed for the fixation of small moving objects. The two systems in Figure 12 (in black and white, respectively) are based on separate neuronal networks. Anatomical and electrophysiological studies have demonstrated the existence of two groups of large interneurons, one with characteristics that would be required for the LF system, and the other for the SF system. In Figure 12, to simplify the situation, the left and right eye are considered to have no overlap in their visual fields. The optical panorama is sampled by an array of local elementary movement detectors (EMDs) of the correlation type. The large-field giant neurons (LFGN) of each eye add up the signals from EMDs that survey a large part of the visual field of that eye, whereas the small-field giant neurons (SFGN) add up EMD signals most likely covering a smaller angular region.

So far, this diagram corresponds to the organizational scheme developed by Egelhaaf et al. (1988) and Egelhaaf and Borst (1993). In their model, however, the LF and SF systems are each provided with an output filter, and differences in the low-pass characteristics of the two filters (the LF system being taken to be slower than the SF system) are considered to bring about a separation of the two systems in the time domain.

The model of Figure 12 does not include these filters. Instead, here both systems are considered to be capable of high temporal resolution, for reasons discussed in more detail elsewhere (Kirschfeld, 1994). The LF and SF systems in Figure 12 are separated functionally from each other by rapid inhibition of the SF system, exerted by the LF system, via the inhibition elements INH. The assumption of this inhibitory interaction is consistent with several experimental findings (e.g., Egelhaaf, 1985, 1990), its consequence is also illustrated in Figure 13.

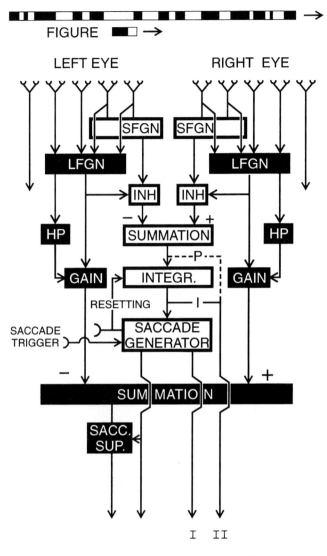

Figure 12. Functional diagram of the interactions among various elements in the large-field (LF) and small-field (SF) systems (see text) governing the optomotor responses of Diptera. Black boxes denote elements of the LF system; white boxes components of the SF system; LFGN, giant neurons of the LF system; SFGN, giant neurons of the SF system; SUMMATION summing elements; HP, high-pass filter; GAIN, gain-control element; SACC.SUP, element in which the signal of the LF system is suppressed during a saccade; INH, element through which the SF system is inhibited by the LF system; INTEGR., integrator; SACCADE GENERATOR, element that generates the signals that trigger a saccade. From Kirschfeld (1994).

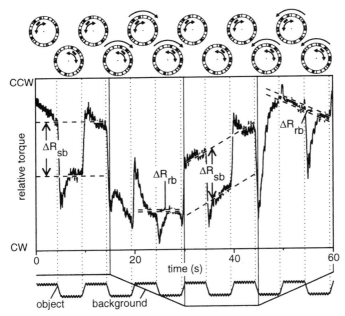

Figure 13. Experimental evidence for inhibitory interactions between the SF and LF systems postulated in Figure 12. Shown is the mean torque generated by female Musca viewing various combinations of movement of object and background (monitor: lower traces; pictograms on top of the figure) (average from 237 sweeps of 12 flies) in open-loop conditions. If only the object (width 3°) is oscillated, first 30° to the left of the animal's long axis and then 30° to the right (in each case at 2 Hz with amplitude ±5°), the fly tries with considerable force to turn toward the object (ccw = counter clockwise, cw = clockwise). The strength of the torque responses (ordinate) are given in relative units. These responses, exept for transients, are largely suppressed when the background is moved uniformly (at 4°/s) to either the right or the left. From K. Kirschfeld et al., in preparation.

The sign convention in Figure 12 is as follows. Positive signals at the LF and SF output correspond to a torque to the right, negative signals to a torque to the left. Consequently, motion of a pattern from front to back of the animal, in the left eye, leads to a negative signal as indicated at the input of the summation elements of the LF and SF systems, respectively. Motion from front to back in the right eye leads to positive signals.

The model sown in Figure 12 is based solely on the detection of the *velocity* of objects moving against a patterned background. This model is sufficient to explain the experimental results described in this chapter. It must be noted, however, that control systems based on *position signals* generated by targets are important as well. The proporties of such signals and their contribution to the animal's performance in orientational tasks are discussed in detail by Land (1992).

How the model is thought to function

The way by which the model shown in Figure 12 is expected to function in several specific theoretical as well as experimentally tested situations will be now discussed in more detail.

Let us first consider a case in which a small object viewed, say, by the left eye, moves to the right in front of an *unstructured ground*. In this case, the small-field giant neurons (SFGN) on the left deliver a positive signal to the summation unit, whereas those on the right are not stimulated and deliver no signal. The summation unit gives a net small-field output (II in Fig. 12) that is positive. This produces a torque to the right, allowing smooth pursuit of the figure. Because the ground is unstructured and the small object hardly activates the large field system (LF), there is no inhibition of the small field system (SF).

In the next hypothetical example, a small object moves again to the right in front of a *textured* background. As in the former case, the SF system will produce smooth pursuit towards the right. But now the "brakes" are applied by the LF system. As soon as the pursuit turn to the right begins, there is a relative rotation of the panorama to the left, which excites the LF system and counteracts rightwards turning for two reasons: (i) the sign of the LF signal is opposite to that of the SF signal induced by the object, and (ii) the LF system inhibits the SF system in the INH elements (see Fig. 12). As the object continues to move to the right, while the eye is largely fixed, the movement elicits a constant response of the SFGN, a steady signal that feeds into the integrator of the SF system. After a certain time, by a still unspecified signal fed into the saccade generator, a saccade is triggered of magnitude $\Delta\phi$. $\Delta\phi$, to a first approximation, is proportional to the duration of the interval Δt between the saccades. (For definition of Δt and $\Delta\phi$, see Fig. 10A).

One mechanism by which the value of $\Delta\phi$ may be determined could be based on position detectors: they could determine the angular position of the target at the end of the saccade and, from this, the necessary magnitude $\Delta\phi$ of the saccade to be triggered. However, as Rossel (1980) has shown, there is at least one further determinant of saccade magnitude, namely the *angular velocity*, which improves the performance of the system. In this case, an element is required that stores the information about velocity during the time Δt, which, in Figure 12, is represented by the "integrator".

Experimental evidence supporting the model

The system illustrated in Figure 12 can thus initiate three forms of activity: (i) an optomotor large-field reaction; (ii) saccadic tracking of small objects in front of a structured background, and (iii) smooth pursuit of a target in front of a homogeneous background. In the latter case, the LFGNs are not activated and can therefore not counteract the continuous rotation characteristic for smooth pursuit. It is only the saccade generator that need be inactivated, as discussed above.

The model (Fig. 12) can explain the unexpected experimental finding that a small object is not detected if object and ground oscillate in counter-phase (Fig. 11B). It actually follows directly from the model that large-field inhibition renders the system more or less blind to the motion of small objects as long as there is significant motion of the eye relative to the background. The finding that the object can be detected when the phase lag between it and the background is 90° is, again, accounted for by the model. The object is detected because the background is stationary before it changes direction, whereas the object is still in motion.

Suppression of the large-field optomotor response during saccadic motion

A closer look at the model shown in Figure 12 reveals that three conditions must be met in order for this system to function.

(i) The input into the integrator of the SF system must be suppressed during every saccade, because, otherwise, signals deriving not only from the moving object, but also from the background, would contribute to the size of the saccade. This rapid-onset inhibition occurs in the element INH (Fig. 12). It has been argued that during a saccade the background could be moving too fast across the eye to be effective (Land, 1992). This, however, can be excluded at least for the input elements of the optomotor system in *Musca* and *Calliphora*. Here the motion-sensitive giant neurons respond strongly and rapidly to the onset of movement in the range of 200–2000°/s (K. Kirscheld et al., unpublished data), typical for saccadic rotation (Wagner, 1986). Therefore, responses of the SF system to saccadic rotation have to be suppressed. Such a suppression need not be realized by an inhibitory interaction like that proposed in the model (Fig. 12), but could also be brought about by an efferent signal from the motor system (corollary discharge). We included the inhibition by the LF system in Figure 12, because there is experimental evidence for its existence (Egelhaaf, 1985, 1990). Furthermore, such an inhibitory mechanism has the advantage that it would work even if body rotation is induced *passively* (for example, by air turbulence).

(ii) The value in the integrator must be reset to zero by each saccade, to guarantee that only the signals entering during the time Δt are effective.

(iii) To ensure rapid rotation during a saccade, the LF output that would otherwise counteract rotation must be temporarily suppressed (SACC.SUP in Fig. 12). The occurrence of this saccadic suppression has already been demonstrated directly in behavioural experiments on *Drosophila* (Heisenberg and Wolf, 1984).

The model (Fig. 12) does not explain in detail how the smooth pursuit of a small object can occur. A simple assumption is that, in this situation (in which the background is homogeneous), because the rotation is not counteracted by the LF system, the signal produced by the integrator in the SF system can be used to form an integrating control system (channel I in Fig. 12). It would also be possible to use, instead of the output from the integrator, the input signal to the integrator (channel indicated by P in Fig. 12), so that a proportional control system would be responsible for smooth tracking. These various possible functional states are unknown and have therefore not been further specified in the model of Figure 12.

Another look at object-ground discrimination

The model presented in Figure 12 immediately inspires an experiment which allows to validate further one of the two hypotheses on how objects are detected by flies: by relative movement between object and background, or by stabilizing the background. The paradigm is as follows: A stripe is oscillated 30° to the left as in the experiment shown in Figure 11B in front of a stationary, textured background. After 5 s, the oscillating stripe is moved 30° to the right, and 5 s later back to the left. As can be seen in Figure 13, the fly's torque is to the left and the right, as expected, with a difference ΔR_{sb} which is a measure of the detectability of the stripe. At this time, the previously stationary background was made to rotate slowly (at 4°/s) around the animal, thus imitating flight on a slightly curved path, whereas the stripe continued to oscillate as before, alternately to the right or the left of the animal's longitudinal axis. It can be seen that there are transients in torque as the stripe is displaced from 30° left to 30° right, and back, probably due to the response of the large-field system to object motion. The difference in the steady state responses (ΔR_{rb}), after the transients have died out, to the oscillating stripes during rotation of the background is, however, almost zero (compare vertical separation is again stationary, ΔR_{sb} is strong again (times 30 to 45 s).

Because the stripe was oscillated with amplitude $\pm 5°$ at 2 Hz, the absolute angular velocity varied between 0 and 40°/s. Because the background rotated only slowly (with 4°/s), the amount of relative movement between object and background remained practically unchanged, regardless of whether the background was rotating (relative motion 36°/s and 44°/s, depending on the direction of oscillation) or was stationary (relative motion 40°/s). The results of this experiment (Fig. 13) show that the fly, during rotation of the eye reltive to the background is, as predicted from the model (Fig. 12), practically blind for the motion of small objects. They favour the hypothesis that object detection is mediated by stabilizing the background. Relative motion between object and background, at least under the experimental paradigm used here, is not a sufficient condition for object detection as defined in Figure 2B.

Based on these arguments, the two subsystems addressed in this chapter, SF and LF, are apparently not a mechanism meant to solve the problem of figure-ground discrimination via relative motion between figure and ground, but rather one meant to enable saccadic tracking.

Conclusion

Cues derived from image motion are involved in several orientational tasks, addressed in further chapters included in this volume (see T.S. Collett and J. Zeil; M.V. Srinivasan and S.W. Zhang; M. Lehrer). The present chapter focuses on two tasks based on motion detection that have one aspect in common, namely the use of mechanisms for image stabilization. We have seen that stabilizing the large-field image on the retina enables the animal to maintain its desired course of locomotion, and that tracking of a small moving target, also accomplished by stabilizing the image of the surround, will enable the animal to precisely follow such an object either with eyes, head or the whole body.

The arguments put forward in the present chapter further suggest that detection of large-field image motion requires a different neural network than that required for detecting small-field image motion, but that the parallel processing of these two types of information includes and also requires interactions between the two neural pathways. Thus, large-field and small-field motion detection act together in guiding the animal to its goal.

We have seen, in addition, that the mechanisms involved in these two types of oriented behaviour are shared by arthropods and vertebrates, suggesting that these mechanisms have either developed independently several times in the course of evolution, or else very early in phylogeny. Indeed, every animal must cope with the problems described in this chapter.

Acknowledgements
I thank Stephanie Schaerer, John Thorson and Christian Wehrhahn for discussion, M.A. Biederman-Thorson for translation. The manuscript has been substantially improved by including suggestions made by Miriam Lehrer. Experiments described in Figures 8 and 13 were performed by Claudia Holt.

References

Bilo, D. (1992) Optocollic reflexes and neck flexion-related activity of flight control muscles in the airflow-stimulated pigeon. *In*: A. Berthoz, W. Graf and P. Vidal (eds): *The Head-Neck Sensory-Motor System.* Oxford University Press, New York, Oxford, pp. 96–100.

Borst, A. and Egelhaaf, M. (1987) Temporal modulation of luminance adapts time constant of fly movement detectors. *Biol. Cybern.* 56:209–215.

Borst, A. and Egelhaaf, M. (1992) *In vivo* imaging of calcium accumulation in fly interneurons as elicited by visual motion stimulation. *Proc. Natl. Acad. Sci.* USA 89:4139–4143.

Briggs, B.H., Phillips, G.J. and Shinn, D.H. (1950) The analysis of observations on spaced receivers of the fading of radio signals. *Proc. Phys. Soc.* 63:106–121.

Buchner, E. (1984) Behavioural analysis of spatial vision in insects. *In*: M.A. Ali (ed.): *Photoreception and Vision in Invertebrates*. Plenum Press, New York, London, pp 561–621.

Collett, T.S., Nalbach, H.O. and Wagner, H. (1993) Visual stabilization in arthropods. *In*: F.A. Miles and J. Wallman (eds): *Visual Motion and its Role in the Stabilization of Gaze*. Elsevier Science Publishers B.V., Amsterdam, pp 239–264.

David, C.T. (1982) Compensation for height in the control of groundspeed by *Drosophila* in a new, "Barber's Pole" wind tunnel. *J. Comp. Physiol*. 147:485–493.

Eckert, H. and Hamdorf, K. (1981) The contrast frequency dependence: A criterion for judging the non-participation of neurones in the control of behavioural responses. *J. Comp. Physiol*. 145:241–247.

Egelhaaf, M. (1985) On the neuronal basis of figure-ground discrimination by relative motion in the visual system of the fly. III. Possible input circuitries and behavioural significance of the FD-cells. *Biol. Cybern*. 52:267–280.

Egelhaaf, M. (1990) Spatial interactions in the fly visual system leading to selectivity for small-field motion. *Naturwissens*. 77:182–185.

Egelhaaf, M. and Borst, A. (1989) Transient and steady-state response properties of movement detectors. *J. Opt. Soc. Am*. A 6:116–127.

Egelhaaf, M. and Borst, A. (1993) Motion computation and visual orientation in flies. *Comp. Biochem. Physiol*. 104A:659–673.

Egelhaaf, M., Hausen, K., Reichardt, W. and Wehrhahn, C. (1988) Visual course control in flies relies on neuronal computation of object and background motion. *TINS* 8:351–358.

Egelhaaf, M., Borst, A. and Reichardt, W. (1989) Computational structure of a biological motion detection system as revealed by local detector analysis in the fly's nervous system. *J. Opt. Soc. Am*. A 6:1070–1087.

Götz, K.G. (1964) Optomotorische Untersuchungen des visuellen Systems einiger Augenmutanten der Fruchtfliege *Drosophila*. *Kybernetik* 2:77–92.

Graaf, B., de, Wertheim, A.H., Bles, W. and Kremers, J. (1990) Angular velocity, not temporal frequency, determines circular vection. *Vision Res*. 30:637–646.

Hassenstein, B. und Reichardt, W. (1956) Systemtheoretische Analyse der Zeit-, Reihenfolgen- und Vorzeichenauswertung bei der Bewegungsperzeption des Rüsselkäfers *Clorophanus*. *Z. Naturf*. 11b:513–524.

Hausen, K. (1984) The lobula-complex of the fly: structure, function and significance in visual behaviour. *In*: M.A. Ali (ed.): *Photoreception and Vision in Invertebrates*. Plenum Press, New York, pp 523–559.

Hausen, K. and Wehrhahn, C. (1990) Neural circuits mediating visual flight control in flies. II. Separation of two control systems by microsurgical brain lesions. *J. Neurosci*. 10:351–360.

Heisenberg, M. and Wolf, R. (1984) *Vision in Drosophila*. Springer-Verlag, Berlin, Heidelberg, New York.

Ilg, U.J., Brenner, F., Thiele, A. and Hoffmann, K.P. (1992) Neuronal coding of retinal slip during smooth pursuit eye movements. *Eur. J. Neurosci*. (Suppl.) 5:253.

Kelly, D.H. (1979) Motion and vision. II. Stabilized spatio-temporal threshold surface. *J. Opt. Soc. Am*. 69:1340–1349.

Kirschfeld, K. (1989) Automatic gain control in movement detection of the fly. *Naturwissens*. 76:378–380.

Kirschfeld, K. (1991) An optomotor control system with automatic compensation for contrast and texture. *Proc. R. Soc. Lond*. B 246:261–268.

Kirschfeld, K. (1994) Tracking of small objects in front of a textured background by insects and vertebrates: phenomena and neuronal basis. *Biol. Cybern*. 70:407–415.

Kunze, P. (1961) Untersuchung des Bewegungssehens fixiert fliegender Bienen. *Z. vergl. Physiol*. 44:656–684.

Land, M. (1992) Visual tracking and pursuit: humans and arthropods compared. *J. Insect Physiol*. 38:939–951.

Lehrer, M. and Srinivasan, M.V. (1992) Freely flying bees discriminate between stationary and moving objects: Performance and possible mechanisms. *J. Comp. Physiol*. A 171:457–467.

Metzger, W. (1975) *Gesetze des Sehens*. Waldemar Kramer Verlag, Frankfurt/M.

Miles, F.A. and Kawano, K. (1987) Visual stabilization of the eyes. *Trends Neurosci*. 10:153–158.

Radl, E. (1903) *Untersuchungen über den Phototropismus der Tiere*. W. Engelmann, Leipzig.

Reichardt, W. and Poggio, T. (1979) Figure-ground discrimination by the relative movement in the visual system of the fly. (Part I: Experimental results). *Biol. Cybern.* 35:81–100.

Reichardt, W. and Varjú, D. (1959) Übertragungseigenschaften im Auswertesystem für das Bewegungssehen (Folgerungen aus Experimenten an dem Rüsselkäfer *Chlorophanus viridis*). *Z. Naturf.* 14b:674–689.

Reichardt, W., Poggio, T. and Hausen, K. (1983) Figure-ground discrimination by relative movement in the visual system of the fly. (Part II: Towards the neural circuitry). *Biol. Cybern.* (Suppl.) 46:1–30.

Rossel, S. (1980) Foveal fixation and tracking in the praying mantis. *J. Comp. Physiol.* 139:307–331.

Ruyter van Steveninck, R.R., de, Zaagman, W.H. and Mastebroek, H.A.K. (1986) Adaptation of transient responses of a movement-sensitive neuron in the visual system of the blowfly *Calliphora erythrocephala*. *Biol. Cyern.* 53:451–463.

Sandeman, D.C. (1978) Eye-scanning during walking in the crab *Leptograpsus variegalus*. *J. Comp. Physiol.* 124:249–257.

Schaerer, S., Feiler, R. and Kirschfeld, K. (1996) Object perception in goldfish. Proc. 24th Göttingen Neurobiology Conf, Vol. II. Thieme-Verlag, Stuttgart, New York, p 386.

Srinivasan, M.V., Lehrer, M., Kirchner, W.H. and Zhang, S.W. (1991) Range perception through apparent image speed in freely flying honeybees. *Vis Neurosci.* 6:519–535.

Türke, W. (1996) *Die Eigenschaften der Eingangselemente des akzessorisch optischen Systems der Taube* (Columba livia). PhD Thesis, Universität Tübingen.

Varjú, D (1959) Optomotorische Reaktionen auf die Bewegung periodischer Helligkeitsmuster (Anwendung der Systemtheorie auf Experimente am Rüsselkäfer *Chlorophanus viridis*). *Z. Naturf.* 14b:724–735.

Varjú, D. and Reichardt, W. (1967) Übertragungseigenschaften im Auswertesystem für das Bewegungssehen II (Folgerungen aus Experimenten an dem Rüsselkäfer *Chlorophanus viridis*). *Z. Naturf.* 22b, 12:1343–1351.

Wagner, H. (1986) Flight performance and visual control of flight of the freeflying housefly (*Musca domestica* L.) III. Interactions between angular movement induced by wide- and smallfield stimuli. *Phil. Trans. R. Soc. Lond.* (Biol) 312:581–595.

Wehrhahn, C. (1985) Visual guidance of flies during flight. *In*: G.A. Kerkut and L.I. Gilbert (eds): *Comprehensive Insect Physiology Biochemistry and Pharmacology*. Pergamon Press, Oxford, New York, pp 673–684.

Wolf, R. and Heisenberg, M. (1990) Visual control of straight flight in *Drosophila melanogaster*. *J. Comp. Physiol.* A 167:269–283.

Wolf-Oberhollenzer, F. and Kirschfeld, K. (1994) Motion sensitivity in the nucleus of the basal optic root of the pigeon. *J. Neurophysiol.* 71:1559–1573.

Orientation and Communication in Arthropods
ed. by M. Lehrer

Visual control of honeybee flight

M.V. Srinivasan and S.W. Zhang

Centre for Visual Science, Research School of Biological Sciences, Australian National University, P.O. Box 475, Canberra, ACT 2601, Australia

Summary. Recent research has uncovered a number of different visual cues which bees use for controlling and stabilising flight. Bees flying through a tunnel maintain equidistance to the flanking walls by balancing the speeds of the images of the two walls. This strategy enables them to negotiate narrow passages or to fly between obstacles. The speed of flight in the tunnel is controlled by holding constant the average image velocity as seen by the two eyes. This mechanism prevents potential collisions by ensuring that the bee slows down when it flies through a narrow passage. Bees landing on a horizontal surface hold constant the image velocity of the surface as they approach it, thus automatically ensuring that flight speed is close to zero at touchdown. The movement-sensitive mechanisms underlying these various behaviours differ qualitatively as well as quantitatively, from those that mediate the well-investigated optomotor response. Flight thus appears to be co-ordinated by a number of visuomotor systems acting in concert.

Introduction

A glance at a fly orchestrating a flawless landing on the rim of a teacup, or at a honeybee finding her way back home effortlessly after foraging for nectar several kilometres away, would convince even the most skeptical observer that many insects are excellent fliers and navigators. Many of the early studies of visual flight control in insects concentrated on the so-called "optomotor response". This is a reflex that helps the insect maintain a straight course by visually detecting and correcting unwanted deviations. More recent studies, carried out primarily on honeybees, have revealed a number of additional reflexes and responses that serve to co-ordinate flight. It now appears, for example, that bees use simple and elegant visual strategies for negotiating narrow gaps, for controlling flight speed, and for executing smooth landings. Here we describe some of these strategies, and attempt to elucidate the properties of the underlying motion-sensitive mechanisms.

Flying through the middle of a gap: The centering response

When a bee flies through a hole in a window, it tends to fly through its centre, balancing the distances to the left and right boundaries of the opening. Most insects, including the bee, possess very small interocular separations, and therefore they cannot rely on stereopsis to measure the distances of objects (Collett and Harkness, 1982; Horridge, 1987; Srinivasan, 1993; but see Rossell, 1983). Even if the bee possessed stereopsis, she could not use it to navigate through the hole because the left rim of the hole would be seen by the left eye alone, and the right rim by the right eye alone. How, then, does the bee gauge and balance the distances to the two rims?

One possibility is that she does not measure distances at all, but simply balances the speeds of image motion on the two eyes, as she flies through the opening. To investigate this possibility, we trained bees to enter an apparatus that offered a reward of sugar solution at the end of a tunnel 40 cm long, 20 cm high and 12 cm wide (Kirchner and Srinivasan, 1989). Each side wall carried a pattern consisting of a vertical black-and-white grating of period 5 cm (alternating black and white bars, each of width 2.5 cm; Fig. 1A). The grating on one wall could be moved horizontally at any desired speed, either towards the reward or away from it. After the bees had received several rewards with the gratings stationary, they were filmed from above as they flew along the tunnel. When both gratings were stationary, the bees tended to fly along the midline of the tunnel, i.e., equidistant from the two walls (Fig. 1A). But when one of the gratings was moved at a constant speed in the direction of the bees' flight – thereby reducing the speed of retinal image motion on that eye relative to the other eye – the bees' trajectories shifted towards the side of the moving grating (Fig. 1B). When the grating moved in a direction opposite to that of the bees' flight – thereby increasing the speed of retinal image motion on that eye relative to the other – the bees' trajectories shifted away from the side of the moving grating (Fig. 1C). These findings demonstrate that when the walls were

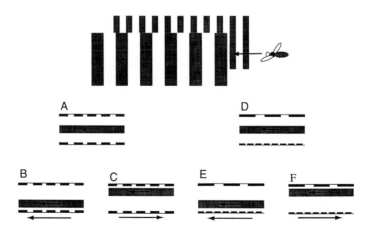

Figure 1. The centering response. Bees are flying through a tunnel flanked by vertical gratings (top panel). The shaded areas in (A–F) represent the means and standard deviations of the positions of the flight trajectories, as inferred from video recordings of several hundred flights. The results demonstrate that flying bees infer range from apparent image speed. Modified from Srinivasan et al. (1991).

stationary, the bees maintained equidistance by balancing the apparent angular speeds of the two walls, or, equivalently, the speeds of the retinal images in the two eyes. A lower image speed on one eye was evidently taken to mean that the grating on that side was farther away, and caused the bee to fly along a trajectory closer to it. A higher image speed, on the other hand, had the opposite effect.

The centering response is independent of contrast frequency

Were the bees indeed measuring and balancing image speeds on the two sides as they flew along the tunnel, or were they simply balancing the contrast frequencies produced by the succession of dark and light bars of the gratings? This question was investigated by analysing the flight trajectories of bees when the two walls carried gratings of different spatial periods (Fig. 1, top panel) (Srinivasan et al., 1991). When the gratings were stationary, the trajectories were always equidistant from the two walls (Fig. 1D), even when the periods of the gratings on the two sides – and therefore the contrast frequencies experienced by the two eyes – differed by a factor of as much as four (10 cm and 2.5 cm). When one of the gratings was in motion, the trajectories, again, shifted towards or away from the moving grating, according to whether it moved with or against the direction of the bees' flight (Fig. 1E, F). These results indicate that the bees were indeed balancing the speeds, and not the contrast frequencies of the retinal images, on the two eyes.

The above findings hold irrespective of whether the gratings possess square-wave intensity profiles (with abrupt changes of intensity) or sinusoidal profiles (with gradual intensity changes), and regardless of whether the contrasts of the gratings on the two sides are equal or considerably different (Srinivasan et al., 1991). Further experiments have revealed that – knowing the velocities of the bee and the pattern – it is even possible to predict the position of a bee's flight trajectory along the width of the tunnel, on the assumption that the bee balances the apparent angular velocities on the two sides of the tunnel (Srinivasan et al., 1991).

Taken together, these results suggest that the bee's visual system is capable of computing the apparent angular speed of a grating independently of its contrast and spatial-frequency content. Indeed, if movement cues are to be exploited to estimate the range of a surface, it is necessary to use a mechanism that measures the speed of the image independently of its geometrical structure. It is this kind of mechanism that would enable an insect to fly through the middle of a gap between, say, two vertical branches of a tree, regardless of the textural properties of the bark on the two sides.

Other work has shown that flying bees use image motion to determine the ranges of objects in the ventral field of view (e.g., Srinivasan et al., 1989), as well as in the frontal field (Lehrer and Collett, 1994).

The centering response is distinct from the optomotor response

The characteristics of the optomotor response (see Introduction) have been investigated extensively in tethered flying insects (e.g., Reichardt, 1969). When an insect is placed inside a rotating striped drum, it tends to turn in the same direction as the drum, thereby stabilising its orientation relative to the surroundings. During free flight, this response helps the insect to maintain a stable orientation and a straight course. The neural basis of the optomotor response is now fairly well understood (e.g., Hausen and Egelhaaf, 1989; K. Kirschfeld, this volume). It is a response to large-field motion driven by a mechanism which is sensitive primarily to the *contrast frequency* of the stimulus, and which therefore confounds the angular velocity of a striped pattern with its spatial period (Reichardt, 1969). The centering response, however, is mediated by a mechanism that is sensitive primarily to the *speed* of the stimulus, regardless of its spatial structure, as demonstrated above. Thus, the centering response is distinct from the optomotor response.

A recent study has investigated further properties of the centering response and compared them with those of the optomotor response (Srinivasan et al., 1993). The experimental procedure is as follows. Bees are trained to fly through a tunnel 60 cm long, 15.5 cm wide, and 30 cm high with textured side walls, to collect a food reward at the far end of the tunnel. The middle section of one wall presents a visual stimulus consisting of a checkerboard (diameter 21 cm, check size 2 cm) generated on a CRT (Fig. 2A). The bees' flight trajectories are filmed from above and, at the same time, from the side (using a mirror, tilted by 45°, viewing the checkerboard through a transparent window). The checkerboard can move horizontally or vertically (Fig. 3). It can also present interleaved motion, wherein adjacent rows (or columns) of the checkerboard move in opposite directions (Fig. 4).

If a bee flies past the stimulus when the checkerboard is moving upwards, she moves laterally away from the stimulus (Fig. 2B), and at the same time upwards (Fig. 2C). The lateral response, defined as shown in Figure 2B, is a measure of the centering response, and the vertical response, defined as shown in Figure 2C, is a measure of the optomotor response.

A comparison of the relative magnitudes of the optomotor and centering responses to a variety of stimulus conditions is shown in Figure 3. At a velocity that produces a contrast frequency of 50 Hz, horizontal backward (front-to-back) motion elicits a strong centering response (shaded bar), but, as expected, virtually no vertical optomotor response (white bar). On the other hand, vertically upward motion elicits strong centering as well as vertical optomotor responses. Thus, the centering response is non-directional, whereas the optomotor response is directionally selective (upward movement deflects a bee upwards, and downward movement deflects her downwards; see also Fig. 5). When the speed of the checkerboard is

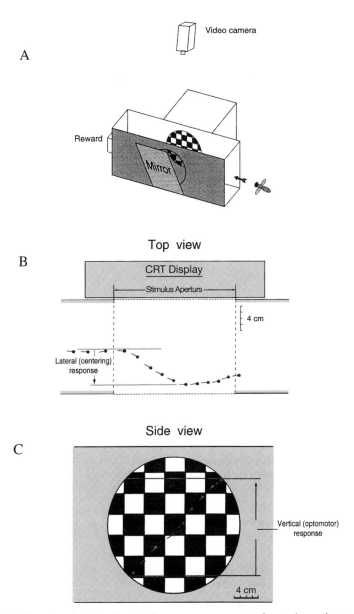

Figure 2. (A) Experimental setup for simultaneous measurement of centering and optomotor responses. (B) An example of lateral (centering) response from the top view. (C) An example of vertical (optomotor) response from the side view. Bee positions are shown every 40 ms in (B) and (C). Modified from Srinivasan et al. (1993).

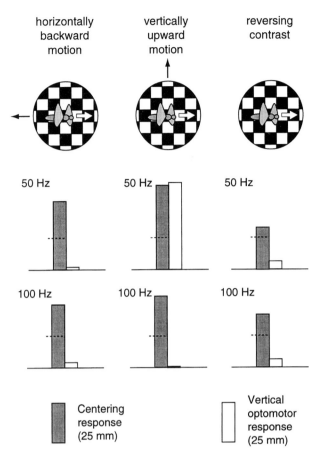

Figure 3. Comparison of magnitudes of centering response (shaded bars) and optomotor response (white bars) elicited by a checkerboard stimulus that moves horizontally backward (left column) or vertically upward (centre column) at speeds that produce contrast frequencies of 50 Hz (upper row) or 100 Hz (lower row). The right column shows, for comparison, the centering and optomotor responses elicited by a checkerboard whose contrast reverses at a rate of 50 Hz (upper row) or 100 Hz (lower row). The dashed horizontal lines depict the magnitude of the centering response elicited by a stationary checkerboard (see Fig. 4, right column).

increased to produce a contrast frequency of 100 Hz, the optomotor response disappears. However, the centering response persists, regardless of the direction of the motion of the checkerboard. Thus, the centering response is sensitive to higher contrast frequencies than is the optomotor response. A reversing-contrast checkerboard (flicker) induces virtually no optomotor response, as one might expect. It induces a centering response which is smaller than that induced by a moving checkerboard, and only slightly greater than that induced by a stationary checkerboard (dotted horizontal lines in Fig. 3). Thus, flicker *per se* makes a measurable contribution

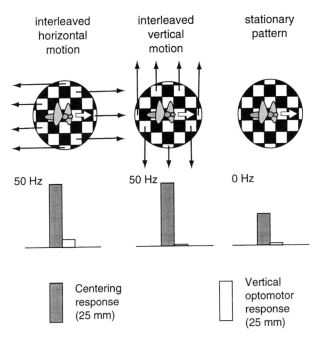

Figure 4. Comparison of magnitudes of centering response (shaded bars) and optomotor response (white) elicited by a checkerboard stimulus that presents interleaved motion in the horizontal direction (left column) or the vertical direction (centre column) at a speed that produces a contrast frequency of 50 Hz to a stationary bee. The right column shows, for comparison, the centering and optomotor responses elicited by a stationary checkerboard.

to the centering response, but it is small compared to the contribution resulting from *motion* of the image.

The non-directionality of the centering response is reaffirmed by experiments using stimuli that present interleaved motion (Fig. 4). Interleaved motion elicits virtually no optomotor response, regardless of whether the motion is horizontal or vertical. This is as expected because, globally, these stimuli present no directional motion. On the other hand, the same interleaved motion stimuli elicit strong centering responses of approximately the same magnitude as the coherent motion stimuli of Figure 3, irrespective of whether the interleaved motion is horizontal or vertical. (Results of statistical analyses of the data are given in Srinivasan et al., 1993.) This result not only supports the notion that the movement-sensitive mechanism mediating the centering response is non-directional, but shows, in addition, that non-directional motion is computed within receptive fields whose diameter is not larger than ca. 20° (which is the visual angle subtended by a single check). Thus, while the optomotor response is generated by pooling the responses of an array of *directionally selective* "elementary movement detectors" with small visual fields (for review, see Reichardt, 1969), the

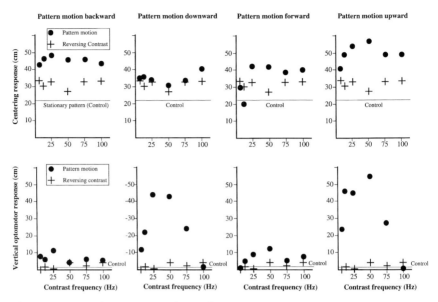

Figure 5. Contrast-frequency dependence of centering response (upper panels, filled circles) and vertical optomotor response (lower panels, filled circles) measured for four different directions of motion of a checkerboard. Also shown are the responses elicited by a checkerboard whose contrast reverses at the same temporal frequencies (+). The response to a stationary checkerboard is depicted by the horizontal lines. Details in text.

centering response appears to be mediated by pooling the responses of an array of *non-directional* motion detectors.

The stationary checkerboard (right-hand panel in Fig. 4), of course, elicits no vertical optomotor response, but it does evoke a centering response, because it is present only on one side of the tunnel. The centering response evoked by this stimulus is used as a baseline (dotted lines in Fig. 3, solid lines in Fig. 5) against which the responses to other stimuli are compared.

The dynamic characteristics of the centering and optomotor responses are compared in Figure 5 for checkerboards moving in four different directions at a range of different speeds, as well as for reversing-contrast checkerboards. In each case, as the bee flies through the tunnel she encounters a checkerboard on one side and a structureless screen on the other. Backward pattern motion induces a strong centering response at all contrast frequencies ranging from 8 Hz to 100 Hz, but virtually no vertical optomotor response at any of these frequencies. Forward (back-to-front) pattern motion generates a centering response comparable to that induced by backward motion, and, again, a negligible vertical optomotor response at all contrast frequencies. Upward pattern motion elicits the strongest centering response – which again persists at all contrast frequencies ranging from 8 Hz to 100 Hz – and a vertical optomotor response that peaks in the vicinity of 25 Hz, and drops to zero at 100 Hz. Downward pattern motion induces a weaker centering response, but a vertical optomotor response similar to that induced by upward motion, which, again, peaks between 25 Hz and 50 Hz and disappears at 100 Hz.

These results indicate that the centering response and the optomotor response possess very different temporal characteristics: the centering response is approximately constant over the range of 8 Hz to 100 Hz (and clearly persists at 100 Hz), whereas the optomotor response exhibits a clear peak in the vicinity of 50 Hz and drops to zero at 100 Hz.

In addition, the vertical optomotor response is directional, whereas the centering response is not. The reversing-contrast checkerboard evokes no vertical optomotor response, as expected. It does elicit a centering response at all contrast frequencies, but this response is only slightly stronger than that induced by a stationary checkerboard.

Two peculiarities of the data on the centering response, evident from Figure 5, require explanation. (i) The centering response is stronger when the pattern moves upwards than when it moves downwards. The reason for this may be that upward pattern motion (which is restricted to one side of the bee) causes the bee to roll and therefore veer slightly toward the other side, thus increasing the apparent magnitude of the centering response. Downward motion has the opposite effect. (ii) The centering response is weaker when the pattern moves in the forward direction at slow speeds than under other conditions. The reason for the diminished response may be that, at low forward speeds, the bee and the pattern move in the same direction and at a similar speed, so that there is relatively little movement of the image of the pattern on the bee's eye.

The results described in this and in the previous section demonstrate that the centering response differs from the optomotor response in three respects: (i) the centering response is sensitive to the *angular speed* of the stimulus, whereas the optomotor response is sensitive to the *contrast frequency* of the stimulus; (ii) the centering response is sensitive to higher contrast frequencies than is the optomotor response, and (iii) the centering response appears to be generated by pooling the outputs of an array of *nondirectional* elementary motion detectors, whereas the optomotor response is generated by pooling the outputs of an array of *directionally-selective* elementary motion detectors (see also K. Kirschfeld, this volume). Thus, it appears that the neural mechanisms underlying the centering response are different from those that mediate the optomotor response.

A neural model of a directionally insensitive speed detector

Why is the centering mechanism sensitive only to the *speed* of the image, and not to direction in which the image moves? We can think of two reasons. Firstly, in neural terms, it may be a lot simpler to build a non-directional speed detector than to build a detector that computes speed as well as direction of motion. In straight-ahead flight, the direction of image motion along each viewing direction is predetermined (Wehner, 1981), and therefore it does not need to be computed. It is the local *speed* that conveys information on range. The insect visual system may thus be adopting a "short-cut" that takes advantage of the fact that the optic flow experienced in straight-ahead flight is constrained in special ways, i.e., it always involves front-to-back motion. Secondly, a non-directional speed detector offers a distinct advantage over a detector that measures speed along a given axis: the later can produce large spurious responses when the orientation of an edge is nearly parallel to the detector's axis. For example, a detector configured to measure speed along the horizontal axis will register large horizontal velocities if it is stimulated by a near-horizontal edge moving in the vertical direction. This "obliquity problem" can be avoided by using either a two-dimensional velocity detector, or a non-directional speed detector – of which the latter offers a simpler, more elegant solution (Srinivasan, 1992). In this context, it is of interest to note that peering locusts also use a non-directional speed-sensitive mechanism to estimate the range of a target towards which they are about to jump (Sobel, 1990).

A simple model of a non-directional speed detector consists of four functional stages of processing (Srinivasan et al., 1991; Srinivasan, 1992), as illustrated in Figure 6. The moving image is first converted to a binary image composed of two levels ("black" and "white") by an array of neurons which possess high sensitivity to contrast and saturate at low contrasts. This neural image, which moves at the same velocity as the original image, is then spatially low-pass-filtered by a subsequent array of neurons, resulting in a moving neural image in which the abrupt edges of the binary image have been converted to ramps of constant slope. The speed of the image can then be monitored by measuring the rate of change of response at the ramps. Accordingly, the neural image at this level is temporally differentiated by an array of phasically-responding neurons, giving a moving neural image composed of a train of pulses, one located at each edge of the binary image. The amplitude of each pulse will then be proportional to the rate of change of intensity at the corresponding ramp, and therefore to the instantaneous speed of the image at that location. A subsequent stage of rectification ensures that the response is positive, regardless of the polarity of the edge (or direction of movement). This (as yet hypothetical) mechanism would measure the local speed of the image, independently of structure, contrast or direction of movement.

In the early stages of processing in the insect visual pathway – the lamina and the medulla – there is an abundance of neurons that exhibit phasic responses which saturate at low contrasts (Laughlin et al., 1987; D. Osorio, unpublished data). It remains to be seen, however, whether such neurons are indeed involved in the speed-based computation of range.

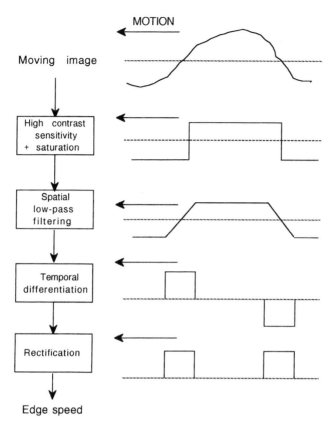

Figure 6. Model of a directionally insensitive motion speed detector. Explanations in text. Redrawn after Srinivasan et al. (1991).

Visual control of flight speed

Do bees control their flight speed by monitoring the apparent velocity of the surrounding environment? An experiment investigating this possibility is illustrated in Figure 7, where bees are trained to fly through a tapered tunnel lined with black-and-white vertical stripes of period 4 cm (top panel in Fig. 7). The bees slow down as they approach the narrowest section of the tunnel, and accelerate when the tunnel widens beyond it (Fig. 7, middle panel). In fact, the variation of flight speed is very close to that expected if the bees were to hold the angular velocity of the image in the lateral eye region constant as they fly through the tunnel (dashed line in Fig. 7, bottom panel). It is evident that the bees are able to hold the angular velocity of the image on the wall constant, despite the changes in the angular period of the stripes that accompany the narrowing and widening of the tunnel. On the other hand, bees flying through a tunnel of constant width do not change

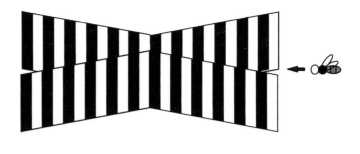

Top view

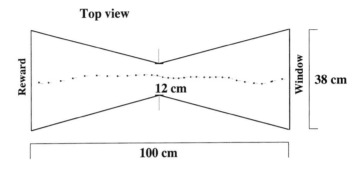

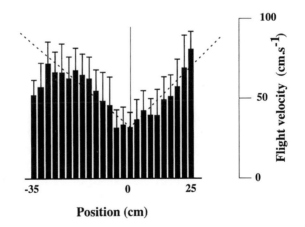

Position (cm)

Figure 7. Flight velocity profiles (bottom panel) of bees flying through a tapered tunnel carrying vertical stripes on either side wall (top and centre panels). The dashed line shows the theoretically expected flight velocity profile if the bees were to hold the angular velocity of the image of the walls constant as they fly through the tunnel. The data represent flight speeds (mean and standard deviations) measured from 18 flights. Details in text. From Srinivasan et al. (1996).

their speed when they encounter an abrupt change in stripe period halfway through the tunnel (Srinivasan et al., 1996). This finding indicates that bees are measuring the angular velocity of the image accurately, irrespective of its spatial structure. Visual control of flight speed is thus achieved by monitoring and regulating the apparent motion of the visual panorama using a movement-sensitive mechanism that is capable of measuring the angular velocity of the image. A similar conclusion was drawn by David (1982) investigating visual control of flight speed in the fruitfly *Drosophila*.

An obvious advantage of controlling flight speed by regulating image speed is that the bee would automatically slow down when negotiating a narrow passage. In addition, such a principle of flight control would provide the bee with a simple, safe strategy for landing, as discussed below.

Executing smooth landings

How does a bee execute a smooth touchdown on a horizontal surface? An approach that is perpendicular to the surface would generate strong looming (image expansion) cues, that could, in principle, be used to decelerate flight at the appropriate moment (Lee and Reddish, 1981; Wagner, 1982; Borst and Bahde, 1988). However, when a bee performs a grazing landing on the surface, looming cues are weak. To investigate how bees make grazing landings, we trained bees to collect a reward of sugar water on a horizontal, textured surface. We then removed the reward and video-filmed from above the landings that the bees made on the surface in search of the food. The experiments were conducted outdoors on a clear day with the sun at an elevation of about 45°. This arrangement allowed us to monitor the height of the bee in terms of the horizontal distance between the bee and its shadow on the wooden surface (Fig. 8A). Height was calibrated in terms of the length of the shadow cast by a vertical rod of a known height (Fig. 8B). This technique, first employed by Zeil (1993), enabled the trajectory of the landing bee to be captured in three dimensions.

A typical trajectory of a landing bee is shown in Figure 8B. Analysis of many such trajectories reveals that the forward speed of the bee decreases steadily as the bee's height above the surface decreases (Fig. 9). In fact, flight speed is approximately proportional to altitude, indicating that the bee is holding the angular velocity of the image of the surface roughly constant as the surface is approached. This mechanism may be a simple way of controlling flight speed during landing, ensuring that its value is close to zero at touchdown. The advantage of such a strategy is that control of flight speed is achieved without explicit knowledge of height. Landing bees maintain angular velocities that range between 400 and 600 deg/s,

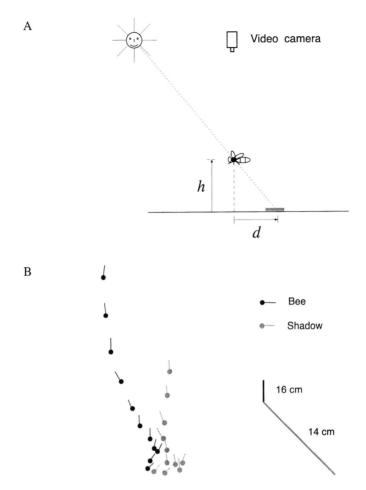

Figure 8. Experimental investigation of landing behaviour. (A) Experimental arrangement for video-filming trajectories of bees landing on a horizontal surface, in three dimensions. The horizontal distance d between the bee and its shadow is proportional to the bee's height h above the surface. (B) Example of a landing trajectory, as filmed from above. Positions of bee and shadow are shown every 40 ms. Also shown are the images of the vertical calibration rod and its shadow. From Srinivasan et al. (1996).

as revealed by analysis of a number of trajectories. It should be noted, however, that this may only be one of several possible mechanisms that could subserve the landing process. It has been shown, for example, that, in the case of landing trajectories that approach the surface in a "head on", rather than a grazing direction, the deceleration and extension of the legs in preparation for landing are triggered by movement-detecting mechanisms that sense *expansion* of the image (Wagner, 1982; Borst and Bahde, 1988).

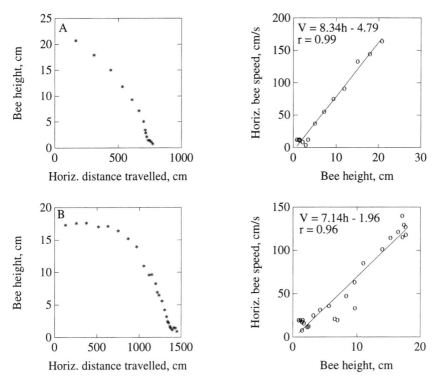

Figure 9. Analysis of flight trajectory whilst landing on a horizontal surface, shown for two bees (A) and (B). In each case, the left-hand panel shows the relationship between height (h) and horizontal distance travelled, whilst the right-hand panel shows the relationship between horizontal flight speed (V) and height (h). The landing bee holds the angular velocity of the image of the ground constant at 8.34 rad/s (480°/s) in (A), and at 7.14 rad/s (410°/s) in (B), as calculated from the slopes of the linear regression lines. Also shown are the values of the correlation coefficient (r). From Srinivasan et al. (1996).

Conclusion

The findings described in the present chapter suggest that visual control of insect flight involves the participation of a number of different motion-sensitive mechanisms, each with a distinctive set of characteristics. The need for several different mechanisms subserving flight control is probably due to the many different requiremens that flight performance imposes on the animal.

The optomotor response stabilises flight by detecting and correcting for unwanted image motion. The optomotor mechanism does not measure image speed reliably: the response at a given speed depends upon the spatial structure of the image. Such a system is adequate, however, for monitoring and correcting for yaw, pitch, roll, or changes in altitude, because all that is needed in these tasks is a signal that conveys the *direction* of image movement reliably. Behavioural studies of the optomotor yaw response in honeybees (Kunze, 1961) indicate that the response peaks at a frequency of about 10 Hz and drops to zero at about 100 Hz. Our measurements of the vertical optomotor response (see Figs 3 and 5) show a similar dependence on contrast frequency.

The centering response, on the other hand, ensures that narrow gaps are negotiated safely by equalising the distances to the walls. The movement-detecting mechanism that mediates this response is sensitive to the *angular speed* of the image on the eye, largely independently of the spatial structure or contrast of the image (Srinivasan et al., 1991). This is precisely what is needed to avoid collisions while flying through narrow gaps or between obstacles, irrespective of the textures of the objects encountered. The system that mediates the centering response has a much broader temporal-frequency bandwidth than does the optomotor system: whereas the opto-motor response drops to zero at a contrast frequency of 100 Hz, the center-ing response shows no signs of abatement at this frequency (see Fig. 5). This property makes sense, because flying through narrow gaps requires sensitivity to high image speeds, and therefore to high contrast frequencies. Unlike the optomotor response, which is directionally sensitive, the center-ing response is mediated by a movement-detecting mechanism that is largely insensitive to the direction of motion. Possible reasons for this directional insensitivity have been discussed above.

The centering response may be partly related to the so-called "movement avoidance" response (MAR) studied by Srinivasan and Lehrer (1984). The MAR is a tendency exhibited by bees to avoid stimuli that generate rapid image movement. Bees trained to find a reward of sugar water at the centre of a sectored, black-and-white disk, find it difficult to approach the disk when it is rotated to produce image motion. We suspect that the MAR is an innate reflex that prevents a flying bee from colliding with nearby objects, which would generate high image velocities. The MAR is strongest at a contrast frequency of about 60 Hz, and drops to zero at about 200 Hz. Thus, the MAR, like the centering response, is sensitive to a broad range of contrast frequencies. Experiments using sectored disks of various periods suggest that, at relatively high contrast frequencies (50 Hz–200 Hz), the MAR depends on contrast frequency, rather than image speed. At lower contrast frequencies (1 Hz–20 Hz), however, the response seems to depend on image speed, rather than contrast frequency (see Fig. 5 in Srinivasan and Lehrer, 1984). Flying bees approaching a rapidly rotating disk do not spin with the disk: rather, they behave as though they are attempting (unsuccess-fully) to penetrate an invisible barrier which prevents them from getting close to the stimulus (Srinivasan and Lehrer, 1984). This suggests that the MAR, like the centering response, is mediated by a movement-detecting system that is not sensitive to the direction of image motion. It is possible, then, that the centering response that bees exhibit whilst flying through a tunnel is a result of equal and oppositely-directed movement-avoidance responses generated by the image motion experienced by the two eyes. In our experiments investigating the centering response (Srinivasan et al., 1991), the bee's flight speeds and the speeds and spatial periods of the gratings lining the tunnel were such that the induced contrast frequencies rarely exceeded 20 Hz. Thus, the centering response may be a manifestation of the movement-avoidance response operating in its speed-sensitive range.

Control of flight speed appears to be accomplished by a system that is, again, sensitive to image speed, rather than contrast frequency. The dynamic characteristics of this system, and its sensitivity to direction of image motion have not yet been investigated.

Finally, visual control of grazing landings also appears to be mediated by a system that is sensitive to image speed. Here again, the dynamic characteristics and sensitivity to direction of image motion remain to be investigated.

Table 1 lists the behaviours described in this chapter and compares their various properties. It now seems clear that visual control of flight is mediated by a variety of special-purpose systems, each dedicated to a specific task. However, it would seem unlikely that all of these systems (and other, as yet undiscovered ones) are operative all of the time. Obviously, the optomotor system would have to be "turned off", or its corrective commands ignored, when the insect makes a voluntary turn or chases a target (Srinivasan and Bernard, 1977; Heisenberg and Wolf, 1993; see also K. Kirschfeld, this volume). Equally, it would be impossible to perform a grazing landing on a surface without first disabling the movement avoidance system. As a third example, there is evidence that the optomotor system is

Table 1. Summary of characteristics of various responses involved in flight control

Type of response	Directionality of response	Bandwidth	Crucial motion parameter
Optomotor response	Directional	Peak: ca. 25 Hz–50 Hz Cuttoff: ca. 100 Hz	Contrast frequency
Centering response	Non-directional	Broad-band; Cutoff: well beyond 100 Hz	Image speed
Movement-avoidance response (MAR)	Non-directional (?)	Cutoff: ca. 200 Hz	Image speed (1 Hz–20 Hz); Contrast frequency (50 Hz–200 Hz)
Flight speed control	?	?	Image speed
Grazing landing	?	?	Image speed (?)

at least partially non-functional when insects fly through narrow gaps (Srinivasan et al., 1993). A major challenge for the future, then, is to discover the conditions under which individual systems are called into play or ignored, and to understand the ways in which these systems interact to co-ordinate flight.

References

Borst, A. and Bahde, S. (1988) Visual information processing in the fly's landing system. *J. Comp. Physiol.* A 163:167–173.

Collett, T.S. and Harkness, L.I.K. (1982) Depth vision in animals. *In*: D.J. Ingle, M.A. Goodale and R.J.W. Mansfield (eds): *Analysis of Visual Behavior.* M.I.T. Press, Cambridge, Massachusetts, pp 111–176.

David, C.T. (1982) Compensation for height in the control of groundspeed by *Drosophila* in a new, "Barber's Pole" wind tunnel. *J. Comp. Physiol.* 147:485–493.

Hausen, K. and Egelhaaf, M. (1989) Neural mechanisms of visual course control in insects. *In*: D.G. Stavenga and R.C. Hardie (eds): *Facets of Vision.* Springer-Verlag, Berlin, Heidelberg, pp 391–424.

Heisenberg, M. and Wolf, R. (1993) The sensory-motor link in motion-dependent flight control of flies. *In*: F.A. Miles and J. Wallman (eds): *Visual Motion and its Role in the Stabilization of Gaze.* Elsevier, Amsterdam, pp 265–283.

Horridge, G.A. (1987) The evolution of visual processing and the construction of seeing systems. *Proc. R. Soc. Lond.* B. 230:279–292.

Kirchner, W.H. and Srinivasan, M.V. (1989) Freely flying honeybees use image motion to estimate object distance. *Naturwissens.* 76:281–282.

Kunze, P. (1961) Untersuchung des Bewegungssehens fixiert fliegender Bienen. *Z. vergl. Physiol.* 44:656–684.

Laughlin, S.B., Howard, J. and Blakeslee, B. (1987) Synaptic limitations to contrast coding in the retina of the blowfly Calliphora. *Proc. R. Soc. Lond.* B. 231:437–467.

Lehrer, M. and Collett, T.S. (1994) Approaching and departing bees learn different cues to the distance of a landmark. *J. Comp. Physiol.*A 175, 171–177.

Lee, D.N. and Reddish, P.E. (1981) Plummeting gannets: a paradigm of ecological optics. *Nature* 293:293–294.

Reichardt, W. (1969) Movement perception in insects. *In*: W. Reichardt (ed.): *Processing of Optical Data by Organisms and by Machines.* Academic Press, New York, pp 465–493.

Rossell, S. (1983) Binocular stereopsis in an insect. *Nature* 302:821–822.

Sobel, E.C. (1990) The locust's use of motion parallax to measure distance. *J. Comp. Physiol.*167:579–588.

Srinivasan, M.V. (1992) How flying bees compute range from optical flow: Behavioral experiments and neural models. *In*: R.B. Pinter (ed.): *Nonlinear Vision.* CRC, Boca Raton, pp 353–375.

Srinivasan, M.V. (1993) How insects infer range from visual motion. *In*: F.A. Miles and J. Wallman (eds): *Visual Motion and its Role in the Stabilization of Gaze.* Elsevier, Amsterdam, pp 139:156.

Srinivasan, M.V. and Bernard, G.D. (1977) The pursuit response of the housefly and its interaction with the optomotor response. *J. Comp. Physiol.* 115:101–117.

Srinivasan, M.V. and Lehrer, M. (1984) Temporal acuity of honeybee vision: behavioural studies using moving stimuli. *J. Comp. Physiol.* A 155:297–312.

Srinivasan, M.V., Lehrer, M., Zhang, S.W. and Horridge, G.A. (1989) How honeybees measure their distance from objects of unknown size. *J. Comp. Physiol.* A 165:605–613.

Srinivasan, M.V., Lehrer, M., Kirchner, W. and Zhang, S.W. (1991) Range perception through apparent image speed in freely-flying honeybees. *Vis. Neurosci.* 6:519–535.

Srinivasan, M.V., Zhang, S.W. and Chandrashekara, K. (1993) Evidence for two distinct movement-detecting mechanisms in insect vision. *Naturwissens.* 80:38–41.

Srinivasan, M.V., Zhang, S.W., Lehrer, M. and Collett, T.S. (1996) Honeybee navigation en route to the goal: visual flight control and odometry. *J. Exp. Biol.* 199:237–244.

Wagner, H. (1982) Flow-field variables trigger landing in flies. *Nature* 297:147–148.

Wehner, R. (1981) Spatial vision in insects. *In*: H. Autrum (ed.): *Handbook of Sensory Physiology, Vol. VII/6C.* Springer-Verlag, Berlin, Heidelberg, pp 287–616.

Zeil, J. (1993) Orientation flights of solitary wasps (Cerceris; Sphecidae; Hymenoptera). I. Description of flight. *J. Comp. Physiol.* A 172:189–205.

Orientation and Communication in Arthropods
ed. by M. Lehrer
© 1997 Birkhäuser Verlag Basel/Switzerland

Honeybees' visual spatial orientation at the feeding site

M. Lehrer

Institute of Zoology, University of Zurich, Winterthurerstrasse 190, CH-8057 Zurich, Switzerland

Summary. Results of behavioural studies on freely flying honeybees (*Apis mellifera*) are reviewed with the aim of summarizing the various visual spatial parameters by which bees orient to the food source. The results show that the accuracy of the bee's performance in recognizing the target depends on the magnitude of intensity-, colour-, and motion contrast produced by the target against its background, as well as on the retinal position onto which the stimulus projects at the eye. Mainly, however, the results reveal that bees use a variety of configurational spatial parameters in this task. These include spatial frequency, geometry, symmetry, angular size, absolute size, distance, different types of edges, and orientation of contours. When viewed jointly, the results demonstrate the bees' remarkable learning capacity and the behavioural flexibility of their orientation performance. They show, in addition, that orientation towards a target does not always require a learning process, i.e., several types of response to the food source are based on hard-wired tendencies. Finally, the results show that colour vision, although not a prerequisite for spatial vision, participates in spatial vision, and that spatial cues extracted from image motion are processed by a colour blind system.

Introduction

Foraging honeybees keep returning to a familiar feeding site, be it several meters, several hundred meters, or several kilometers away from the hive. They navigate by using spatial visual cues, i.e., the skylight pattern (von Frisch, 1965; R. Wehner, this volume), far and near landmarks (Menzel et al., 1996; Wehner et al., 1966; T. S. Collett and J. Zeil, this volume), and optic flow (Esch and Burns, 1996; M. V. Srinivasan and S. W. Zhang, this volume). A forager's task, however, is not restricted to this in many cases amazing route-finding performance. Once it has arrived at its goal, the bee's orientation will be aimed at recognizing the target, which is, under natural conditions, a flower of a particular species that has, on previous visits, been found to be profitable. In this task, the bee uses visual as well as olfactory cues.

The bee's capacity to recognize the profitable flower on every foraging trip improves foraging efficiency and ensures, at the same time, species-specific pollination. Thus, there is much reason to believe that the diversity of flower scents, colours, and shapes developed in coevolution with the pollinators' capacity to discriminate among different signals within each of these sensory modalities.

A large amount of work has been devoted to the bee's performance in discriminating among colours (e.g., von Frisch, 1915; Daumer, 1956; Menzel, 1967, 1969; von Helversen, 1972; Menzel et al., 1986; Menzel and Backhaus, 1989) and scents (e.g., von Frisch, 1919; Opfinger, 1949; Couvillon and Bitterman, 1987; Fonta et al., 1991). Colour and scent were found to be very effective signals not only for attracting potential pollinators, but also as cues for discriminating among flowers. Still, the bee's well-known flower constancy (e.g., Manning, 1957; Waser, 1986; Gross, 1992) cannot be accounted for by the use of colour and scent alone. The large variety of *spatial* parameters by which flowers differ would not have evolved in the absence of any selection pressure. It is probably the *combination* of scent, colour, and various spatial parameters that enables flower discrimination (see also Leppik, 1953).

Earlier workers on the bee's spatial vision suggested that the main spatial parameter that bees use for pattern discrimination is spatial frequency, i.e., the amount of contours, or of on- and off-stimulation (flicker), per area of the pattern (e.g., Hertz, 1930, 1933; Wolf, 1933; Zerrahn, 1934; Wolf and Zerrahn-Wolf, 1935; Free, 1970; Anderson 1977a). This "flicker theory" of insect vision, already proposed by Exner (1876) based on the anatomy of the compound eye, persisted for several decades. It was not until the late 1960s and early 1970s that an alternative theory was proposed. Using patterns presented on a vertical plane, Wehner and Lindauer (1966) and Wehner (1969, 1972a, b) demonstrated that bees discriminate between patterns that differ in no other parameter than their spatial orientation. Based on these and further results, it was proposed that bees memorize an eidetic ("photographic") image of the rewarded pattern, and that pattern discrimination depends on the degree to which a novel test pattern matches the stored "template" (Wehner 1972a, b; Cruse, 1972). This conclusion was later supported by results of studies using cinematographic methods that revealed a fixation phase in which the bee, prior to landing, hovers on the spot in front of the pattern, thus maintaining a constant spatial relation between eyes and pattern (Wehner and Flatt, 1977).

The image-matching theory was soon embraced by workers studying the bee's use of landmarks in navigational tasks (e.g., Anderson, 1977b; Cartwright and Collett, 1983; Wehner et al., 1996; Collett, 1996; T. S. Collett and J. Zeil, this volume). But even for investigators who remained sitting at the food source, observing the bee's performance at recognizing their target, Wehner's findings have initiated a new era in work on spatial vision. Much of this work has already been reviewed (Wehner, 1981). Here we add results of more recent studies, with the aim of demonstrating the bee's use of a large variety of spatial cues in the task of detecting and recognizing the food source.

General methods

In all of the experiments to be described below, an artificial food source placed in the laboratory is used. The bees are flying freely between their hive and the experimental set-up, just as they do when visiting natural flowers.

The details of the experimental procedures employed differ among the various studies. They will be described in due context. The present section deals with those methodological aspects that are common to all or to many of the studies.

Training and test procedures

In all of the studies, the food source presents a visual stimulus that bees learn to associate with a reward of sucrose solution.

Even at an artificial food source, there is no way of presenting one particular visual parameter in complete isolation from others. For example, a stimulus of a particular shape or geometry possesses also a particular colour and size, a particular spatial frequency, a particular spatial relation to neighbouring visual marks, and often also a particular predominant orientation of contours. Still, it is posible to encourage bees to use mainly one parameter in the orientation task. One method to do so is to keep the parameter under consideration constant during the training, but randomize the others. For example, to investigate the use of size-independent cues to distance, the target's distance is kept constant, but its size is varied between rewarded visits. The second is to offer, simultaneously, two different stimuli during the training, one rewarded, the other not, that differ from each other only in the parameter under consideration. For example, to examine the use of shape, the rewarded and unrewarded stimuli must differ in shape, but not in colour, size, or spatial frequency. To prevent the bees from using positional cues, the rewarded and unrewarded stimuli interchange places throughout the experiment.

In subsequent tests, the rewarded stimulus is tested against stimuli that differ from it in the parameter under consideration (but sometimes other types of tests are used, as will be specified below). In most cases, the target offers no reward during the tests, to obtain a large number of choices. The criterion for a choice is usually the bee's landing on the target, but sometimes choices are recorded at some distance from it, depending on the question to be investigated. Between tests, bees are rewarded several times in the training situation.

The bee's use of olfactory cues is excluded in all cases. The various methods employed to eliminate such cues will not be specified.

Definition of intensity contrast, colour contrast, and receptor-specific contrasts

Spatial vision is based on partitioning the image into contrasting pixels, the size of which (and therefore the fineness of the image) being dependent on the spatial resolution power (see review by Wehner, 1981) of the visual system under consideration. Three types of contrast can be used for spatial resolution: intensity contrast, colour contrast, and receptor-specific contrasts.

Intensity contrast, also termed intensity modulation (m), is defined as the ratio between the modulation amplitude and the mean intensity of the two adjacent stimuli that produce the contrast. The ratio can also be expressed as

$$m = |I_1 - I_2| / |I_1 + I_2| \tag{1}$$

where I_1 is the intensity of one of the stimuli, and I_2 that of the other. Intensity contrast can have any value betwen 0 % (when $I_1 = I_2$) and 100 % (when $I_2 = 0$). Intensity contrast values, whenever such are given in the present review, are bee-subjective, i.e., the intensities of Stimuli (1) and (2) are calculated using the spectral sensitivity curve of the bee's three photoreceptor types (UV, blue, and green) (Menzel et al., 1986) and the spectral composition of each of the two stimuli. These calculations render relative excitation values evoked in each receptor type by each of the two stimuli. The total intensity of each stimulus is calculated as the sum of the three excitation values obtained. The total intensity contrast m between the two stimuli is then calculated using Equation (1).

Receptor-specific contrasts (UV-contrast, blue-contrast, and green-contrast) are calculated from the excitation values obtained for each receptor type separately, using Equation (1). For more details, see Srinivasan and Lehrer (1988). Note that receptor-specific contrasts are pure intensity contrasts, because a single spectral type of receptor cannot encode colour.

Colour contrast is produced by two adjacent stimuli that differ in colour, regardless of whether or not they produce, in adition, intensity contrast. Quantification of colour contrast is not as simple as is that of intensity contrast. Broadly speaking, colour contrast is the higher, the greater the distance of the two colour loci is in the bee's colour space. (For more details, see Menzel and Lieke, 1983, and Menzel and Backhaus, 1989). A good measure for the magnitude of colour contrast is the bees' discrimination between the two colours that produce the contrast.

Experiments and results

The experimental results (Sections 1–10) will be presented according to the spatial cue under consideration. Section 1–3 are concerned with spatial

parameters that are expected to quantitatively influence the bee's performance in resolving, detecting, and recognizing the target, namely various
types of contrast, and the position of the stimulus in the visual field. Sections 3–9 deal with the bee's use of specific spatial parameters, namely
spatial orientation, geometry, symmetry, spatial frequency, edges, distance,
depth, angular size and absolute size. Finally, Section 10 provides examples
of the bee's capacity to actively acquire visual information that is useful in
coping with particular spatial tasks.

Spatial parameters influencing the accuracy of target recognition

Intensity and colour contrast (1)
The role of intensity contrast in a spatial task was first demonstrated
exploiting the optomotor response (Götz, 1965; Hengstenberg and Götz,
1967; Kaiser and Liske, 1974): the strength of the yaw response to a
moving grating decreases as the contrast contained in the grating is reduced. The role of *colour* contrast, however, could not be investigated using
this method, because the optomotor system is colour blind, being driven
exclusively by intensity contrast (Schlieper, 1927; Kaiser, 1975).
   Lehrer and Bischof (1995) investigated the role of intensity contrast, as
well as of colour contrast, in a task involving detection of a single object
presented against a contrasting background. The experimental set-up
(Fig. 1A) consisted of two parallel channels whose back walls were covered by a white or a yellow paper that served as a background. A tube penetrating the centre of the wall led to a dark reward box placed behind the
back wall, containing a feeder with sucrose solution (Fig. 1B). In one of the
channels (alternately the right of the left one), a contrasting disc of a variable size was fixed in front of the tube. Bees choosing this channel had
access to the reward box through an incision in the tube, immediately
behind the disc. The tube in the alternative channel carried a disc constructed out of the same paper as that used for the background, thus providing
no contrast. Here, the entrance to the tube leading to the reward box was
blocked.
   When a white background (W) was used, the contrasting disc was either
black (B) or grey, using two different shades of grey (G1 and G2). When a
yellow wall was used, the disc was violet, using two different combinations
of yellow and violet pigment papers (V1/Y1 and V2/Y2). These two colour
combinations do not differ in the amount of colour contrast (see *General
methods*), but they differ strongly in the magnitude of intenstiy contrast
(34% and 2%, respectively). On each visit, each bee's first decision
between the two channels was recorded at the channel entrance.
   The results obtained with the achromatic stimuli are shown in Figure 1C.
With the highest intensity contrast (combination B/W), a disc subtending
only 3° at the bee's eye (as viewed from the channel entrance) is still

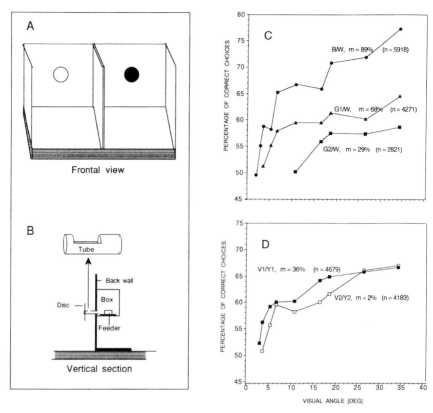

Figure 1. The role of contrast in object detection. (A, B) Experimental set-up. "Correct choice" is defined as the bee entering the positive channel (see text). In (C) and (D), percentage of correct choices is shown as a function of the visual angle subtended by the disc at the bee's eye, as viewed from the channel entrance. (C) Achromatic stimuli. B/W, black disc against a white background, G1/W and G2/W, grey disc against a white background, G1 being darker than G2. (D) Coloured stimuli. V1/Y1 and V2/Y2, violet disc against a yellow background, using two different hues of violet and yellow. m, total intensity contrast; n, total number of visits. Modified from Lehrer and Bischof (1995).

detected. With the decrese of intensity contrast (combinations G1/W and G2/W), detectability deteriorates markedly. Consequently, intensity contrast is crucial in this task.

The results obtained using the coloured combinations, on the other hand, were independent of intensity contrast (Fig. 1D), showing that colour contrast is sufficient in this task. However, recent experiments using colour combinations that differ in the amount of colour contrast (M. Lehrer and B. Stuber, in preparation) revealed that the magnitude of colour contrast (similar to the magnitude of intensity contrast in achromatic stimuli) influences the detectability of the contrasting disc.

Using a slightly different method for investigating the same question, Giurfa et al. (1996) confirmed the finding that detection of a coloured disc depends on the colour contrast, and not on the intensity contrast that it produces against the background. However, detectability is strongly enhanced when the stimulus contains, in addition to colour contrast, achromatic contrast perceived by the green receptor. The minimum detectable angular size obtained with coloured stimuli was similar to that reported by Lehrer and Bischof (1995).

Motion contrast (2)

Apart from chromatic and achromatic contrasts, another type of contrast is conceivable by which an object may become visible against its background. When the bee's target is nearer to the bee than is the background (which is usually the case), a bee flying above or in front of the target will perceive a faster speed of image motion from the target than from the background. The thus created relative motion (motion parallax) between object and

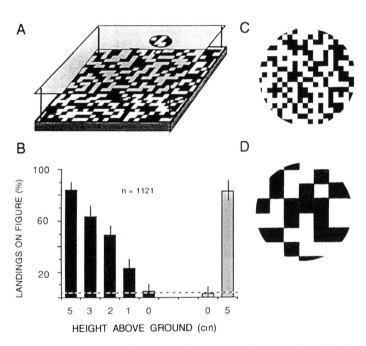

Figure 2. The use of motion parallax in an object-ground discrimination task. (A) Experimental set-up. A randomly patterned disc was placed on a Perspex sheet raised above a similarly patterned ground. In the tests, landings on the disc as well as elsewhere on the Perspex sheet were recorded. (B) Percentages of landings on the disc as a function of its height above the ground. Mean values and SD are shown as calculated from several tests conducted at each height. Dashed line denotes random-choice level. Black bars show results obtained with a disc (C) that carried the same pattern as the ground. Shaded bars are results obtained with a disc (D) that had a much coarser pattern than the ground. n, total number of landings. Modified from Srinivasan et al. (1990).

background might be used in an object detection task even when object and background do not differ in brightness or in pattern.

Srinivasan et al. (1990) investigated the bee's use of motion parallax in an object-ground discrimination task by training bees to collect sugar water from a randomly patterned disc (6 cm in diameter) placed above a similarly patterned ground (Fig. 2A). In the tests, the bees did not detect the disc when it was placed flat on the ground (0 cm in Fig. 2B). However, when raised above the ground, the disc was the better detected, the higher it was placed, i.e., the larger the amount of motion parallax between it and the ground (Fig. 2B, black bars). Similar results (Fig. 2B, shaded bars) were obtained using a disc that carried a much coarser pattern (Fig. 2D) than did the ground. Thus, an object may be detected, regardless of pattern, simply on the basis that it is nearer than is the ground. In analogy to intensity- and colour contrast (see above), detectability is crucially influenced by the *amount* of motion contrast.

Position in the visual field (3)

Using patterns presented on vertical planes, Wehner (1972a, b) demonstrated that eidetic memory (see *Introduction*) is not space-invariant, i.e., specific elements of the visual pattern are mapped topographically on the ommatidial array and are accordingly stored in memory. The strongest evidence for eye-region-specific pattern learning was provided by results of experiments in which the eye region onto which the pattern has projected during the training was occluded by light-tight paint prior to the test (Wehner, 1972a; Wehner and Flatt, 1977). Bees treated in this way did not recognize the pattern, although other eye regions were free to view it.

The eye-region specificity of pattern learning made it possible to compare the accuracy of pattern recognition among different frontal regions of the bee's eye. Wehner (1972a) trained bees to a white disc and then offered them a choice between that disc and another white disc that had a black sector (30°) inserted in it in different positions. The sector was detected best when it projected onto the fronto-ventral eye region. Because the bee does not posses a high-acuity zone in this particular eye region, Wehner (1972a,b) concluded that eye-region-specific weighting factors are involved in the bee's spatial vision (see also Giurfa et al., 1995). Position-weighting effects were also found to be involved in colour learning (Menzel and Lieke, 1983).

The dominant role of the fronto-ventral visual field in spatial vision was demonstrated in yet another study concerned, this time, with pattern discrimination (M. Lehrer, unpublished data). The stimuli were white discs with a sector of 90° displaying a black-and-white pattern inserted in them in either the ventral, the lateral, or the dorsal position (Fig. 3, insets), training a fresh group of bees to each position. In one set of experiments (Fig. 3A), the rewarded pattern had a spatial period (λ) of 5.5° (high fre-

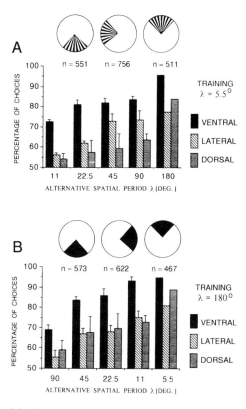

Figure 3. Position-weighted pattern discrimination. The rewarded stimulus was a patterned sector projecting onto either the ventral, lateral, or dorsal eye region (insets), using a fresh group of bees in each training. n, number of choices. In (A), bees were trained to a high-frequency pattern that was then tested against lower-frequency ones (abscissa). In (B), it was *vice versa*. Mean values and SD are shown, as calculated from three tests conducted at each test frequency. M. Lehrer, unpublished data.

quency), and the unrewarded one had $\lambda = 180°$ (low frequency). In another set of experiments (Fig. 3B), training was reciprocal (i.e., 180° rewarded, 5.5° unrewarded). In the tests, the bees were offered a choice between the previously rewarded pattern and a series of patterns of lower or higher frequencies, respectively. In either set of experiments, best discrimination between the trained pattern and each of the alternative patterns was obtained when training and tests were conducted with the sectors presented in the ventral position, showing that pattern discrimination, similar to pattern detection, is best in the fronto-ventral visual field.

Spatial parameters used for target recognition

Spatial orientation (4)

The role of spatial orientation of patterns could not be investigated in the earlier studies using patterns presented on a horizontal plane, because in this situation pattern orientation depends on the bee's direction of approach. Using patterns placed on a *vertical* plane, Wehner (1972a, b) trained bees to a half-white and half-black cardboard disc with the orientation of the edge being 45° with respect to the vertical. In subsequent dual-choice tests, bees discriminated the trained disc from a series of test discs that deviated from it in the orientation of the edge. However, discrimination was found to be asymmetrical, depending on whether the test disc was rotated clockwise or counter-clockwise. Thus, the crucial parameter was the spatial distribution of the black and white areas in the visual field, rather than the orientation of the edge *per se*. Menzel and Lieke (1983), investigating the combined effect of colour and position using two-coloured circular glass filters, obtained a similar asymmetry when the orientation of the edge during training was 45°, but not when training was with the edge oriented horizontally. In the latter case, rotation of the test disc by +45° and by −45° had the same effect, which is as expected, because the amount of deviation (from the trained pattern) was the same in the two test patterns with respect to both the edge orientation and the colour distribution.

M. Lehrer (unpublished data) trained bees to a two-coloured (blue/yellow) carboard disc with the edge oriented horizontally (Fig. 4A, B, insets). In the tests, the bees had to choose between the trained pattern and one that was rotated by either 45° or 135°. The angular deviation of the contour from the trained contour orientation was 45° in either pattern, but the spatial distribution of the two colours strongly differed between the two. If contour orientation is the crucial parameer, then the two tests are expeced to render similar results. If, however, the distribution of colours is crucial, then the trained pattern is expected to be discriminated from the 135° pattern better than from the 45° pattern. The test results (Fig. 4A, B) demonstrate that discrimination depends on the amount of deviation in colour distribution, and not on the angular deviation in edge orientation. Thus, when a single contour between two contrasting areas is used, the role of contour orientation is masked by the role of the spatial distribution of contrasting areas, i.e., by the dominant role of eidetic memory.

This difficulty was overcome in an extensive series of studies training bees to randomized gratings, rather than to a single edge between two

---➤

Figure 5. Example demonstrating the use of contour orientation regardless of pattern. Bees were trained using randomized gratings of a constant contour orientation (left panel). In subsequent tests, novel patterns were used (right panel). In each panel, the percentage of the bees' choices for each of the two test patterns is shown. Data from van Hateren et al. (1990).

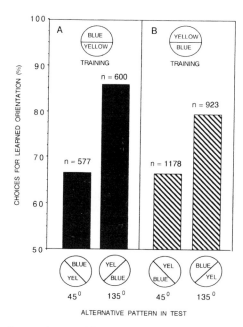

Figure 4. The role of eye-region specific distribution of contrasting coloured areas in a task involving orientation discrimination. Bees were trained to a half-yellow and half-blue disc, blue being presented in either the upper (A) or the lower visual field (B). In the tests, the orientation of the edge was pitted against the distribution of colours. Shown is the percentage of choices in favour of the trained disc. n, number of choices. M. Lehrer, unpublished data.

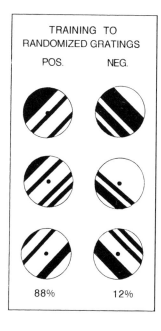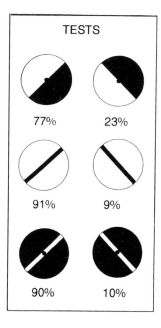

contrasting areas, thus excluding the participation of eidetic memory (e.g., van Hateren et al., 1990; Srinivasan et al., 1993; Srinivasan, 1994; Zhang and Srinivasan, 1994; Srinivasan et al., 1994; Geiger and Srinivasan, 1995). Bees trained in this way were shown to use the orientation of contours in the pattern discrimination task even when presented with entirely novel patterns (Fig. 5).

Geometry (5)
In most of the studies investigating pattern discrimination, the criterion for a bee's choice was its landing on a pattern. At this close range, however, the bee can perceive only local cues contained in the pattern. Therefore, to test the role of geometry, which is a global spatial parameter, a choice criterion other than landing is needed.

Lehrer et al. (1994) investigated the bees' pattern preferences by measuring their choices at some distance from the patterns. The experimental apparatus (Fig. 6A) consisted of 12 identical compartments opening onto a central arena. The bees entered the arena from above through a circular opening in the transparent Perspex sheet that covered the whole apparatus. During training, one of six checkerboard patterns (Fig. 6B) was presented, in a semi-random succession, on the back wall of one of the compartments, the access to the reward box being through a tube penetrating the center of the pattern. Because the six checkerboards were randomized with respect to spatial frequency and orientation of contours, bees could not rely on these parameters. In subsequent tests, each of the 12 compartments had a

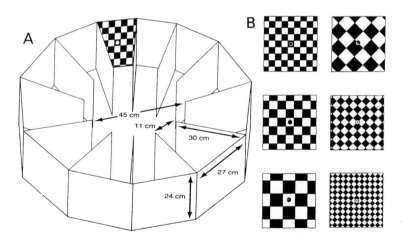

Figure 6. Set-up used for investigating the bees' pattern preferences. (A) View of the arrangement of 12 compartments. (B) Checkerboard patterns used during training, one at a time, in a semi-random succession. In the tests, the bees were given a choice among four patterns, each repeated three times in the different compartments. The criterion for a choice was the bee's entering a compartment. Modified from Lehrer et al. (1994).

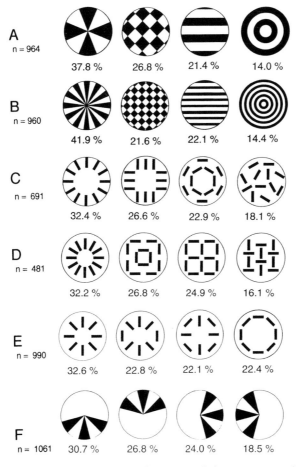

Figure 7. Pattern preferences. Percentages of the bees' choices among novel patterns after training to the checkerboard patterns (see Fig. 6). In each set, the four test patterns differ either in type (A, B), in the arrangement of bars (C–E), or in orientation (F), but not in spatial frequency. n, total number of choices. Modified from Lehrer et al. (1994).

novel pattern placed on its back wall, using four different patterns, each repeated in three different compartments. The criterion for a choice was the bee's entering a compartment.

   In the tests shown in Figure 7A and B, bees preferred the sectored pattern, the least attractive pattern being the one with concentric rings. When the four test patterns contained a constant number of bars arranged in different ways, bees expressed a preference for radiating bars (Fig. 7C–E). Circular arrangements of bars (Fig. 7E), as well as random arrangements (Fig. 7C), were unattractive. In addition, symmetrical patterns were more attractive than les symmetrical or asymmetrical ones (Fig. 7D), and bilaterally symmetrical patterns with a vertical plane of symmetry were preferred over such with a transverse plane of symmetry (Fig. 7F). In other words, bees prefer flower-like patterns, despite the fact that the checkerboard patterns to which they have been trained bear no resemblance to natural flowers. The possible neural mechanisms that might underlie such a hard-wired recognition system are discussed by Srinivasan et al. (1993) and by Horridge (1994).

Spatial frequency (6)
Bees trained to randomized checkerboards as described above (see Fig. 6) were now tested with sets of patterns that differed in their spatial frequency, rather than in their geometry (Lehrer et al., 1994). The test results (Fig. 8A–D) show that, regardless of the type of pattern used, the pattern of the lowest spatial frequency is the most attractive one. These results are in contrast to the results of earlier workers, who found a spontaneous preference for high-frequency patterns when decisions were judged by the bees' landings on a pattern (e.g., Hertz, 1930; Free, 1970; Anderson, 1977). Thus, at a close range, bees rely on subtle, local cues, whereas at some

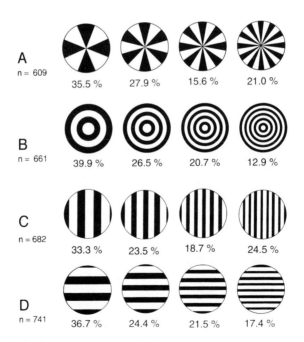

Figure 8. Discrimination among patterns of different spatial frequencies. As in the tests shown in Figure 7, bees were previously trained to randomized checkerboard patterns (see Fig. 6). Modified from Lehrer et al. (1994).

distance they use coarser, more global cues (see also Zhang et al., 1992). Nonetheless, in agreement with the earlier findings, the results show that spatial frequency is a reliable discrimination criterion (see also Fig. 3 above, and Fig. 67 in Wehner, 1981).

Depth (7)
With very few exceptions (e.g., Rossel et al., 1992), all arthropods, including the bee, possess stereoscopic (binocular) vision that is restricted to a range of only a few mm, due to the small separation of the two eyes (Burkhardt et al., 1973; Collett and Harkness, 1982). Still, even at ranges at which stereoscopic vision cannot work, bees were found to be adept at several spatial tasks that require information on distance, such as selecting a particular flight height (Levin and Kerster, 1971; Faulkner, 1976), and estimating the range of a landmark (Cartwright and Collett, 1979). How does the bee measure the distance of an object?

One way would be by using the objects' angular size: an object of a given size subtends a larger visual angle at the eye when it is near than when it is farther away. The bee's capacity to estimate distance by using angular size has been demonstrated by Cartwright and Collett (1979) who trained bees to a feeding site with a cylinder of a constant size placed at a constant distance from it. When the cylinder was replaced by a smaller or a larger one, bees searched for the food source at a nearer or a farther site, respectively.

However, bees can rely on angular size only if they are familiar with it. A bee recruited to a novel feeding site, where sizes of objects are unknown, cannot rely on this cue. In this situation, the only reliable cue to distance is the speed at which the contours of an object move on the bee's eye during flight: The contours of a near object, irrespective of its size, move faster than do those of a more distant object (see also Section 2).

To investigate the bee's use of image speed in a distance estimation task, Lehrer et al. (1988) trained bees to collect a reward of sugar water from a black disc placed on a stalk 70 mm above a white ground (Fig. 9A, top panel). Six further discs of different sizes were placed flat on the ground and each carried a drop of plain water. The positions of all seven discs were varied after every rewarded visit, and, at the same time, the size of the rewarded disc was altered. The only parameter that always remained constant was its height above ground.

In subsequent tests, five discs, each of a different size, were placed at five different heights (Fig. 9A, botton panel). In these tests, the proportions of the bees' landings on the five discs were strictly correlated with the discs' height (Fig. 9B), showing that bees discriminate range irrespective of size. Similar results were obtained with blue discs on a yellow ground, using a colour combination that offered exclusively green-contrast (see *General methods*) (Fig. 9C). However, in the absence of green-contrast, range discrimination broke down (Fig. 9D). Consequently, the bee's per-

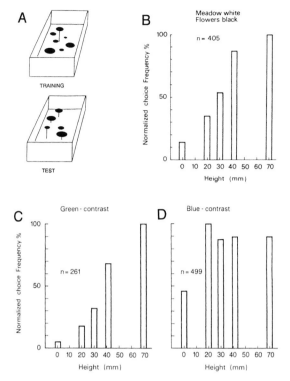

Figure 9. The use of image motion for distance estimation. The rewarded dummy flower was placed at a constant height (70 mm) above the ground (A, top panel), but its size and position were randomized between rewarded visits. Tests were conducted using five dummy flowers of different sizes placed at different heights (A, bottom panel). (B–D) Distribution of the bees' landings among the five test flowers as a function of flower height. (B) Black discs on a white ground. (C) and (D) Blue discs on a yellow ground. In (C), contrast was detectable by the green-sensitive receptors only. In (D), green-contrast was absent. n, number of landings. Modified from Lehrer et al. (1988).

formance in this task is colour blind, as are all of the bee's motion-induced behaviours investigated previously (reviewed by Lehrer, 1987).

Using a modified set-up, Srinivasan et al. (1989) were able to train bees not only to the highest disc, but also to the lowest one, and to one placed at an intermediate height, showing that the bees' preference for the highest disc manifested in the previous experiment is not simply a consequence of a spontaneous attraction to the fastest-moving object, but rather of the bee's capacity to discriminate among different speeds of image motion. The same conclusion was drawn from results of experiments involving the bee's frontal, rather than the ventral visual field (Horridge et al., 1992). The use of motion speed was demonstrated also in studies on flight control (Srinivasan et al., 1996; M. V. Srinivasan and S.W. Zhang, this volume).

Size (8)

The role of flower size in the bee's choice preferences has been demonstrated in many field observations (reviewed by Dafni et al., 1966). In training experiments, bees were shown to discriminate between a trained target and targets of other sizes placed at the same distance by using the visual angle subtended by the target at the eye (e.g., Schnetter, 1972; Mazokhin-Porshnyakov et al., 1977; Lehrer and Collett, 1994), but not between a learned target and one of the same angular size placed at a different distance (Wehner and Flatt, 1977; Collett and Cartwright, 1979; Lehrer and Collett, 1994). Thus, angular size cannot serve as a cue for discriminating among targets that are at different ranges. Only knowledge of absolute size would be useful in this task.

Based on the finding that bees measure distance irrespective of size (see previous section), Horridge et al. (1992) set out to investigate the bee's performance in learning the absolute size of a target. The idea was that, at least theoretically, when the distance of an object is known, its absolute size can be inferred from its angular size.

The experimental apparatus (Fig. 10A and B) consisted of two sheets of transparent Perspex. One was placed 60 mm above a white ground, the other (serving to control for the bees' height of flight) 60 mm above the lower one. The targets were two black discs of different sizes, one rewarded, the other not. A drop of sugar-water was placed on the top sheet over the projection of the rewarded disc, and a drop of plain water over the other one. Because a drop of water and one of sugar-water look alike and lack scent, the drops could not serve for accomplishing the discrimination. On

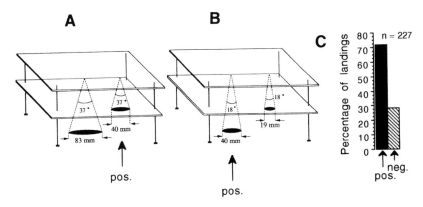

Figure 10. Learning of absolute size. Bees were trained alternately in the two situations shown in (A) and (B). The positive (rewarded) disc had a constant absolute size, but its distance from the bee and therefore its angular size differed between the two training situations. In either situation, the unrewarded disc had the same angular size as had the rewarded one. (C) Proportion of the bees' choices between the positive and the negative disc, situations (A) and (B) taken together. (A) and (B) modified from Horridge et al. (1992). (C) Data from Horridge et al. (1992).

each visit to the apparatus, each bee scored a positive or a negative point, depending on whether its first landing was on the drop associated with the rewarded or the unrewarded disc, respectively.

In the experiment shown in Figure 10, the rewarded disc had a diameter of 40 mm and was placed either on the middle level, where it subtended a visual of 37° (Fig. 10A), or on the ground, where it subtended 18° (Fig. 10B), as viewed by a bee flying at the level of the uppermost sheet. The unrewarded disc subtended the same visual angle as did the rewarded one, i.e., a disc of 83 mm diameter was placed on the ground when the rewarded disc was on the middle level (Fig. 10A), and a 19 mm diameter disc was placed on the middle level when the rewarded disc was on the ground (Fig. 10B). Thus, angular size alone could not serve for accomplishing the discrimination. The situations (A) and (B) alternated at regular intervals throughout the experiment.

The bees' decisions between the two discs (Fig. 10C) show that they recognize the rewarded disc despite the fact that it is presented at two different heights and therefore it subtends two different visual angles. From these and further results it was concluded that bees infer absolute size by combining distance and visual angle. The same conclusion was drawn from results obtained using stimuli presented on vertical planes (Horridge et al., 1992).

Edges (9)

The attractiveness of edges has already been noted by Hertz (1930) and Free (1979) using black figures placed flat on a white ground. They reported that bees preferentially land on the borders between the figure and its background. One explanation of this preference might be the conspicuous contrast perceived at the edge, eliciting a perigrammotactic response (see R. Campan, this volume). Could it be that this response is due to the retinal motion perceived from the edge as the bee arrives at it?

To examine this question, Lehrer et al. (1990) video-recorded bees trained to black discs placed on a white background and found that landings occur mainly on the boundaries, facing the inner area of the disc (Fig. 11A). Repeating the experiment using blue discs placed on a yellow ground, the results were the same as before when green-contrast was present (Fig. 11B), but not when green-contrast was absent (Fig. 11C), as indeed expected in a motion-dependent task. Using discs that produced against the background green-contrast without additional intensity contrast, landings were again at the boundaries, facing the inner area of the discs (Fig. 11D). Thus, green-contrast is sufficient in the edge detection task.

Bees were shown to be guided by edges in yet another spatial task, involving localization of a frontal target with the help of lateral marks. Lehrer (1990) trained bees to collect sugar water from a small box placed behind a vertical board containing an array of 27 holes, arranged in nine

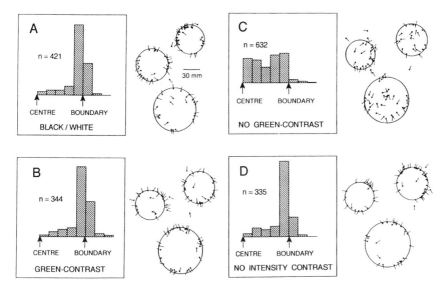

Figure 11. The use of absolute image motion in an edge detection task. Bees were trained to collect sugar water from three discs placed on a horizontal plane, and were then video-recorded from above. Histograms show the normalized distribution of landings along the radii of the discs. The stimuli used are specified in each panel (A–D). n, total number of landings. Panels to the right of each histogram show, as an example, positions of the bees' landings during one test; dots denote the bee's head position, dashes denote the orientation of its longitudinal axis, at the instance of landing. Modified from Lehrer et al. (1990).

rows and three columns (Fig. 12, top left panel), the entrance to the box being through the central hole of the array. To reach the reward, the bees had to fly between two white walls placed perpendicularly to the board. During training, a black horizontal stripe (width 3 cm) was placed on each of the lateral walls at the height of the reward hole. In the tests, the bees' choices among the 27 holes were recorded. The percentage of choices was then calculated for the upper, central and lower subarray of holes, comprising nine holes each. When the stripes were presented at the height of the reward hole, as during training, the bees preferred the central array. However, when the stripes were displaced to a lower or a higher position, the same bees preferred the lower or the higher array of holes, respectively (Fig. 12A). Consequently, orientation towards the frontal goal is guided by the retinal position of the laterally viewed contours.

Repeating the experiment using yellow stripes on blue lateral walls, bees trained with green-contrast performed as well as before (Fig. 12B), whereas in the absence of green-contrast, the marks were ineffective in guiding the bees to the target (Fig. 12C). These results suggest that motion cues are involved in the localization performance, i.e., the bee keeps the mark in the trained lateral retinal position by correcting for any drift of the

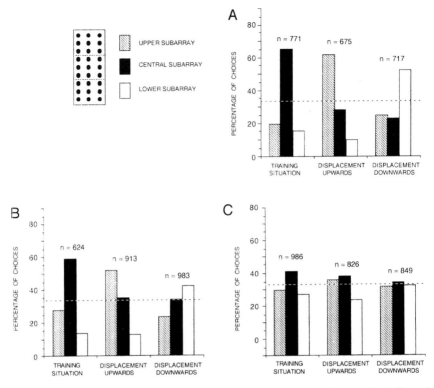

Figure 12. The use of laterally positioned contours for localizing a frontal target. Top left panel shows definition of upper, central and lower subarray of holes viewed frontally. To reach the central (rewarded) hole, bees had to fly between two lateral walls each carrying a stripe placed at the height of the central hole. Tests were conducted with the stripe at the training height, as well as with the stripe displaced to either a higher or lower position. Histograms show the distribution of the bees' choices among the three subarrays. Dashed lines denote random-choice level. (A) Black stripe on a white background; (B) and (C) yellow stripe on a blue background, using two different colour combinations: (B) Green contrast, (C) no green contrast. n, number of landings. Modified from Lehrer et al. (1990).

mark from this position during flight. Similar results (not shown) were obtained using a single edge between a yellow and a blue area, rather than a stripe (Lehrer, 1990). In this case, the polarity of the edge was randomized (i.e., blue and yellow were presented alternately in the lower and the upper visual field), to prevent bees from simply using the distribution of colours in the lateral visual field (see Section 4).

The particular role of edges is evident not only in the bees' choice behaviour, but, in addition, in their flight behaviour. Lehrer et al. (1985) trained bees to various patterns presented on a vertical plane. The trained bees were then video-recorded while they were flying in front of the patterns. The evaluation of the bees' flight trajectories revealed that they follow, in free

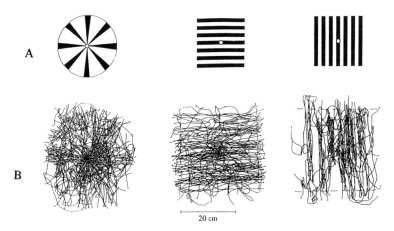

Figure 13. The bees' scanning behaviour. Bees were trained to one of the patterns shown in (A). In (B), frame by frame analysis of flight trajectories of bees video-recorded after 30–50 rewarded visits to the pattern in front of which they are flying is shown. Lines connect the bee's head position in successive frames of the video tape. Bees video-recorded on their visit to a novel pattern displayed the same contour-following behaviour (not shown). Modified from Lehrer et al. (1985).

flight, the contours contained in the pattern (Fig. 13). This behaviour, which the authors termed "scanning", is independent of a learning process and is not subject to habituation, i.e., bees follow the contours regardless of whether they fly in front of the learned pattern, even after 50 and more rewarded visits, as in Figure 13, or are presented with a totally novel pattern. Using coloured patterns (not shown), it was found that bees scan the pattern if it contains green-contrast, but not if green-contrast is absent (Lehrer et al., 1985).

The attractiveness of edges is manifested even when motion contrast (see Section 2), rather than intensity contrast, is present at the edge. Lehrer and Srinivasan (1993) trained bees to collect food at the edge between two randomly patterned surfaces, one placed raised above the other (Fig. 14A). in the tests, the bees were video-recorded from above. Frame-by-frame analyses of the video tapes revealed that the majority of the landings occur at the edge. Furthermore, bees flying in the direction of the raised surface usually land as soon as they arrive at the edge, whereas bees flying in the opposite direction usually cross the edge without landing on it (see ratios landings:crossings in Fig. 14B). It was concluded that the edge elicits landing when the bee perceives a local increase in the speed of image motion, but not when the speed of image motion decreases as the bee arrives at the edge. This mechanism automatically determines the direction in which the insect would land on an edge, namely facing the nearer object.

Active acquisition of motion information (10)

Because image motion is an inevitable consequence of locomotion, the bee's use of motion cues for distance estimation, object-ground discrimination, and edge detection (see above) may be considered as being a by-product of the bee's flight activity. However, in some cases bees acquire motion information *actively* by selecting flight patterns that create the particular type of motion cues needed to cope with the task at hand. For example, in the tests described in Figure 9A, bees were found to approach the discs perpendicularly to the boundaries (Srinivasan et al., 1989), thus maximizing the amount of image motion perceived from the edge.

Another, even more impressive example for active acquisition of motion information is the following. When the experiment described in Figure 14 is repeated using striped, rather than randomly patterned surfaces (Fig. 15, left-hand panel), the bees initially follow the contours (Fig. 15A), based on their spontaneous scanning tendency. Flight parallel to the contours, however, produces no image motion and thus it prevents the bees from detecting the edge and finding the reward. Although normally the scanning

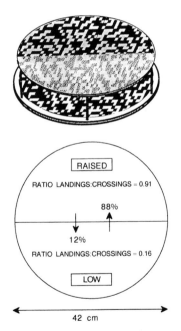

Figure 14. The use of speed of image motion in an edge detection task, and for selecting direction of landing on the edge. Bees were trained to an edge between two randomly patterned surfaces, one raised 5 cm above the other (A). During training, the reward of sugar water was placed at a random position along the edge. (B) Results of evaluation of video recordings done from above during tests. Percentage of landings on the edge in the direction of the raised and the low pattern (arrows) are shown. Interpretation of the ratios landings:crossings is given in the text. Modified from Lehrer and Srinivasan (1993).

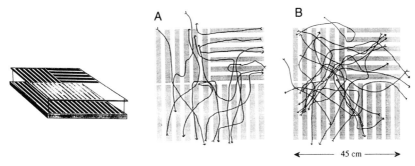

Figure 15. Active acquistion of motion information. Bees were trained to collect food at the edge between a low and a raised black-and-white striped surface (left-hand panel). (A, B) Examples of frame-by-frame analyses of video recordings done from above. Lines connect the position of the bee's head in successive frames of the video tape. Shaded bars (not to scale) denote the orientation of the black stripes. During their initial visits to the experimental set-up (A), the bees display the scanning behaviour (see Fig. 13): 75% out of the 850 single frames evaluated show flight parallel to the contours. As training continues (B), the bees abandon the scanning behaviour and select oblique flight patterns (74% out of 4000 frames evaluated) that produce image motion suitable for detecting the edge between the low and the raised surfaces. Modified from Lehrer and Srinivasan (1994).

behaviour does not habituate even after 50 and more rewarded visits (see Fig. 13), the bees now, after only 5–10 visits, abandon the scanning behaviour and select oblique flight paths (Fig. 15B) that produce the image motion needed for detecting the edge (Lehrer and Srinivasan, 1994).

Further examples for active acquisition of motion cues are the stereotyped fight manoeuvres performed by bees upon departing from a novel feeding site (Lehrer, 1991b, 1993), and by wasps upon departure from the nest (Zeil, 1993a, b). These so-called learning flights were shown, in both bees and wasps, to serve for acquiring size-independent cues to distance (Brünnert et al., 1994; Lehrer and Collett, 1994; Zeil et al., 1996; Collett and Zeil, this volume).

Discussion

The purpose of the present chapter was to highlight one of the many aspects of orientation performance, namely the use of spatial cues for recognizing the food source, which is, under natural conditions, a flower species that has been proved to be profitable on the basis of previous experience. Behavioural work on honeybees has provided the majority of the data available on this topic, manily due to this social insect's excellent learning capacity that enables to examine the use of particular cues vers specifically.

Most of the studies reviewed above employed training methods designed so as to encourage bees to use a particular spatial parameter, rather than

others. Therefore, nothing can be said about the hierarchy of the effectiveness of the various cues that have been investigated. Still, it is noteworthy
that, when the dominant learned parameter is absent in the tests, bees
are able to use others, also present in the previously rewarded target (see
Fig. 5 in Lehrer et al., 1985).

The use of eidetic memory

Several spatial parameters, such as the positon of the stimulus in the visual
field (Section 3), contour orientation (Section 4), and distribution of contrasting areas (Section 4 and 9) are only useful if the image of the target has
been stored retinotopically, such that it forms a template that can serve for
future recognition. In other words, eidetic memory, because it is linked to
spatial constancy, can only be used for target recognition if the target is
always approached from a constant direction and projects, on every
approach, onto the same eye region, as has also been suggested for landmark learning (Zeil, 1993b; Wehner et al., 1996; Collett, 1996; T. S. Collett
and J. Zeil, this volume). At an artificial food source presenting the visual
stimulus on a vertical plane (see Fig. 1 and Figs. 3–7), these conditions are
met. As yet, however, no results of field observations have been reported
to document that they are met at the bee's natural food source as well.
Natural flowers, even of one and the same species, may present themselves
in any plane, and may be approached from any direction.

Thus, to use eidetic memory in the natural task, bees should be capable
of selecting a particular direction of approach relative to the flower's
orientation in space on every visit to every particular flower, or, alternatively, to store several snapshots of the flower from several vantage points,
as do wasps (*vespula vulgaris*) upon departing from an artificial food
source (Collett and Lehrer, 1994). The question of whether or not bees
employ these strategies to acquire template-like memories of a flower
remains to be investigated.

*The role of colour vision in spatial tasks, and the colour blindness
of motion vision*

Most of the studies concerned with the bee's close-range spatial vision
were conducted using black-and-white patterns (see review by Wehner,
1981). The combination black/white produces the highest possible intensity
contrast, because the modulation amplitude is high, and mainly, because
the intensity of black is close to zero (see Equation (1) in *General methods*).
Natural flowers produce a much lower intensity contrast against the background, because neither the intensity of the flower nor that of the background (foliage, soil, rock, or the sky) is ever close to zero. However,

flowers produce, in all cases, colour contrast against their background. We have seen that colour contrast is sufficient for coping with several spatial tasks (Sections 1, 3, 9). Thus, colour vision, although it is not a prerequisite for spatial vision, participates in spatial vision.

We have seen, in addition, that in tasks that require the use of cues derived from image motion, the bee's performance is colour blind (Sections 7 and 9). The bee's motion vision is mediated exclusively by intensity contrast perceived by the green-sensitive photoreceptor. Further evidence for this conclusion is given in Lehrer (1991a), and in Zhang and Srinivasan (1993). The advantage of colour blindness in motion vision, and the particular role that the green receptor plays in it, are discussed in Srinivasan (1985) and in Lehrer (1987), respectively. Recently, the green receptor was shown to be crucial also in the analysis of contour orientation (Geiger and Srinivasan, 1996), although results using stroboscopic illumination have suggested that bees are adept in this task even in the absence of image motion (Srinivasan et al., 1993).

Learning processes and hard-wired orientation behaviours

Some of the spatial cues examined must be learned before they can be used by the bee in the discrimination task. These include position in the visual field (Section 3), contour orientation (Section 4), angular size (Section 7), absolute size (Section 8), and even spatial frequency (Section 3 and 6), although the latter parameter also plays a role independently of a learning process (see Introduction and Section 6).

All of these parameters are actually displayed by natural flowers. It should, indeed, be noted that, even at an artificial food source, bees cannot be trained to use spatial cues that are never provided by natural flowers. For example, although bees use directional information derived from the polarization pattern of the sky in the task of navigation (e.g., Rossel and Wehner, 1984; R. Wehner, this volume), they cannot be trained to a particular E-vector direction when it is used to mark the food source (Lau, 1976). Another example is the inability of bees to associate a magnetic vector with a food reward, although they use magnetic directional information in the context of recruitment dances and of comb building (see M.M. Walker, this volume). And finally, at the food source, bees cannot learn to discriminate between a steady and a flickering light (Srinivasan and Lehrer, 1984), although they clearly perceive flicker (Autrum and Stöcker, 1950). As has already been put forward by Lauer and Lindauer (1971) and Menzel (1985, 1990), learning capacity is genetically fixed. It would indeed make little sence to provide a small-brained forager with a learning capacity for cues that will hardly ever be encountered at the feeding site.

Similar considerations may apply to the finding that some of the responses to spatial parameters reviewed here are spontaneous, and are thus altogether independent of a learning process. Such responses include the preference for high-frequency patterns at close range and for low-frequency patterns at a farther range (Section 6), the preference for radiating elements and symmetrical shapes (Section 5), the preference for edges while landing (Section 9), and the contour-following tendency (Sections 9 and 10). It indeed makes a lot of sense for small-brained animals to come equipped with hard-wired neural mechanisms that require neither learning nor storage capacity and are nevertheless perfectly suitable for coping with natural tasks (see also R. Campan, this volume). Still, it must be emphasized that many of these hard-wired mechanisms are modifiable by experience (see, again, R. Campan, this volume). For example, bees can be trained to prefer either a low or a high spatial frequency (Fig. 3; see also Fig. 67 in Wehner, 1981), although spontaneously they would not (Section 6). They even learn to suppress the innate optomotor response when it prevents them from landing on their target (Lehrer and Srinivasan, 1992), and to abandon the otherwise innate scanning behaviour when it prevents them from pin-pointing the goal and thus from collecting the reward (see Fig. 15).

Acknowledgements
I wish to thank Raymond Campan for his inspiring comments on the manuscript and for many helpful suggestions. The studies conducted at the Centre for Visual Sciences in Canberra (Sections 2, 5–7, part of Section 8, and Section 9) were supported by grants to G. Adrian Horridge and Mandyam V. Srinivasan.

References

Anderson, A.M. (1977a) Parameters determining the attractiveness of stripe patterns in the honey bee. *Anim. Behav.* 25:80–87.
Anderson, A.M. (1977b) A model for landmark learning in the honey-bee. *J. Comp. Physiol.* 114:335–355.
Autrum, H. and Stöcker, M. (1950) Die Verschmelzungsfrequenzen des Bienenauges. *Z. Naturforsch.* 5b:38–43.
Brünnert, U., Kelber, A. and Zeil, J. (1994) Ground-nesting bees determine the location of their nest relative to a landmark by other than angular size cues. *J. Comp. Physiol.* A 175:363–370.
Burkhardt, D., Darnhofer-Demar, B. and Fischer, K. (1973) Zum binokularen Entfernungsmessung der Insekten. I. Die Struktur des Sehraums von Insekten. *J. Comp. Physiol.* 87:165–188.
Cartwright, B.A. and Collett, T.S. (1979) How honey-bees know their distance from a near-by visual landmark. *J. Exp. Biol.* 82:367–372.
Cartwright, B.A. and Collet, T.S. (1981) Landmark learning in bees: experiments and models. *J. Comp. Physiol.* 151:521–543.
Collett, T.S. (1996) Honeybee navigation *en route* to the goal: multiple strategies for the use of landmarks. *J. Exp. Biol.* 199:227–235.
Collett, T.S. and Harkness, L.I.K. (1982) Depth vision in animals. *In*: D.J. Ingle, M.A. Goodale and R.J.W. Mansfield (eds): *Analysis of Visual Behavior*. M.I.T. Press, Cambridge, Massachusetts, pp 111–176.
Collett, T.S. and Lehrer, M. (1993) Looking and learning: a spatial pattern in the orientation flight of Vespula vulgaris. *Trans. Phil. Roy. Soc. Lond.* B 252:129–134.

Couvillon, P.A. and Bitterman, M.E. (1987) Discrimination of color-odor compounds by honeybees: Tests of a continuity model. *Anim. learn. Behav.* 15:218–227.

Cruse, H. (1972) Versuch einer quantitativen Beschreibung des Formensehens der Honigbiene. *Kybernetik* 11:185–200.

Dafni, A., Lehrer, M. and Kevan, P. (1996) Flower spatial parameters and insect spatial vision. *Biol. Rew.* 72:239–282.

Dafni A., Lehrer, M., Kevan, P. (1997) Flower spatial parameters and insect spatial vision. *Biol. Rev.* 72:239–282.

Daumer, K. (1956) Reizmetrische Untersuchung des Farbensehens der Biene. *Z. vergl. Physiol.* 38:413–478.

Esch, H.E. and Burns, J.E. (1996) Distance estimation by foraging honeybees. *J. Exp. Biol.* 199:155–162.

Exner, S. (1876) Über das Sehen von Bewegung und die Theorie des zusammengesetzten Auges. *Sitz. Ber. Kaiserl. Akad. Wiss., Mat.-Nat. Wiss.* C 172:156–191.

Faulkner, G.J. (1976) Honeybee behaviour as affected by plant height and flower colour in Brussels sprouts. *J. Apicult. Res.* 15:15–18.

Fonta, C., Sun, X.J. and Masson, C. (1991) Cellular analysis of odour integration in the honeybee. *In*: L.J. Goodman and R.C. Fisher (eds): *The Behaviour and Physiology of Bees*. CAB International, Wallingford, UK, pp 185–202.

Free, J.B. (1970) Effect of flower shapes and nectar guides on the behaviour of foraging bees. *Behav.* 37:269–285.

Frisch, K. von (1915) Der Ferbensinn und Formensinn der Bienen. *Verhandl. d. Deutsch. Zool. Ges.* 35:1–182.

Frisch, K. von (1919) Über den Geruchssinn der Bienen und seine blütenbiologische Bedeutung. *Zool. Jb. (Physiol.)* 37:1–238.

Frisch, K. von (1965) *Tanzsprache und Orientierung der Bienen*. Springer-Verlag, Berlin, Heidelberg, New York.

Geiger, A.D. and Srinivasan, M.V. (1995) Pattern recognition in honeybees: Eidetic imagery and orientation discrimination. *J. Comp. Physiol.* A 176:791–795.

Geiger, A.D. and Srinivasan, M.V. (1996) Pattern recognition in honeybees: chromatic properties of orientation analysis. *J. Comp. Physiol.* A 178:763–769.

Giurfa, M., Backhaus, W. and Menzel, R. (1995) Color and angular orientation in the discrimination of bilateral symmetry patterns in the honeybee. *Naturwissenschaften* 82:198–201.

Giurfa, M., Vorbyev, M., Kevan, P. and Menzel. R. (1996) Detection of coloured stimuli by honeybees: minimum visual angles and receptor specific contrasts. *J. Comp. Physiol.* A 178:699–710.

Götz, K.G. (1965) Die optischen Übertragungseigenschaften des visuellen Systems einiger Augenmutanten der Fruchtfliege *Drosophila*. *Kybernetik* 2:215–221.

Gross, C.L. (1992) Floral traits and pollinator constancy: foraging by naive bees among three sympatric legumes. *Austral. J. Ecol.* 17:67–74.

Hateren, H.J., van Srinivasan, M.V. and Wait, P.B. (1990) Pattern recognition in bees: orientation discrimination. *J. Comp. Physiol.* A 167:649–654.

Helversen, O. von (1972) Zur spektralen Unterschiedsempfindlichkeit der Honigbiene. *J. Comp. Physiol.* 80:439–472.

Hertz, M. (1930) Die Organisation des optischen Feldes bei der Biene II. *Z. vergl. Physiol.* 11:107–145.

Hertz, M. (1933) Über figurale Intensitäten und Qualitäten in der optischen Wahrnehmung der Biene. *Biol. Zbl.* 54:10–40.

Hengstenberg, R. and Götz, K.G. (1967) Der Einfluss des Schirmpigmentgehalts auf die Helligkeits- und Kontrastwahrnehmung bei *Drosophila*-Augenmutanten. *Kybernetik* 3:276–285.

Horridge, G.A. (1994) Bee vision of pattern and 3D. *BioEssays* 16:877–884.

Horridge, G.A., Zhang, S.W. and Lehrer, M. (1992) Bees can combine range and visual angle to estimate absolute size. *Phil. Trans. R. Soc. Lond.* B 337:49–57.

Kaiser, W. (1975) The relationship between visual movement detection and colour vision in insects. *In*: G.A. Horridge (ed.): *The compound eye and vision of insects*. Clarendon Press, Oxford, pp 359–377.

Kaiser, W. and Liske, E. (1974) Die optomotorischen Reaktionen von fixiert fliegenden Bienen bei Reizung mit Spektrallichtern. *J. Comp. Physiol.* 80:391–408.

Lau, D. (1976) Reaktionen von Honigbienen (*Apis mellifica* L.) auf Polarisationsmuster an der Futterquelle. *Zoologischer Garten, Neue Folge*, Jena 46:34–38.

Lauer, J. and Lindauer, M. (1971) Genetisch fixierte Lerndispositionen bei der Honigbiene. *Abhanl. Akad. Wissensch. u. Literatur, mathematisch-naturwissenschaftliche Klasse*, 1:1–87.

Lehrer, M. (1987) To be or not to be a colour-seeing bee. *Isr. J. Entomol.* 21:51–76.

Lehrer, M. (1990) How bees use peripheral eye regions to localize a frontally positioned target. *J. Comp. Physiol.* A 167:173–185.

Lehrer, M. (1991a) Locomotion does more than bring the bee to new places. *In*: L.J. Goodman and R.C. Fisher (eds): *The Behaviour and Physiology of Bees*. CAB International, Wallingford, UK, pp 185–202.

Lehrer, M. (1991b) Bees which turn back and look. *Naturwissens.* 78:274–276.

Lehrer, M. (1993) Why do bees turn back and look? *J. Comp. Physiol.* A 172:544–563.

Lehrer, M. (1996) Small-scale navigation in the honeybee: active acquisition of visual information about the goal. *J. Exp. Biol.* 199:253–261.

Lehrer, M. and Bischof, S. (1995) Detection of model flowers by honeybees: the role of chromatic and achromatic contrast. *Naturwissens.* 82:145–147.

Lehrer, M. and Collett, T.S. (1994) Approaching and departing bees learn different cues to the distance of a landmark. *J. Comp. Physiol.* A 175:171–177.

Lehrer, M. and Srinivasan, M.V. (1992) Freely flying bees discriminate between stationary and moving objects: Performance and possible mechanisms. *J. Comp. Physiol.* A 171:457–467.

Lehrer, M. and Srinivasan, M.V. (1993) Object-ground discrimination in bees: Why do they land on edges? *J. Comp. Physiol.* A 713:23–32.

Lehrer, M. and Srinivasan, M.V. (1994) Active vision in honeybees: task-oriented suppression of an innate behaviour. *Vesion Res.* 34:511–516.

Lehrer, M., Wehner, R. and Srinivasan, M.V. (1985) Visual scanning behaviour in honeybees. *J. Comp. Physiol.* A 157:405–415.

Lehrer, M., Srinivasan, M.V., Zhang, S.W. and Horridge, G.A. (1988) Motion cues provide the bee's visual world with a third dimension. *Nature* 332:356–357.

Lehrer, M., Srinivasan, M.V. and Zhang, S.W. (1990) Visual edge detection in the honeybee and its spectral properties. *Proc. R. Soc. Lond.* 238:321–330.

Lehrer, M., Horridge, G.A. Zhang, S.W. and Gadagkar, R. (1994) Shape vision in bees: innate preference for flower-like patterns. *Phil. Trans. R. Soc. Lond.* 347:123–137.

Leppik, E.E. (1953) The ability of insects to distinguish number. *Am. Naturalist* 87:229–236.

Levin, D.A. and Kerster, H.W. (1971) Pollinator flight directionality and its effect on pollen flow. *Evolution* 25:113–118.

Manning, A. (1957) Some evolutionary aspects of the flower constancy of bees. *Proc. Roy. Phys. Soc.* 25:67–71.

Mazokhin-Porshnyakov, G.A., Semyonova, S.A. and Milevskaya, I.A. (1977) Characteristic features of the identification by apis mellifera of objects by their size (in Russian). *J. Obsch. Biol.* 38:855–962.

Menzel, R. (1967) Untersuchungen zum Erlernen von Spektralfarben durch die Honigbiene, *Apis mellifica. Z. vergl. Physiol.* 56:22–62.

Menzel, R. (1969) Das Gedächtnis der Honigbienen für Spektralfarben. II. Umlernen und Mehrfachlernen. *Z. vergl. Physiol.* 63:290–309.

Menzel, R. (1985) Learning in honey bee in an ecological and behavioural context. *In*: M. Lindauer and B. Hölldobler (eds): *Experimental & Behavioural Ecology*. Fischer, Stuttgart, pp 55–74.

Menzel, R. (1990) Learning, memory and "cognition" in honeybees. *In*: R.P. Kesner and D.S. Olton (eds): *Neurobiology and comparative cognition*. Erlbaum, Hillsdale NJ, pp 237–292.

Menzel, R. and Backhaus, W. (1989) Colour vision in honeybees: phenomena and physiological mechanisms. *In*: D.G. Stavenga and R.C. Hardie (eds): *Facets of Vision*. Springer-Verlag, Berlin, Heidelberg, pp 281–297.

Menzel, R. and Lieke, E. (1983) Antagonistic color effects in spatial vision of honeybees. *J. Comp. Physiol.* 151:441–448.

Menzel, R., Ventura, D.F., Hertel, H., de Souza, J.M. and Greggers, U. (1986) Spectral sensitivity of photoreceptors in insect compound eyes: comparison of species and methods. *J. Comp. Physiol.* A 158:165–177.

Menzel, R., Geiger, K., Chittka, L., Joerges, J., Kunze, J. and Müller, U. (1996) The knowledge base of the bee navigation. *J. Exp. Biol.* 199:141–146.

Opfinger, E. (1949) Zur Psychologie der Duftdressuren bei Bienen. *Z. vergl. Physiol.* 31: 441–453.

Rossel, S. and Wehner, R. (1984) How bees analyse the polarisation patterns in the sky. *J. Comp. Physiol.* A 154:607–615.

Rossel, S., Mathis, U. and Collett, T.S. (1992) Vertical disparity and binocular vision in the praying mantis. *Vis. Neurosci.* 8:165–170.

Schlieper, C. (1927) Farbensinn der Tiere und optomotorische Reaktionen. *Z. vergl. Physiol.* 6:453–472.

Schnetter, B. (1972) Experiments on pattern discrimination in honey bees. *In*: R. Wehner (ed.): *Information Processing in the Visual Systems of Arthropods.* Springer-Verlag, Berlin, Heidelberg, New York, pp 195–200.

Srinivasan, M.V. (1985) Shouldn't directional movement detection necessarily be "colour-blind"? *Vision Res.* 25:997–1000.

Srinivasan, M.V. (1994) Pattern recognition in the honeybee: recent progress. *J Insect Physiol.* 40:183–194.

Srinivasan, M.V. and Lehrer, M. (1984) Temporal acuity of honeybee vision: behavioural studies using flickering stimuli. *Physiol. Entomol.* 9:447–457.

Srinivasan, M.V., Lehrer, M., Zhang, S.W. and Horridge, G.A. (1989) How honeybees measure their distance from objects of unknown sizes. *J. Comp. Physiol.* A 165, 605–613.

Srinivasan, M.V., Lehrer, M. and Horridge, G.A. (1990) Visual figure-ground discrimination in the honeybee: The role of motion parallax at boundaries. *Proc. Roy. Soc. Lond.* 238: 331–350.

Srinivasan, M.V., Zhang, S.W. and Rolfe, B. (1993) Is pattern vision in insects mediated by "cortical" processing? *Nature* 362:539–540.

Srinivasan, M.V., Zhang, S.W. and Whitney, K. (1994) Visual discrimination of pattern orientation by honeybees. *Phil. Trans. Roy. Soc. Lond.* B 343:199–210.

Srinivasan, M.V., Zhang, S.W., Lehrer, M. and Collett, T.S. (1996) Honeybee navigation *en route* to the goal: visual flight control and odometry. *J. Exp. Biol.* 199:237–244.

Waser, N.M. (1986) Flower constancy: definition, cause and measurement. *Am. Naturalist* 127: 593–603.

Wehner, R. (1969) Der Mechanismus der optischen Winkelmessung bei der Biene (*Apis mellifica*). *Zool. Anz., Suppl.* 33:586–592.

Wehner, R. (1972a) Pattern modulation and pattern detection in the visual system of Hymenoptera. *In*: R. Wehner (ed.): *Information Processing in the Visual Systems of Arthropods.* Springer-Verlag, Berlin, Heidelberg, New York, pp 183–194.

Wehner, R. (1972b) Dorsoventral asymmetry in the visual field of the bee, *Apis mellifera. J. Comp. Physiol.* 77, 256–277.

Wehner, R. (1981) Sptial vision in arthropods. *In*: H. Autrum (ed.): *Handbook of Sensory Physiology VII/6C.* Springer-Verlag, Berlin, Heidelberg, New York, pp 287–616.

Wehner, R. (1992) Homing in arthropods. *In*: F. Papi (ed.): *Animal Homing.* Chapman and Hall, London, pp 45–144.

Wehner, R. and Flatt, I. (1977) Visual fixation in freely flying bees. *Z. Naturforsch.* 32:469–471.

Wehner, R. and Lindauer, M. (1966) Die optische Orientierung der Honigbiene (*Apis mellifica*) nach der Winkelrichtung frontal gebotener Streifenmuster. *Zool. Anz., Suppl.* 30:239–246.

Wehner, R., Michel, B. and Antonsen, P. (1996) Visual navigation in insects: coupling of egocentric and geocentric information. *J. Exp. Biol.* 199:129–140.

Wolf, E. (1933) Das Verhalten der Bienen gegenüber flimmernden Feldern und bewegten Objekten. *Z. vergl. Physiol.* 20:151–161.

Wolf, E. and Zerrahn-Wolf, G. (1935) The effect of light intensity, area, and flicker frequency on the visual reactions of the honeybee. *J. Gen. Physiol.* 18:853–863.

Zerrahn, G. (1934) Formdressur und Formunterscheidung bei der Honigbiene. *Z. vergl. Physiol.* 20:117–150.

Zhang, S.W. and Srinivasan, M.V. (1993) Behavioural evidence for parallel information processing in the visual system of insects. *Jap. J. Physiol. 43, Suppl.* 1:247–258.

Zhang, S.W. and Srinivasan, M.V. (1994) Pattern recognition in honeybees: analysis of orientation. *Phil. Trans. R. Soc. Lond.* B 346:399–406.

Zhang, S.W., Srinivasan, M.V. and Horridge, G.A. (1992) Pattern recognition in honeybees: local and global analysis. *Proc. R. Soc. Lond.* B 248:55–61.

Zeil, J. (1993a) Orientation flights of solitary wasps (*Cerceris*, Sphecidae; Hymenoptera): I. Description of flight. *J. Comp. Physiol.* A 172:189–205.

Zeil, J. (1993b) Orientation flights of solitary wasps (*Cerceris;* Sphecidae; Hymenoptera): II. Similarity between orientation and return flights and the use of motion parallax. *J. Comp. Physiol.* A 172:209–224.

Zeil, J., Kelber, A. and Voss, R. (1996) Structure and function of lerning flights in groundnesting bees and wasps. *J. Exp. Biol.* 199:245–252.

Orientation and Communication in Arthropods
ed. by M. Lehrer
© 1997 Birkhäuser Verlag Basel/Switzerland

The ant's celestial compass system: spectral and polarization channels

R. Wehner

Department of Zoology, University of Zürich, Winterthurerstrasse 190, CH-8057 Zürich, Switzerland

Summary. Ants as well as bees derive compass information not only from the direct light of the sun, but also from the scattered light in the sky. In the present account, the latter phenomenon is described for desert ants, genus *Cataglyphis*. Due to the scattering of sunlight by the air molecules of the earth's atmosphere, spatial gradients of polarization, spectral composition and radiant intensity extend across the celestial hemisphere. All of these optical phenomena are exploited by the *Cataglyphis* navigator. Here I concentrate on the use *Cataglyphis* makes of the polarization and spectral skylight gradients. Either type of information is neurally processed by a separate sensory channel receiving its input from a separate part of the retina. These channels are characterized and their possible interactions are analyzed in a variety of behavioural experiments, in which ants, whose compound eyes are partially occluded by light-tight caps, are presented with spatially restricted and spectrally altered parts of the celestial hemisphere. It is discussed whether skylight patterns are used by the insect navigator simply to read a reference direction (e.g., the azimuthal position of the solar meridian) from the sky, or whether they are used to determine any particular point of the compass. Different approaches to examine these questions – behavioural and neuro-physiological analyses, computer simulations and robotics – are described, and results obtained by these approaches are reported. New ways of portraying the pattern of polarized light in the real sky are presented in Figures 2 (lower part) and 3, and Figure 22 introduces an autonomous agent navigating by polarized skylight. Conceptually, the last paragraph of this chapter provides my most general conclusions drawn from the analyses of the insect's skylight compass.

Introduction: scattered skylight as a compass cue

In the afternoon of May 16, 1914, the Swiss physician Felix Santschi visited an isolated stretch of desert habitat outside the ramparts of the North African city of Kairouan to perform an ingenious biological experiment. By the time darkness fell, he had obtained an intriguing result, but the interpretation of this result remained obscure for nearly half a century. What Santschi had observed – and described at length in his 1923 treatise – was a harvester ant, *Messor barbarus*, walking directly back home after a successful foraging trip. To his surprise, the ant maintained its homeward course even after he had surrounded it with a cardboard cylinder screening off the sun and providing the ant with only a small patch of sky. Santschi could even move the cylinder along with the ant as it walked, without disturbing its homeward course. Only after he had covered the opening of the cylinder with a ground-glass plate, did the ant stop homing and start walking in random directions. He repeated the experiment several times, and always got the same result.

A few years earlier, in 1911, he had found that ants – actually members of the same *Messor* species – could use the sun as a compass cue. In the present experiment, however, the ants could not see the sun. Santschi hypothesized that they might have been able to infer the position of the sun from an intensity gradient potentially visible to the ants in the top opening of the cylinder; but he immediately dismissed this interpretation, because he could perform this type of experiment even when the sun was close to the horizon. At this time of day, he himself was unable to discern any intensity differences within the small, zenith-centred patch of sky, which the ants were able to see.

Santschi's seminal experiment showed for the first time that animals can derive compass information not only from the direct light of the sun, but also from the scattered light in the sky. The scattering of sunlight by the air molecules in the earth's atmosphere provides spatial gradients of both colour and polarization. The latter is invisible to the unaided human eye, and to humans even the former is not very spectacular.

As far as Santschi's harvester ants are concerned, we can conclude that in determining their homeward courses they had exploited either the spectral or the polarization gradients, or both. The fact that insects – and as we now know, arthropods in general (Waterman, 1981) – can perceive the polarization gradients and use them for navigation, has first been established by v. Frisch (1949). Not knowing of Santschi's classical work in ants, he performed a similar experiment in bees – with the only crucial difference being that he presented the bees with Polaroid sheets which had become available just at that time (and that he tested them by evaluating their recruitment dances in the hive, rather than their actual routes between the hive and the feeding site). Since the early 1970s we have tried to unravel the mechanisms underlying the bee's and ant's "polarization compass" (for reviews see Wehner, 1994a, b). The use of spectral cues was discovered much later (for *Apis mellifera* see Brines and Gould, 1979; Edrich et al., 1979; Rossel and Wehner, 1984b). In the following we show that *Cataglyphis* ants exploit spectral skylight information as well, and that they process this information within a visual channel separate from the polarization channel. Furthermore, we investigate whether they can transfer information from one channel to another. Finally, we show how they behave when they are left alone with gradients detectable by only one spectral type of photoreceptor. In the present account I briefly summarize our results, inferences and hypotheses.

Let us first have a quick look at the physical parameters available in the sky, i.e., at the spatial structure of the polarization and spectral gradients which bees and ants have been shown to exploit. Light radiated by the sun is "unpolarized", e.g., its electric (E-) vector, which vibrates perpendicularly to the direction of propagation, changes its plane of vibration randomly every 10^{-8} s (Clarke and Graininger, 1971). However, on its way through the earth's atmosphere sunlight is scattered by the air molecules, i.e., by particles much smaller than the wavelength λ of light (diameter $d \ll \lambda$). This scattering results in the polarization of light: at any particular point in the sky the plane within which the E-vector oscillates is oriented in a particular direction. In general, it is oriented perpendicularly to the plane of scattering, i.e., perpendicularly to the great circle passing through the sun and the point observed. Consequently, the E-vectors in the sky form concentric circles around the sun.

This general rule can be easily read in Figure 1A, which portrays the celestial E-vector patterns as seen by an earthbound observer positioned in the centre of the celestial hemisphere. From these two figures the main geometrical feature common to all possible E-vector patterns in the sky is immediately apparent: the symmetry plane formed by the *solar meridian*

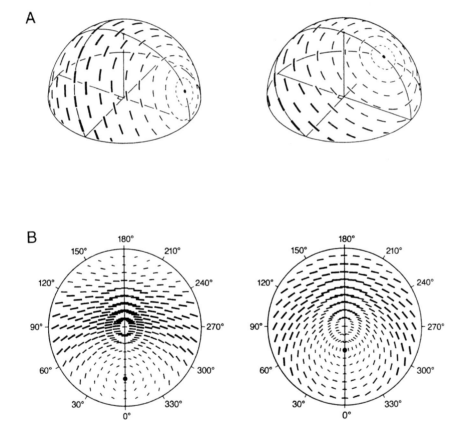

Figure 1. Three- (A) and two-dimensional (B) representations of the pattern of polarized light in the sky depicted for two different elevations of the sun (filled circle; left, $\mu_s = 25°$; right, $\mu_s = 60°$). The orientation and the sizes of the black bars mark the direction (χ) and degree (d) of polarization, respectively. Combined from Wehner (1982, 1994a).

and the *antisolar meridian*. Along this *solar vertical* light is invariably polarized parallel to the horizon. At all other points in the sky the E-vector orientation changes as the sun changes its elevation. While the sun is at the horizon, the pattern exhibits a twofold symmetry insofar as a second line of symmetry passes through the zenith at an angle of 90° to the solar vertical. Hence, in this special case there are not only left and right, but also solar and antisolar halves of the sky that form mirror images of each other.

Figure 1 reveals yet another skylight phenomenon, depicted by the size of the black bars. Apart from the orientation (χ) of the E-vector, it is also the degree (d) – or percentage – of polarization that varies across the celestial hemisphere. As this parameter (d) is proportional to some sine-square function of the scattering angle η (the angle formed by the sun and the point observed), it varies from 0 ($\eta = 0°$; direct, unscattered and hence unpolarized light from the sun) to 1.0 ($\eta = 90°$; great circle of maximally polarized light extending across the sky at an angular distance of 90° from the sun). When the sun is at the horizon (elevation of sun $\mu_s = 0°$), this circle of maximum polarization passes through the zenith, and tilts down within the antisolar half of the sky as the sun rises ($\mu_s > 0°$). Hence, the antisolar half of the sky is always more strongly polarized than the solar one.

The story told so far refers to an ideal atmosphere within which each ray of sunlight is scattered only once – primary (Rayleigh) scattering as first described by the English physicist and Nobel Prize laureate John William Strutt (1871), the later Baron Rayleigh – and which is free of particles with diameters d $> \lambda$. In the real atmosphere, however, particles of the latter size almost always occur. Multiple scattering, haze, dust and clouds, as well as reflections from the ground decrease the degree of polarization so that d hardly ever reaches its theoretical maximum of 1.0. In the natural sky it rarely exceeds values of 0.75 (Brines and Gould, 1982). Even though clouds depolarize skylight, E-vector patterns are often undisturbed by patchy cloud cover as long as the sun is visible. In this case, direct solar rays can illuminate the air between the clouds and the observer, so that polarization occurs due to light scattering within this volume of air. Because the polarization pattern is geometrically related to the position of the sun in the same way as it is in the unobscured sky, the E-vector pattern tends to be continuous over the sky even if parts of it are covered by clouds (Stockhammer, 1959; Brines and Gould, 1982).

Apart from some entoptic phenomena, such as Haidinger's and Boehm's brushes (Haidinger, 1844; Boehm, 1940), humans have only one way to visualize celestial E-vector patterns, namely by using polarization filters as

Figure 2. Skylight polarization at sunset. The upper two pictures were taken with a 180° fish-eye lens (Nikkor-Auto 1 : 2.8, f = 8 mm) equipped with a linear polarizer (HNP'B Polaroid). The transmission axis of the polarizer runs either parallel (upper left figure) or at a right angle (upper right figure) to the solar/antisolar meridian. In the lower figure, the same photographic

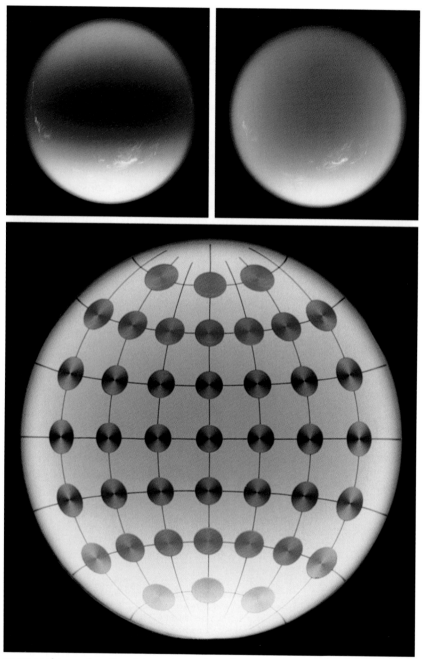

set-up used to produce the pictures shown above (camera and 180° wide-angle lense) was mounted in the centre of a Plexiglass hemisphere provided with a set of 41 circumferentially polarizing filters, axis finders. Celestial E-vector orientations χ are marked by the orientation of the dark hour-glass shaped figures. The two upper pictures are taken from Wehner (1982).

optical aids. This kind of visualization might date back to the times of the Vikings, who used cordierite crystals as polarization filters (Ramskou, 1969), but it was only in 1809 that the French physicist Etienne-Louis Malus, while looking at a glass through a calcite crystal, discovered and correctly interpreted the phenomenon of the polarization of light. However, the polarization he observed by looking at a glossy surface was produced by the reflection rather than the scattering of light. The first to describe the latter phenomenon was Dominique Arago (1811). Looking at the sky through a rotating dichroic (polarization) filter, he perceived the alternating appearance and disappearance of an impressive dark band extending across the sky at a distance of 90° from the sun. This phenomenon is portrayed in the upper part of Figure 2, which presents two photographs of the sunset sky, taken through a 180° fish-eye lens equipped with a linear polarizer. The dark band representing the area of maximum polarization shows up when the transmission axis of the polarizer is oriented parallel to the solar vertical, and vanishes when the filter is rotated by 90°. In order to fully

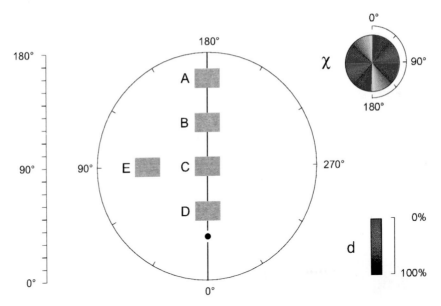

Figure 3. Direction (χ) and degree (d) of skylight polarization measured in the ultraviolet (UV) range at five locations (A–E) in the celestial hemisphere. The sky window C is centred about the zenith. 0° and 180° denote the solar and antisolar meridian, respectively. The elevation of the sun (filled circle) is 35°. The pictures were taken in southern Tunisia, Chott-el-Djerid (34.0°N, 8.4°E) by using a video polarimeter composed of a UV-sensitive Hamamatsu camera (Beam Finder III), a Video Walkman (Sony, Video 8), a UV-filter (Hamamatsu) and a UV-transmitting linear polarizer (HNP'B, Polaroid). Each picture resulted from a series of three recordings with the polarizer oriented at 0°, 45° and 90° relative to the vertical (for more details about the videopolarimetric method, see Horvàth and Varjú, 1997). False colour images of the distribution of χ and d are given on p. 151 (for colour conventions see inset figures on this page). The false colour images are accompanied by histograms depicting the frequency distributions of χ and d. G. Horvàth and R. Wehner, in preparation.

appreciate this phenomenon, the reader is invited to compare the photographs of Figure 2 with the schematic drawings of Figures 1 A and B (left panels), in which the sun is close to the horizon. A more detailed visualization of the celestial E-vector patterns is obtained if the fish-eye picture is taken through a set of so-called axis-finders. In these filters the absorption axes are aligned not linearly (as in the Polaroid sheets used in the upper part of Fig. 2), but circumferentially, i.e., along circles around the centres

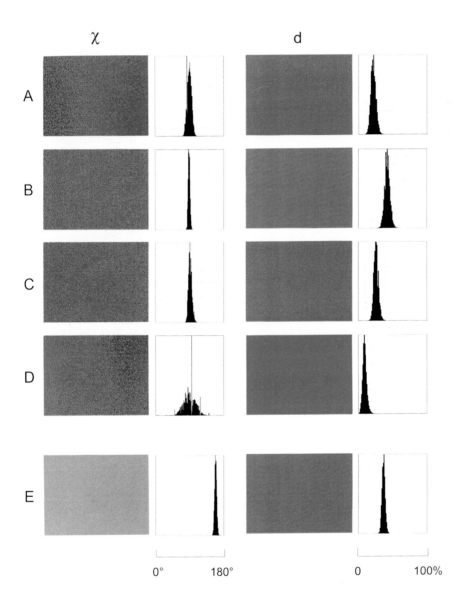

of the filters. Hence, the axes of the dark hour-glass shaped figures indicate the E-vector direction at the pixel in the sky at which the axis-finder is directed. The lower part of Figure 2 provides a view of the sunset sky taken through a Plexiglass hemisphere which was equipped with 41 evenly spaced axis-finders. Even more detailed information about skylight polarization can be obtained by taking more elaborate optical measurements in the sky and deriving from these measurements independent data sets about the direction and degree of polarization (see "false colour" images in Fig. 3).

The atmospheric scattering of sunlight does not only create polarization gradients, but spectral and intensity gradients as well. These are coarser and less reliable than the former, but as they are linked to the position of the sun in the same way as the polarization gradients are centred about the sun, they may provide additional – either supplementary or independent – compass information. Of course, in physical terms, "colour gradients" is not the proper term. What we mean is the intensity (radiance) gradients in the sky that vary with wavelength, so that the colour vision system – the dichromatic system of ants (Mote and Wehner, 1980; Labhart, 1986) or the trichromatic system of bees (Menzel, 1979) – is able to exploit the wavelength-dependent spatial distributions of radient intensity. (In fact, it is also the direction of polarization that varies with the wavelength of light, but these variations are very small and, for all practical purposes, can be neglected.) The principal feature of the spectral gradients in the sky can simply be stated as follows: with increasing angular distances from the sun skylight is increasingly dominated by short-wavelength radiation. Hence, one can distinguish not only between a weakly polarized solar half and a highly polarized antisolar half of the sky (see above), but also between a long-wavelength dominated solar and a short-wavelength dominated antisolar half of the sky. To an earthbound observer, the sun itself appears as the point that is characterized by the highest absolute radiant intensity, by the highest percentage of long-wavelength radiation, and – as outlined above – by zero per cent polarization. Furthermore, the solar vertical forms the symmetry plane not only of the polarization, but also of the spectral gradients of scattered skylight.

There are many more aspects referring to the scattering of light in the sky that could be discussed in physical terms, such as scattering by large particles (Mie scattering), multiple scattering, absorption and underground reflections. However, they are important to insect navigation only insofar as they might deteriorate the insect's ability to read compass information from the polarization and spectral gradients. For example, in the vicinity of the sun and the anti-solar point of the sky, anomalous ("negative") polarization occurs, in which the intensity component within the scattering plane is greater than that perpendicular to it. As behavioural experiments performed in bees show, the orientation of the E-vector cannot be inferred from pixels of sky within which the degree of polarization is lower than $d = 0.1$ (v. Frisch, 1967; Edrich and v. Helversen, 1987). Interestingly, polarization-sensitive interneurons in the optic lobes of crickets exhibit threshold responses at about half that value (d = 0.05; Labhart, 1996). These threshold values imply that anomalous polarization will not affect the insect's compass mechanism, because in the region of the sky in which it occurs the value of d is less than 0.1, and mostly even less than 0.05.

All details of the patterns of scattered skylight that have been mentioned – or could be mentioned in addition – should not distract from the more general aspects of these patterns. The coarse-grain spectral gradients and the finer-grain polarization gradients are highly uniform global phenomena, strictly linked to the position of the sun and characterized by a distinct plane of symmetry formed by the solar and antisolar meridians.

The use of polarization gradients

As shown in desert ants (Wehner, 1982; Fent, 1985) and honey bees (Wehner and Strasser, 1985), a specialized part of the insect's visual system is necessary and sufficient for the detection of polarized skylight. The "polarization channel" receives its input from a small fraction of the retinal photoreceptors (2.5 per cent in *Apis mellifera* and 6.6 per cent in *Cataglyphis bicolor*) positioned at the uppermost dorsal rim of the eye (POL area; for *Cataglyphis* see Fig. 4). Owing to optical, anatomical and neuro-

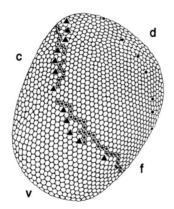

Figure 4. Regionalization within the compound eye of *Cataglyphis bicolor* based on anatomical characteristics of the retina (especially the structure of the rhabdom; see Räber, 1979; Wehner, 1982). The dorsal rim area (POL area) at the dorso-frontal rim of the eye occupies the area delineated by the dorsal rim of the eye and the black dot signatures. The asterisks mark the anatomical border between the dorsal and ventral retina. This border coincides with the row of ommatidia that looks at the horizon (indicated by black triangles) as determined by pseudopupil measurements under antidromic illumination (Räber, 1979; Zollikofer et al., 1995), after the head mounted on a goniometer stage had been adjusted in the normal angular pitch position recorded in running ants (Duelli, 1975; Wehner, 1975). c, caudal (posterior); d, dorsal; f, frontal (anterior); v, ventral.

physiological specializations, these photoreceptors are characterized by high polarization sensitivities. Structural specializations ensure that in the remainder of the eye the polarization sensitivity is markedly decreased (*Cataglyphis*) or even lost (*Apis*) (for review see Wehner, 1994a, b). Furthermore, in both ants and bees it is only the ultraviolet type of receptor that mediates the animal's E-vector responses.

At this juncture, this short physiological characterization of the polarization channel might suffice to make the point that marked functional specializations occur within the insect's visual system (for further details see p. 176 f). In the present context the crucial question is: what information does the polarization channel convey to the navigator's brain? Laborious sets of behavioural experiments showed that, while navigating, ants (Wehner, 1982; Fent, 1986) and bees (Rossel and Wehner, 1982, 1984a), employ a rather simple, stereotyped internal representation of the external E-vector patterns. The rationale of the experiments on which this conclusion is based can be briefly described as follows. The insect is trained to forage in a particular compass direction (under the natural sky or in situations in which the skylight parameters have been changed artificially) and later tested for its ability to recall this compass direction under particular experimental conditions. It is in *Cataglyphis* ants that the most extensive set of such experimental paradigms has been applied.

Paradigm I

During training, the full E-vector pattern is available, but in the test the insect's field of view is restricted to an individual E-vector in the sky. In this case, systematic navigational errors occur. The insects still exhibit finely tuned directional preferences, but the directions preferred by the animals deviate systematically from the compass direction to which the animals have previously been trained, i.e., with the full E-vector pattern available (Fig. 5). This means nothing else than that, in the insect's internal representation of the sky, the particular E-vector presented in the experiment does not occur at its actual position, but is shifted in azimuthal position by the error angle observed in the experiment (Fig. 6). A full description and exemplification of this rationale is given in Wehner and Rossel (1985).

Following this line of argument we presented hundreds of bees and ants with different E-vectors at different elevations above the horizon and at different times of day (i.e., at different elevations of the sun). The error angles exhibited by the insects allowed us to reconstruct the ant's and bee's internal representation of the celestial E-vector pattern which they use as a celestial compass. The most striking result is that this internal representation depends neither on the elevation at which a given E-vector is presented above the horizon, nor on the elevation of the sun. Each point of the compass is characterized invariably by a particular E-vector orientation

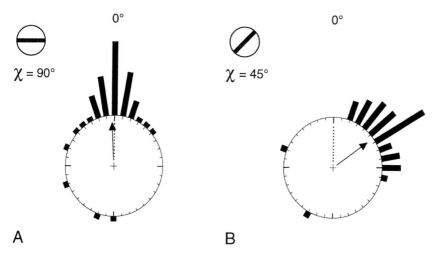

Figure 5. Experimental paradigm I. *Cataglyphis bicolor* trained under the full E-vector pattern is presented in the test period with individual E-vectors, $\chi = 90°$ (A) and $\chi = 45°$ (B) both displayed at an elevation of $\mu = 51°$ above the horizon. Elevation of sun, $\mu_s = 63°$. The correct home (training) direction is denoted by 0°. The angular distribution of the ants' homebound courses are strictly unimodal (length of mean orientation vector r = 0.85 and 0.84 in A and B, respectively; $P_R < 0.001$ in either case; Rayleigh test). In A, the mean orientation angle ($\alpha_m = 357.1°$, n = 55) is not significantly different from the 0° direction ($CI_{S, 0.99} = \pm 13°$, Confidence Interval for the mean angle, for Q = 0.99), but in B ($\alpha_m = 54.0°$, n = 37) it clearly is ($CI_{S, 0.99} = \pm 14°$). For statistical methods see Batschelet (1965, 1981). Data from Fent (1985).

(Fig. 7A, B). In the real sky, the E-vector pattern varies with the elevation of the sun (Fig. 7C), but in the insect's compass it does not. This stereo-typed E-vector compass is used under all conditions tested in bees and ants. Even if beams of artificially polarized light are used to present the animals with E-vectors that, at a given time of day, do not occur at the particular elevation tested in the experiment, the insects are not disturbed at all. They expect any artificially presented E-vector to occur exactly where it should according to their celestial map. Finally, the degree (percentage) of polarization is not encoded in the insect's internal E-vector representation. In the celestial map all E-vectors are of equal importance.

The insect uses its simple internal representation of the sky even if the experimental paradigm I is extended by *presenting the animal with more than one pixel of sky or large parts of the natural skylight pattern*. A short description of one experiment will suffice to make the point (Fig. 8). In this experiment, bees are presented, during the test period, with a large patch of blue sky positioned either symmetrically (Fig. 8A) or asymmetricallay (Fig. 8B) with respect to the solar and antisolar meridian. In the latter case, the bees deviate considerably from the direction to which they have been trained under the full blue sky. Furthermore, the orientation error induced by the skylight window is the arithmetic mean of the orientation errors

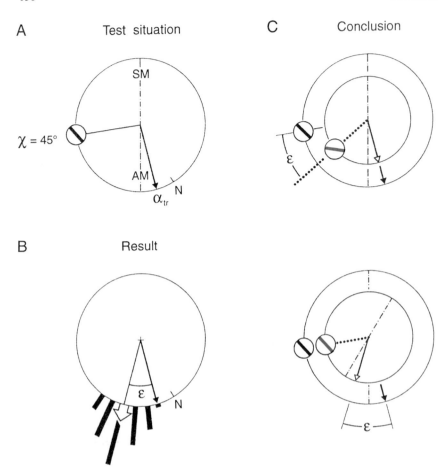

Figure 6. Conclusions drawn from experimental paradigm I. (A) After training under the full E-vector pattern, *Apis mellifera* while performing its recruitment dances on a horizontal comb, is presented with an individual E-vector, $\chi = 45°$, $\mu = 60°$, $\mu_s = 60°$ (see Fig. 5 for explanation). SM and AM, solar and antisolar meridian, respectively; α_{tr}, training direction; N, north. (B) The dancing bees deviate significantly from α_{tr} ($\alpha_m = 193.6°$, $r = 0.98$, $P_R < 0.001$, Rayleigh test) by the error angle $\varepsilon = 30°$ ($CI_{S, 0.99} = \pm 2°$). (C) In the insect's internal representation (inner circle, red bars) of the external E-vector distribution (outer circle, black bar), the $\chi = 45°$ E-vector deviates by the azimuthal distance ε from its actual position in the sky (upper figure). If the insect tried to match its internal representation with the outside world, it would deviate by ε from the training direction, as indeed observed (lower panel). Adopted from Wehner and Rossel (1985).

induced by each E-vector alone (taken in 10° intervals with respect to both azimuth and elevation). Orientation errors do not occur whenever the patch of sky is positioned symmetrically to the solar and antisolar meridian. In this case the errors induced by the left and the right half of the pattern are equal in amount, but opposite in sign, and thus they cancel each other out.

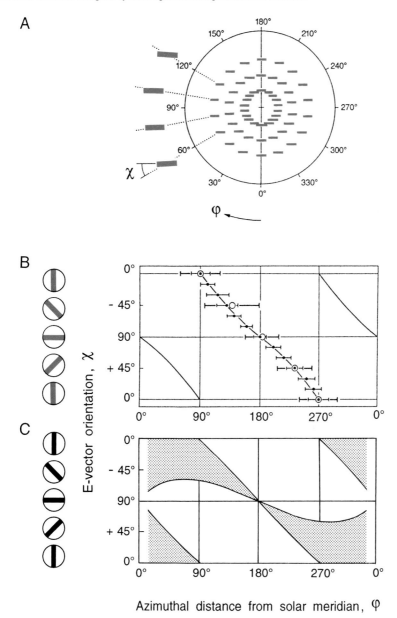

Azimuthal distance from solar meridian, φ

Figure 7. Result of paradigm-I experiments: the ant's and bee's internal representation (red bars in A, B) of the celestial E-vector patterns. (A) Two-dimensional representation of the insect's "internal map" of the sky. 0°, solar meridian; 180°, antisolar meridian; χ, E-vector orientation; φ, azimuthal distance from solar meridian. (B) The insect's standardized χ/φ function, as derived from the experimental data (test paradigm I, see Fig. 6 for rationale; Rossel and Wehner, 1982, 1984a; Fent, 1985). Mean angles and standard deviations are given for *Cataglyphis bicolor* (open symbols) and *Apis mellifera* (filled symbols). (A) is a map-like representation of the χ/φ function. (C) Distribution of E-vectors (black bars) as they occurred in the sky during the periods of time in which the experiments were performed (celestial χ/φ functions). Combined from Wehner and Rossel (1985), Fent (1985) and Wehner (1994a).

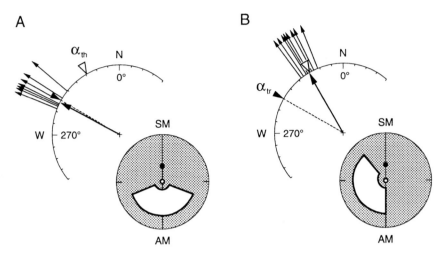

Figure 8. Test paradigm I in which *Apis mellifera* was presented, during the test period, with large skylight windows ($135° \times 50°$) positioned either symmetrically (A) or asymmetrically (B) with respect to the antisolar meridian, AM (insets bottom right: filled circle, sun at elevation $\mu_s = 61°$; open circle, zenith). In the data plots at the upper left, the thin arrows mark the mean directions of orientation of 15 bees tested individually. Total means are represented by heavy arrows. In A, the bees do not deviate from the training direction $\alpha_{tr} = 300°$ (filled arrow head and dashed line), but in B they do. The direction of orientation observed in B coincides with the one to be expected theoretically (α_{th}, open arrow head) if the bee tried to match its internal representation of the sky (see Fig. 7A) with the E-vector pattern displayed within the skylight window. SM and AM, solar and antisolar meridian, respectively; α_{th}, theoretical (predicted) direction of orientation if the bee treated all E-vectors displayed within the skylight window alike; α_{tr}, training direction; N, north; W, west. Modified from Wehner (1983); for further data see Rossel and Wehner (1984a).

This type of experiment finally confirms that the degree of polarization is not used as an additional compass cue. In the experiment described in Figure 8B, the degree of polarization reaches its maximum near the antisolar side of the window, and decreases steadily towards the solar side. If the bees referred to the degree of polarization in addition to the E-vector orientation, the E-vectors should dwindle in importance on the way from the antisolar to the solar side; but this is not the case.

Paradigm II

Methodologically, paradigm II is the reverse of paradigm I. The insect – in this case *Cataglyphis* – is presented with *a spatially restricted part of the skylight pattern during training* and with the *full pattern during the test period*. In particular, the ants are trained to run in narrow channels in a given direction (training direction α_{tr}). Within these channels the view of the sky is restricted to slit-like aerial windows (Fig. 9A). Later the ants are tested under the full skylight pattern (Fig. 9B). Again, navigation errors occur which can be fully explained on the basis of the ant's celestial map. This result is especially intriguing because, in this test, the insect has all skylight information available, but yet it makes mistakes. These mistakes are due to the fact that the insect's celestial compass has been set inappropriately during the training period.

To exemplify this point let us have a closer look at the particular experiment described in Figure 9. In Figure 9A, the long axis of the strip-like celestial window is oriented in the 30°/210° direction, i.e., roughly northeast/south-west, with $\alpha_{tr} = 210°$. As the experiments are performed throughout the whole course of the day during which the celestial E-vector pattern rotates about the zenith, different parts of this pattern are displayed at different times of day. Correspondingly, the errors exhibited by the ants in the test situation vary with the time of day, i.e., with the relative position of the symmetry plane of the skylight pattern and the long axis of the skylight window (Fig. 9C). The error angles observed are in good agreement with the errors predicted from the ant's celestial map. This prediction (represented by the solid curve) is again based on the arithmetic mean of the errors induced by individual E-vectors presented every 10° (with respect to both azimuth and elevation). The correspondence between the predicted values and the experimental data is striking. In Figure 9C, two additional series of experiments are shown in which the long axis of the training channel was oriented in other directions: $\alpha_{tr} = 180°$ (upper graph) and $\alpha_{tr} = 270°$ (lower graph). In these situations the errors the ants are expected to make at different times of day should differ from those observed in the $\alpha_{tr} = 210°$ situation, and this is indeed the case. In all these series of experiments – as well as in others not reported here – there are two conditions under which navigational errors do not occur: When the solar and antisolar meridians are oriented in parallel or at right angles to the long axis of the channel. In these cases the errors induced by the left and right halves of the slit-like aerial window neutralize each other. In the graphs of Figure 9C this is documented by the zero crossings of both theoretical curves and experimental data.

The results of the experiments performed according to paradigms I and II clearly rule out the possibility that the insect remembers the E-vector pattern seen last, and later compares it with whatever fraction of the pattern is available in a given test situation. It is an internal template of the sky that the insect seems to try to match with the external pattern. According to this hypothesis, the best possible match is achieved whenever the animal is aligned with the symmetry plane of the full E-vector pattern. (The distinction between the solar and antisolar meridian can be made by other means, e.g., on the basis of spectral cues; see below.) If parts of the sky are available that are positioned asymmetrically with respect to the solar and antisolar meridian, the best possible match between internal template and external pattern only occurs when the animal deviates from the symmetry plane of the sky. In experimental paradigms I and II, the latter was the case in the test and training situation, respectively.

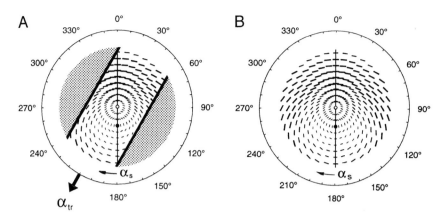

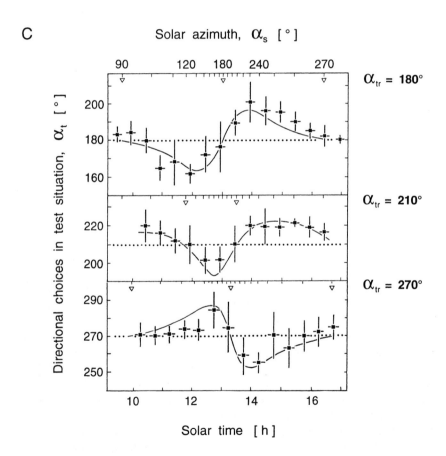

Paradigm III

The kind of reasoning set forth in the previous paragraph leads directly to experimental paradigm III, in which the animals are presented with *the same patch of sky during both the training and test period*. In this situation, experimental errors do not occur even if the patch of sky is positioned asymmetrically to the solar vertical (Tab. 1). It is as if the animals had adjusted their celestial compass to the same zero (reference) direction, i.e., the direction that is characterized by the best possible match between template and skylight pattern. It does not matter what this reference direction actually is: the solar or antisolar meridian or any other point of the compass. It is only important that the same reference direction is used during the animal's entire foraging excursion.

Table 1. Systematic navigational erros exhibited by ants, *Cataglyphis fortis*, when thested under the experimental paradigms described in the text. The ants were either trained or tested when provided with either the total (T) or a partial (P) pattern of polarized light in the sky. Numbers represent the angular deviations (means and standard deviations, given in degrees) from the 0° (home) direction. Significant deviations from the 0° direction ($p < 0.001$) occur in the experimental paradigms I and II (bold lettering), but not in paradigm III and the controls. Deviations to the right and to the left are denoted by plus and minus signs, respectively. 139 individuals were tested. Data from Wehner (1991).

Training	Test	
	T	*P*
T	Control − 0.9 ± 2.7	**Paradigm I** **− 7.3 ± 7.1**
P	**Paradigm II** **+ 12.9 ± 6.1**	Paradigm III − 1.5 ± 5.2

Figure 9. Experimental paradigm II. When ants, *Cataglyphis fortis*, are presented with a skylight window during training (A), but with the full E-vector pattern during the test period (B), they exhibit systematic navigational erros (C). These errors depend on the relative position of the skylight window with respect to the symmetry plane of the sky (0°/180°, solar/antisolar meridian). During the course of the day, the latter rotates about the zenith. The three graphs in C refer to three different training directions. They depict the experimental data (square and bar signatures, means and standard deviations), as well as the hypothetical error functions (continuous lines). These functions are computed as the arithmetic means of the errors the ants would exhibit to each E-vector displayed individually, taken every 10° in both the horizontal and the vertical direction. (The same procedure was used in computing α_{th} in Fig. 8). α_{tr}, training direction (homeward course), parallel to the long axis of the skylight window; α_t, direction chosen by the ants in the test situation; α_s, solar azimuth (azimuthal position of solar meridian); filled circle, sun; open circle, zenith. The open triangles in C mark the zero crossings of the hypothetical error functions.

At this point it might be worth mentioning how the experimental paradigm III was actually implemented in experimental terms. During training the ants walked in narrow channels and were thus presented with slit-like aerial windows in the same way as they were in the former situation (paradigm II). In the critical test, however, they did not walk under the full blue sky, but underneath a flat trolley, a rolling optical laboratory (see Wehner, 1982, 1994a), which presented them with the same slit-like area of the celestial E-vector pattern that they had experienced during training. For the experiment it is amazing to see how freely running animals can be exposed to such drastic changes in their visual world, how they can be restricted in their locomotor activities by walking in channels and within optical laboratories that move along with them, and still perform nearly with the same accuracy as they do when walking unrestrictedly within their natural habitat. The experimental restrictions imposed on *Cataglyphis* become even more impressive, if we now consider behavioural tests in which another sensory channel will be unravelled within the insect's visual system.

The use of spectral and intensity gradients

If those parts of the eyes of bees and ants that provide the input to the polarization channel are covered with light-tight paint, the insects are no longer able to make use of polarized skylight, but they can still infer compass information from the sky – even if the sun is obscured by clouds or screened off experimentally. In this case they are able to refer to other straylight parameters: the spectral and/or intensity gradients in the sky.

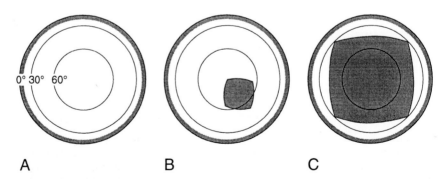

0° 30° 60°

A   B   C

Figure 10. Stimulus conditions used during the test period in examining whether and how spectral and intensity gradients in the sky are used by *Cataglyphis* ants as compass cues. The figures represent the zenith projections of the celestial hemisphere as seen by an ant walking underneath a "rolling optical laboratory" (filter vehicle, in the following referred to as the "trolley") described in Wehner, 1982, 1994a. (A) Sky unobscured (above an elevation of 20°), sun visible. (B) Sun obscured, sky visible. (C) Sun obscured, sky view restricted to a narrow annular region.

The basic type of experiment on which this conclusion is based is described in Figure 10. *Cataglyphis* ants in which the POL areas (Fig. 4) of both compound eyes (and the ocelli; see Fent and Wehner, 1985) have been painted out, are tested under the experimental trolley, in which the sun and the surrounding area of the sky are screened off in one way or another (Fig. 10B, C). Under these conditions, in which the ants can refer only to spectral and/or intensity gradients, they are still able to compute their 0° (homeward) course, but deviate towards the azimuthal position of the sun (Figs 12A and 13A, for control see Fig. 11). Note that, due to the experimental procedures applied in these tests, the sun itself is not visible. This means that the ants must be able to determine the centre of gravity of the celestial intensity distribution. This holds true even in the experimental situation of Figure 10C (for results see Fig. 13), in which the ants are provided with only a narrow annulus of scattered skylight.

The deviation from the 0° direction towards the solar azimuth might reflect some kind of phototactic flight response. It is stronger the smaller the aerial windows with which the ants are presented, i.e., the less precise the compass information is which the animals can gain from the spectral gradients in the sky.

To demonstrate and isolate this phototactic response most clearly, all ants are tested under two mirror-image like conditions: with the sun positioned

0°

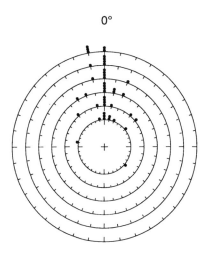

Figure 11. The ants' use of spectral gradients as compass cues. Angular distribution of the positions of ants, *Cataglyphis bicolor*, in which the POL areas of both eyes had been painted out (using Testors Enamel, black). Stimulus condition as depicted in Figure 10A. The trolley was provided with a transparent UV transmittable Plexiglass disc (Roehm and Haas, no. 218). Data are presented at distances of 2, 3, 4, 5, 6, and 7 m from the ant's starting point. Training direction: 0°; training distance: 15 m. Statistical evaluation (4-m circle): mean direction of orientation $\alpha_m = 359.0°$, $n = 9$, length of orientation vector $r = 0.97$, $P_R < 0.001$ (P according to Rayleigh test), $CI_{S, 0.99} = \pm 17°$. For statistical conventions see Fig. 5.

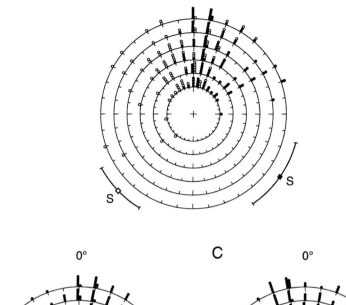

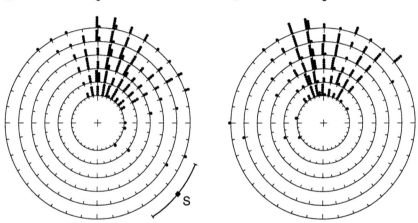

Figure 12. Same as in Figure 11, but with ants, *Cataglyphis fortis*, tested under stimulus condition depicted in Figure 10B, i.e., with no view of the sun. (A) In two experimental series (open and filled symbols), the azimuthal position of the sun (S) was either to the left or to the right of the ants' training direction (0°). Statistics (4-m circle), filled symbols: $\alpha_m = 17.3 \pm 21.6°$ (s.d.), $n = 20$, $r = 0.93$, $P_R < 0.001$, $CI_{S, 0.99} = \pm 14°$, $S = 126.1 \pm 19.6°$; open symbols: $\alpha_m = 339.2 \pm 23.1°$, $n = 16$, $r = 0.92$, $P_R < 0.001$, $CI_{S, 0.99} = \pm 19°$, $S = 225.1 \pm 14.6°$. (B) Experimental data for $S > 180°$ (open symbols) mirrored at the $0°–180°$ axis. Statistics (4-m circle): $\alpha_m = 18.9 \pm 22.3°$, $n = 36$, $r = 0.92$, $P_R < 0.001$, $CI_{S, 0.99} = \pm 11°$, $S = 130.1 \pm 18.1°$. (C) Experimental data depicted in panel A shifted by $\pm 19°$ (α_m in panel B) away from the azimuthal position of the sun. Statistics (4-m circle): $\alpha_m = 358.3 \pm 22.3°$, $n = 36$, $r = 0.92$, $P_R < 0.001$, $CI_{S, 0.99} = \pm 11°$. For conventions see Figure 11.

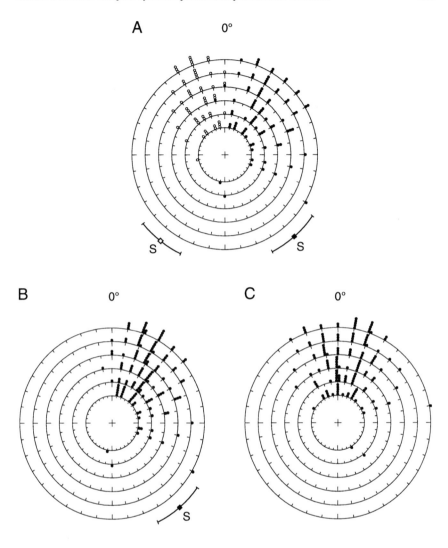

Figure 13. Same as in Figure 11, but with ants, *Cataglyphis fortis*, tested under stimulus conditions depicted in Figure 10C. As shown in (A), again two experimental series were run with the solar azimuth (S) either to the left (open symbols) or to the right (closed symbols) of the ants' training direction (0°). Statistics (4-m circle), filled symbols: $\alpha_m = 44.8 \pm 27.8°$, n = 15, r = 0.88, $P_R < 0.001$, $CI_{S, 0.99} = \pm 23°$, S = 139.2 ± 14.6°; open symbols: $\alpha_m = 333.3 \pm 15.2°$, n = 9, r = 0.97, $P_R < 0.001$, $CI_{S, 0.99} = \pm 25°$, S = 217.6 ± 12.8°. The distributions in (B) and (C) are designed according to the procedures used in Figures 12B and C. Statistics (4-m circle), B: $\alpha_m = 37.6 \pm 25.3°$, n = 24, r = 0.90, $P_R < 0.001$, $CI_{S, 0.99} = \pm 14°$, S = 140.4 ± 14.1°; C: $\alpha_m = 7.38°$, n = 24, r = 0.91, $P_R < 0.001$, $CI_{S, 0.99} = \pm 14°$. For conventions see Figure 11.

either to the left or to the right of the ants' zero (0°) direction (Figs 12 A and 13 A). Later all data points obtained for S > 180° (S, solar azimuth) are mirrored on the 0°/180° axis (Figs 12 B and 13 B). Now, the mean deviation of the data from the 0° direction towards the sun directly reflects the influence of the ants' phototactic responses. To correct for this influence, all data points of Figures 12 B and 13 B are shifted by this angular amount (18.9° and 37.6°, respectively) away from the sun's position. Hence, in the resulting Figures 12 C and 13 C, the ants' systematic deviations towards the sun have been eliminated, and the basic component of orientation in the (correct) 0° deviation becomes evident.

An inspection of the fine structure of the ants' trajectories reveals an important point: the smaller the area of the sky with which the ants are presented, the more curved the ants' trajectories. Furthermore, the mean directions which the ants have chosen (and which are plotted in Figs 12 and 13) are the result of the behavioural strategies: to walk alternatively in the correct 0° compass direction or in the direction of the sun's azimuthal position. One example of such a trajectory exhibiting the repeated switch from one behavioural strategy to the other is given in Figure 14.

In a next step, the experimental situation is refined even more. In the experiments described so far, spectral as well as intensity gradients could have contributed to the ants' responses. We can exclude the use of the spectral gradient by providing the experimental trolley with a short-wavelength cut-off filter transmitting only long-wavelength light that stimulates exclusively the ant's green receptors. Unfortunately, the reverse experiment cannot be performed, because the sensitivity curves of both ultraviolet and green receptors extend into the short-wavelength range (Mote and Wehner, 1980; Labhart, 1986), so that there is no way of stimulating exclusively the ant's ultraviolet receptors. In the former case, with the ultraviolet receptors blocked out, the ants tested under the experimental condition of Figure 10 C show only phototactic responses towards the sun's position (Fig. 15). Hence, the long-wavelength intensity distribution perceived by only one (the green) spectral type of receptor suffices to determine the azimuthal position of the sun.

In discussing the results referring to the use ants make of spectral and intensity gradients in the sky, a caveat is needed. Consulting Figures 12–15, one might get the impression that the ants use the spectral gradients to determine their compass courses, and that the centre of gravity of the spatial intensity distribution, when seen by only one (the long-wavelength) type of photoreceptor, elicits merely phototactic responses. Such, of course, is not the case. If ants are presented with large parts of the celestial hemisphere including the sun, they are able to determine their 0° direction even if their short-wavelength (ultraviolet) receptors are blocked out by means of UV-tight lacquer sheets (Fig. 16). As an illustrative example, the homing trajectory of an individual ant treated in this way is shown in Figure 17. Prior to release at site R the ant's eyes were covered with UV-blocking

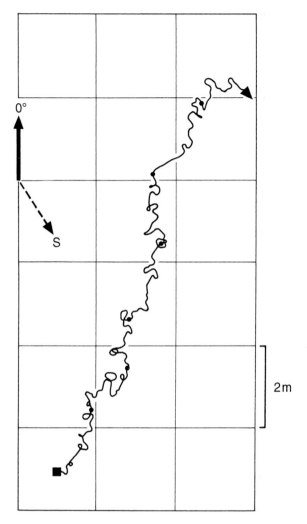

Figure 14. Trajectory of an individual ant, *Cataglyphis fortis*, tested under experimental condition of Figure 10C (see Fig. 13 for experimental design and full data set). Training direction: 0°, solar azimuth S = 146°. The ant's starting position is indicated by the black square. The ant alternatively selects either its proper compass course or deviates towards the azimuthal position of the sun.

lacquer sheets, so that the ant was visually exposed to nothing but long-wavelength signals. While "homing", the ant walked underneath the experimental trolley. During the first part of its homeward run, the ant could not see the sun, which became visible only at position P. On its way from R to P the ant exhibited phototactic responses towards the sun's position, which it must have computed as the centre of gravity of the intensity dis-

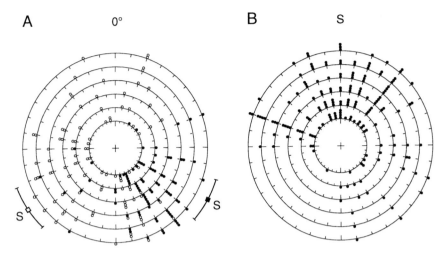

Figure 15. The ants' use of intensity gradients. Angular distribution of the position of ants, *Cataglyphis fortis*, in which the POL areas of both eyes had been painted out. The ants were tested under the stimulus condition shown in Figure 10C. The trolley was provided with a short-wavelength cut-off filter (Plexiglass Roehm and Haas, No. 478), thus transmitting light detectable exclusivley by the ant's green receptors. In this situation the ants can neither rely on polarization patterns nor on colour gradients. (A) Data of two experimental series in which the sun was either to the left (open symbols) or to the right (closed symbols) of the training direction (0°). Statistics (4-m circle), filled symbols: $\alpha_m = 142.2 \pm 27.8°$, n = 15, r = 0.88, $P_R < 0.001$, $CI_{S, 0.99} = \pm 21°$, S = 119.4 ± 12.0°; open symbols: $\alpha_m = 226.5 \pm 66.8°$, n = 17, r = 0.32, $P_R = 0.322$, S = 237.1 ± 12.5°. (B) Same data plotted relative to the solar azimuth S = 0° (rather than to the training direction as shown in A). Statistics (4-m circle): $\alpha_m = 9.6 \pm 54.7°$, n = 32, r = 0.55, $P_R < 0.001$, $CI_{S, 0.99} = \pm 35°$. For conventions see Figure 11.

tribution perceived by the green receptors (Fig. 15). However, when it could see the sun at P (and during the later stages of its homeward run), it immediately used it as a compass cue and moved in the proper home direction (see Figs 16 and 17). Still, in this case, in which colour gradients are abolished as a compass cue, phototactic responses occur in addition to compass responses (see also Lanfranconi, 1982). Cautiously, one may conclude that if the sun, due to experimental manipulations, is part of only an intensity rather than a spectral gradient, it can be used as both a compass cue (when presented within a larger celestial environment) and the source of phototactic responses (when presented under spatially restricted experimental conditions). In the experimental situation depicted in Figure 10C, in which the ants' monochromatic view of the sky is severely restricted and the sun screened off, the centre of gravity of the remaining intensity distribution which the ants are still able to compute elicits merely phototactic responses.

In any event, these results unequivocally show that visual areas outside the POL area are able to derive compass information from skylight cues

0°

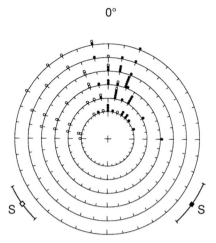

Figure 16. Ants, *Cataglyphis fortis*, are able to use the sun as a compass cue even if only their long-wavelength (green) receptors are stimulated. In this situation, they can neither rely on polarization nor colour contrast. Technically, the eyes of ants previously trained to the 0° direction were covered with laquer sheets (consisting of transtilbene, 4-nitroaniline, 2-nitrodiphenylamine and shellac) which blocked the ants' ultraviolet receptors (transmission $T < 10^{-4}$ for $\lambda < 425$ nm). The animals were tested under the stimulus condition of Figure 10A, i.e., with the sky and the sun fully visible. Statistics (4-m circle), filled symbols: $\alpha_m = 17.0 \pm 22.8°$, n = 11, r = 0.92, $P_R < 0.001$, $CI_{S, 0.99} = \pm 23°$, S = 126.2 ± 12.4°; open symbols: $\alpha_m = 322.8 \pm 29.7°$, n = 6, r = 0.87, $P_R < 0.01$, $CI_{S, 0.99} = \pm 45°$, S = 232.5 ± 11.2°. For conventions see Figure 11.

other than E-vector gradients. Spectral cues of the scattered light in the sky provide such information, and the sun, if displayed merely within a spatially extended long-wavelength intensity gradient, does as well. In the latter case, however, phototactic responses are heavily superimposed on the proper compass orientation.

How large is the non-POL area used as a spectral channel for skylight navigation? Experiments with differently sized eye caps clearly show that only the dorsal half of the ant's compound eye can make use of this information. If the dorsal halves of both eyes (down to the equator; compare Fig. 4) are painted out, but all polarization, spectral and intensity cues of the sky are available, the ants behave as though lost, or exhibit phototactic responses (Fig. 18). This finding is all the more astounding, as the ants tested under these conditions perform strong roll and pitch movements of their heads and thus, time and again, look at the sky with the ventral halves of their eyes. In this situation, however, they perform only phototactic responses towards the sun, but do not use the sun or any other celestial cue for proper compass response. Hence, the ventral part of the ant's visual system seems to lack the neural machinery that mediates compass orientation by skylight cues.

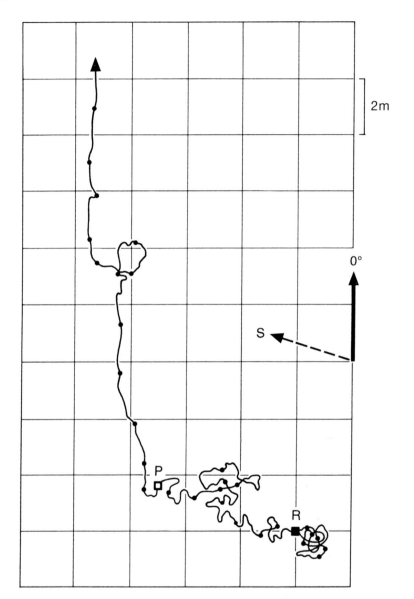

Figure 17. Trajectory of an individual ant, *Cataglyphis bicolor*, trained and tested as specified in Figure 16. Training direction: 0°, solar azimuth S = 289°. Filled circles mark the positions of the ant in 10-s intervals. The ant was released at R (position indicated by the filled square) and first tested with the sun obscured (experimental condition of Fig. 10B). Then, at P (position indicated by the open square), the screen shielding the sun was removed.

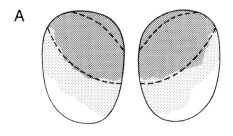

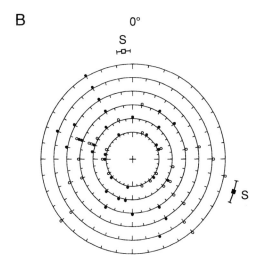

Figure 18. The behaviour of ants, *Cataglyphis bicolor*, which could use only the ventral halves of their compound eyes (see Fig. 4). The dorsal halves including the POL areas – as well as the ocelli – had been painted out. (A) Boundaries of lacquer sheets fitted to the ants' eyes. The dark grey areas had been covered in all 17 ants tested. In some ants, the eye caps extended into the light grey area, i.e., into the ventral half of the eye. The positions of the eye caps were recorded using SEM pictures. (B) Angular distribution of the ants' positions. S, azimuthal position of sun in two experimental series (open and filled symbols). In the latter series, the ventral borders of the eye caps were positioned as closely as possible to the equator for the eye. Filled and open symbols refer to two experimental series which differed in the azimuthal position of the sun (S). Statistics (4-m circle): $\alpha_m = 292.2°$, n = 14, r = 0.16, $P_R = 0.706$. The ants are neither oriented in the training direction (0°) nor towards the solar azimuth (S).

Information transfer between sensory channels

We now ask whether information gained under one stimulus condition – say, with the sun as the only compass cue – can be used under other stimulus conditions, e.g., when spectral or polarization gradients are available. Methodologically, the polarization channel can be blocked in one of three ways: by (i) painting out the POL areas of the compound eyes as well as the ocelli, (ii) covering compound eyes and ocelli with lacquer sheets, which act as cut-off filters for short-wavelength radiation and hence stimulate only the long-wavelength receptors (transmision smaller than 10^{-4} for $\lambda < 425$ nm), (iii) letting the ants walk underneath the experimental trolley equipped with an orange Plexiglass filter (transmission smaller than 10^{-5} for $\lambda < 506$ nm). Methods (ii) and (iii) are based on the fact that in ants (Duelli and Wehner, 1973) and bees (von Helversen and Edrich, 1974) E-vector detection is exclusively mediated by the ultraviolet receptors. It follows from the results mentioned above that methods (i) and (ii) cannot be applied in studying the transfer from the sun-compass to the polarization-compass system, because in either situation spectral skylight gradients are available that could be used by the ants as compass cues (see previous Section). Therefore, in the experiments described below, method (iii) was used to switch off the polarization channel and any system that relies on colour gradients. On the other hand, the sun can be excluded as a compass cue by screening off substantial parts of the sky including the sun's position.

Let us first ask whether the ants are able to transfer information from their polarization-compass to their sun-compass system. The animals are trained in narrow channels as described earlier, which shield most of the sky including the sun and leave open only a narrow slit-like aerial window. In order to prevent the ants from misaligning their compass system due to an asymmetric presentation of parts of the celestial E-vector pattern, the long axis of the channel is always oriented at right angles to the solar and anti-solar meridian. (The other error-free situation – the alignment of channel and solar/antisolar meridian – cannot be used in the present context, because then the position of the sun would be close to the border of the screen used for hiding the sun.) After training, the ants are tested under the orange cut-off filter, and hence left alone with the sun as their only compass cue. They exhibit a combination of compass orientation and phototactic responses (Fig. 19). Hence, they seem to be able to transfer information from their polarization to their sun-compass system. However, there is still another possibility of interpreting the experimental results. In the test period – with the sun visible but with polarization and spectral gradients unavailable – the ants might have referred to information acquired during the training period from the spectral gradient in the sky. To clarify this point note that, during training, ants may have inferred the (invisible) sun's position from the spectral gradient in the sky, and could have used this information when the sun (but not the spectral gradient) was visible.

Next we ask whether in *Cataglyphis* the reverse transfer – from the sun-compass to the polarization-compass system – also occurs. In this case, the eyes of the foragers are covered with lacquer sheets that act as filters optically cutting off the ultraviolet receptors (method ii). Hence, compass information can be gained exclusively from the light emanating directly from the sun. Prior to the critical test, the eye caps (lacquer sheets) are

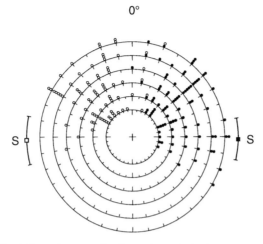

Figure 19. Test for information transfer from the polarization-compass and/or spectral-compass to the sun-compass system. For experimental design see text. The angular distribution of the ant's positions deviates from the solar azimuth (S) towards the training direction (0°). Statistics (4-m circle), filled symbols: $\alpha_m = 56.6 \pm 27.4°$, $n = 15$, $r = 0.89$, $P_R < 0.001$, $CI_{S,0.99} = \pm 22°$; open symbols: $\alpha_m = 320.7 \pm 22.6°$, $n = 14$, $r = 0.92$, $P_R < 0.001$, $CI_{S,0.99} = \pm 20°$, $S = 268.1 \pm 12.7°$. For conventions see Figure 11.

removed, and the animals are tested while they walk underneath the experimental vehicle equipped with a Plexiglass screen that transmits ultraviolet radiation (and hence stimulates the polarization channel), but shields the sun (stimulus situation as depicted in Fig. 10C). For technical reasons, a control experiment must precede the final test. In this control, the ants still wearing their orange eye caps are tested in the same situation as later in the critical test (with the eye caps removed). If the caps were intact and fitted the eyes tightly, the ant's ultraviolet receptors should not respond. Left alone with an intensity gradient in the long-wavelength range, the ants should be disoriented or exhibit only phototactic responses (see above). This is exactly what occurs (Fig. 20). Hence, the eye caps can now be removed, and the animals can be used for the final test. Even though only very few animals could be tested (n = 8), the data (Fig. 21) indicate that sun-compass information might be transferred to the polarization-compass system. Given the complex series of treatments which the animals had to undergo (application and removal of eye caps, test under the experimental trolley with only small parts of the polarized sky available), the large scatter in the data should not be surprising.

Again, however, it cannot be excluded that the compass information gained from the sun's position during training was later used to read compass information from the spectral rather than polarization gradient, as the

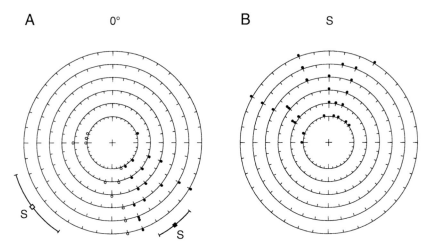

Figure 20. Test for information transfer from sun-compass to polarization-compass system. Angular distribution of the position of homing ants, *Cataglyphis fortis*, provided with eye caps that allowed only the green receptors to be stimulated. Consequently, polarization and colour contrast could not be used as compass cues. The ants had no direct view of the sun (experimental condition of Figure 10C). However, as the data show, they could infer the azimuthal position of the sun from the long-wavelength distribution within the narrow annular window through which they could see the sky, but did not use it to steer the compass course to which they had been trained. The data are presented relative to either the compass direction (0° in A) or to the azimuthal position of the sun (S in B). Statistics (3-m circle), A, filled symbols: $\alpha_m = 130.4 \pm 17.6°$, n = 3, r = 0.95, S = 142.0 ± 10.6°; open symbols: $\alpha_m = 203.5 \pm 38.9°$, n = 3, r = 0.77, S = 231.3 ± 20.0°; B, $\alpha_m = 336.4 \pm 32.7°$, n = 6, r = 0.84, $P_R < 0.01$, $CI_{S, 0.99} = \pm 50°$. For conventions see Figure 11.

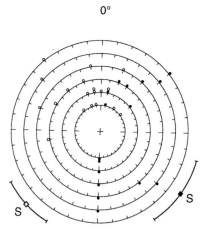

Figure 21. Test for information transfer from the sun-compass to the polarization-compass and/or spectral compass system. For experimental design see text. The two series (filled and open circles) differ in the azimuthal position of the sun (S). Statistics (3-m circle), filled symbols: $\alpha_m = 52.0 \pm 64.4°$, n = 3, r = 0.37, S = 149.6 ± 12.9°; open symbols: $\alpha_m = 354.0 \pm 10.5°$, n = 5, r = 0.98, $P_R < 0.001$, $CI_{S, 0.99} \approx 60°$, $P_S < 0.01$ (deviation from sun, Stephens test, Batschelet 1965), S = 236.6 ± 17.9°. For conventions see Figure 11.

spectral cues were still available in the test situation. To decide between the two possibilities, one could block – during the test period – the POL areas of the eyes. If this treatment prevented the ants from choosing their proper compass course, the information transfer from the sun-compass to the polarization-compass system could be taken for granted. It could also be taken for granted if the ants, when tested after training, were provided with eye caps that left the POL areas open but covered the remainder of the eyes. If information transfer occurred between the two channels under consideration, then the ants should select their compass courses. Either experiment, however, is extremely difficult to perform. At this stage, we probably reach the limit of what a freely walking ant would tolerate in terms of sensory manipulations. Note that prior to training the ant's eyes would have to be provided with orange eye caps covering the entire eyes (to switch off any system relying on polarization and colour gradients). After training, these caps had to be exchanged for small light-tight caps which guaranteed that the POL areas of the eyes were either exclusively covered or exclusively left open. Finally, the ants manipulated that way had to walk underneath the experimental trolley that shielded the sun and provided the animals with a rather narrow annular skylight window.

Inferences and hypotheses: a global skylight compass

I ended the *Introduction* by pointing out that the polarization and spectral gradients in the sky are global phenomena based on simple geometrical rules. Now I can add that, by evolutionary experience, bees and ants have discovered this global simplicity by using an extremely uniform internal representation of the celestial straylight patterns. The "E-vector map" resembles the E-vector pattern actually present in the sky when the sun is at the horizon, but differs in detail from the actual E-vector patterns for all other elevations of the sun (for $\mu_s > 0°$). However, what it shares with all skylight patterns is their principal geometric characteristic – the symmetry plane formed by the solar vertical.

At this point, a short discussion about what we mean by E-vector "map" might be in order. In neurophysiological parlance, the term neural map is used in at least two different ways. On the one hand, it describes the neural representation of the animal's sensory – e.g., visual, somatosensory, auditory – or motor space. This type of map can result either from topic – e.g., retinotopic, somatotopic – projections of the sensory layers onto higher-order neuropiles (Hartline et al., 1978; Murphey, 1983; Sparks, 1988) or – as in auditory maps – from neural computations along the sensory pathway (Knudsen and Konishi, 1978; Konishi et al., 1988). On the other hand, the term applies to neural representations of particular characteristics rather than the spatial positions of sensory stimuli. Telling examples of this type of map (so-called feature maps) are provided by the ordered neural representations of sound-frequencies (tonotopic maps), sound-intensities or sound-source distances (Konishi, 1986; Oldfield, 1988; Römer and Rheinlaender, 1989; Suga, 1990), and by the spatially organized encoding of visual aspects of the world, as it has been unravelled in the primate cortex (e.g., Zeki, 1993). In the insect's E-vector map, any particular E-vector orientation is linked to a particular spatial position and is not encoded as a position-invariant feature in itself. This strong linkage between E-vector direction and position in visual space indicates that, for the insect navigator, E-vectors have no meaning other than labelling spatial (azimuthal) positions in the sky. Hence, E-vector "map" is a descriptive term used to specify the azimuthal position at which the animal expects any particular E-vector to occur, but it does not necessarily imply that the animal relies on a map-like neural representation of the skylight patterns. As will be discussed below, the insect could "read its map" – again a colloquial expression – in ways quite different from what a human navigator might usually mean by reading maps.

The conclusion that insects refer to the global aspects of the rather uniform skylight patterns is in accord with the physiological finding of strong neural convergence occurring within the polarization channel. Large populations of peripheral analyzers (the polarization-sensitive photoreceptors of the POL area) converge onto polarization-sensitive interneurons within the insect's second visual neuropile, the medulla. Three classes of such neurons have been found in crickets (Labhart, 1988; Petzold and Labhart, 1994), each characterized by an extremely wide visual field and a particular E-vector orientation to which it is maximally sensitive (its tuning angle χ_{max}). Most likely, the tuning angle of each of the three classes of polarization-sensitive interneurons[1] ("macro-analyzers") is the arithmetic – or a somewhat weighted – mean of the tuning angles of the individual photoreceptors ("micro-analyzers") converging onto that particular interneuron. However, any further discussion of how the macro-analyzers might be used as part of a neural skylight compass is hampered by an experimental chimera problem. Until now, all electrophysiological and neuroanatomical data concerning the macro-analyzers have been collected in orthopterans, whereas the behavioural phenomena we want to explain by these data have been analyzed in hymenopteran species.

At the level of the retina, the POL areas of both groups of insects share several common properties (for review of neurophysiological data see Wehner, 1994a). First, the micro-analyzers all belong to one spectral

---

[1] In this terminology of micro- and macro-analyzers it does not matter at all whether the antagonistic interaction between opponent analyzers located within a particular visual unit (a particular ommatidium) occurs prior to the convergence step or at the convergence node (the dendrites of the wide-field interneuron) itself. In any event, the local micro-analyzers convey high-contrast E-vector signals to the global macro-analyzers.

type of photoreceptors (the ultraviolet type in bees and ants, the blue type in crickets and locusts) and thus render the E-vector detecting system homochromatic. Second, the polarization-sensitive photoreceptors within each ommatidium, i.e., the photoreceptors that look at the same point in space, are arranged orthogonally, so that two pairs of micro-analyzers form intra-ommatidial "opponent analyzers". This stuctural arrangement enhances polarization contrast already optically – and later, as found in orthopterans, by antagonistic interactions in the first synaptic level, also neuronally. Third, the photoreceptors within the POL areas are spatially arranged in a fan-like way, so that the tuning angles of the micro-analyzers form a marked gradient along the insect's fronto-caudal (anterio-posterior) body axis. Fourth, the POL areas exhibit contralateral fields of view.

Here, however, the similarities might end. The individual fields of view of the micro-analyzers are small in ants, but extremely large in crickets. The POL areas of *Cataglyphis* ants – even of the large *bicolor* species – are composed of only 50–70 pairs of crossed micro-analyzers (depending on the size of the animal and hence the eye), whereas in crickets 500–600 such analyzers contribute to the polarization channel. In ants, the small-field polarization-sensitive photoreceptors form a rather regular fan array providing the animal with a steep spatial gradient of tuning angles. In contrast, the POL retina of crickets exhibits a much less orderly arrangement of photoreceptors and tuning angles. Each of these properties is expected to have critical consequences on the way the polarization-sensitive interneurons sample the output of receptor populations, the way they interact, and thus the mechanism employed in extracting compass information – or any other information used in visual course control – from the skylight patterns. One might argue that the differences in the structural organization of the POL areas found in, say, crickets and ants represent different ways of accomplishing the same task. More likely, however, they are functional adaptations to particular modes of behaviour. In crickets, for example, a whole set of specializations including the optics of the eye, the distribution of screening pigment, the retinal arrays of analyzers and, possibly, the degree of neural convergence forms a functionally related assemblage of derived characters. Most probably, in the course of evolution this set of characters has been selected for fulfilling a specific task. What this task is, remains to be elucidated. In any way, visual stimuli positioned at infinity – as skylight patterns effectively are – provide an earthbound navigator with information about his rotatory (angular) rather than translatory (linear) components of movement. This information could be used in various ways of navigation and course control.

In spite of the obvious differences between the orthopteran and hymenopteran polarization channels, it is tempting to speculate about how the ant's (and bee's) celestial compass system might work; but in following this temptation we must be cautious. We do not even know yet whether skylight patterns are used by these hymenopteran species simply to read a reference direction – e.g., the azimuthal position of the solar meridian – from the sky (Hypothesis I), or whether they are used to determine any particular point of the compass (Hypothesis II).

Hypothesis I
The E-vector patterns are used to read a reference direction (e.g., the solar or antisolar meridian) from the sky. Take, for example, the walking ant. While integrating its path, it might well rely primarily on proprioceptive information derived from cuticular mechanoreceptors of its locomotor apparatus. Such information is assembled and used in an egocentric system of reference. If this self-centred open-integrator ran for extended periods of time, cumulative errors would build up rapidly. As shown by Benhamou et al. (1990) in a theoretical analysis, within a purely egocentric system of computation, random noise added to the assessment of the angular components of movement has a much more deleterious effect on the accuracy of path integration than has noise introduced into the system that is responsible for measuring linear displacements. Consequently, the use of any internal compass reference largely decreases the scatter of the home vector. Nocturnal spiders have been shown to integrate their paths by receiving information exclusively from mechanosensitive proprioceptors (*Cupiennius*: Seyfarth and Barth, 1972; F. G. Barth, this volume), but *Cataglyphis* ants travelling in the desert over distances that are larger by two orders of magnitude would be severely handicapped if they did not use allocentric (e.g., external compass) information as well. In this respect, *Cataglyphis* might refer to skylight information simply for calibrating and, every now and then, recalibrating its internal compass scale. This could be achieved by determining, for example, the azimuthal position of the solar meridian. The simplest way of accomplishing this task is to balance the total outputs of the left and right POL areas. The animal could rotate about its vertical body axis and thus scan the sky with its array of polarization analyzers. Whenever a balance between the outputs of the two POL areas is reached, the animal is oriented parallel to the symmetry line of the skylight pattern, the solar vertical. (To discriminate between the two branches of this line – the solar and the antisolar meridian – the animal could refer to information provided by its spectral channel or, if the receptor array were not completely symmetrical with respect to the animal's transverse body axis, to the absolute values of summed receptor output. Of course, the ambiguity can be resolved by any system that spatially segregates the outputs of the receptor array, e.g., samples the anterior and posterior halves of the array separately.)

In this scanning model, the accuracy with which the solar meridian can be determined depends directly on how strongly the responses of the macro-analyzers are modulated as the animal scans the sky. It is clear from the outset that the modulations get weaker, the larger the populations of micro-analyzers (photoreceptors) are across which the macro-analyzers (interneurons) sample the retinal output. But how large should these sampling areas be? And how should the micro- and macro-analyzers be arranged and connected to maximize response modulations?

As nothing is known yet about polarization-sensitive interneurons and connectivity patterns in the supreme insect navigators, bees and ants, let us approach these problems from a theoretical perspective. In this model approach we start with an ideal fan array of micro-analyzers based on completely hemispherical eye geometry, and then systematically vary a number of parameters such as the fine structure of the fan, its position within the animal's field of view (i.e., the azimuth and elevation of the fan centre), the sampling areas of the macro-analyzers, and the excitatory and inhibitory interactions of these wide-field units. Before we take a cursory glance at this approach (G.D. Bernard and R. Wehner, in preparation), an epistemological remark might be useful. What do such theoretical considerations, and the computer simulations based on them, finally tell us? First, the assumptions underlying the models are not completely up in the air. To a large extent, they are based on what we already know about both the physics of polarized skylight and the optical, anatomical and physiological characteristics of the ant's polarization-sensitive photoreceptors (their spatial, polarization and angular sensitivities, as well as the antagonistic interactions between opponent analyzers). The parameters that are varied – and mentioned above – are those about which experimental evidence is still lacking. If this computational exercise yields solutions to particular tasks, then the theoretical predictions can be used by the neurophysiologist as some kind of conceptual search image. It might well be that later the biological mechanisms, once unravelled, do not correspond in detail to what the simulations have indicated as the theoretically best solutions. Such discrepancies between the observed and the expected raise the stimulating question of why this is so. What are the biological constraints that might have been involved in shaping the particular mechanism the animal employs?

In our theoretical approach, let us first assume that all micro-analyzers of ideal left and right fan arrays converge onto a single large-field macro-analyzer centred about the animal's vertical axis of rotation (Subhypothesis IA). Such a macro-analyzer characterized by a tuning angle $\chi_{max} = \pm 90°$ (as measured relative to the animal's longitudinal body axis) would exhibit peak responses whenever the animal is aligned with the solar vertical. However, as in the neighbourhood of this position yielding maximum responses the response gradient is rather flat, the accuracy with which the solar vertical can be determined is limited. It can be improved by employing more than one type of macro-analyzer, for example by adding a pair of macro-analyzers with tuning angles of $\chi_{max} = +30°$ and $\chi_{max} = -30°$ to the 90°-analyzer (Subhypothesis IB). In this case, the 90°-analyzer reaches maximum responses at a position at which the $(+30°)$- and $(-30°)$-analyzers exhibit steep and opposite response gradients (see, for instance, Fig. 10C in Rossel, 1993). As one can easily imagine, the three types of macro-analyzer can be interconnected in a way that yields sharply tuned peak responses whenever the animal faces the solar vertical. The obvious next step in such model simulations is to systematically vary the sampling and connectivity parameters mentioned above, in order to find solutions that are most robust against changes in the skylight patterns as they occur when the sun changes its elevation above the horizon.

Hypothesis II

The E-vector patterns in the sky are used to determine any particular point of the compass. Let us start with the biologically realistic assumption made above in introducing subhypothesis IB that the array of micro-analyzers is sampled by three macro-analyzers differing in their tuning angles (e.g., $\chi_{max} = \pm 90°, +30°, -30°$). If this set of analyzers shared the same field of view, and looked at the zenith (see again Fig. 10C in Rossel, 1993), then the response ratios of the three analyzers would encode any particular E-vector direction present in the zenith (Kirschfeld, 1972; Bernard and Wehner, 1977). To be exact, as the analyzers are wide-field units they would not encode individual E-vector directions, but rather azimuthal positions of the animal relative to the solar vertical. This point is important, because it emphasizes a general aspect of the insect's "E-vector detecting" system: In the insect navigator, polarization vision has not been designed by evolution to determine individual E-vector directions, but rather to use the global celestial E-vector patterns as some means to read compass information from the sky. To accomplish this task the insect could possess a neural look-up table specifying any particular compass direction by the response ratios of a set of three macro-analyzers. Then, one can enquire theoretically about what the sampling areas and directions of view of the macro-analyzer should look like to accomplish this task as robustly as possible. Robustness, in this context, means that the compass information provided by the system is largely invariant against changes in the elevation of the sun and in the clouding that might render certain parts of the E-vector pattern invisible. Although, in principle, scanning movements are not required in the kind of three-analyzer system discussed here (the measurement could be taken instantaneously by simultaneous recording of separate analyzers), they will become necessary within an unpredictable skylight world in which the parts of the E-vector pattern available to the navigator vary from one foraging trip to another. If this were the case, the animal would have to recalibrate the look-up table of its three-analyzer system every time it starts on a new foraging excursion. In any case, it might be useful not to invoke too strong a dichotomy between a "simultaneous" and "successive" mode of compass orientation.

Let us end the discussion of potential compass mechanisms by adding some more general remarks. First and foremost, the neurophysiological bottom-up approach, i.e., the analysis of polarization-sensitive photoreceptors and interneurons (see also Petzold et al., 1995, and Homberg et al.,

1996, for additional types of polarization-sensitive interneurons in crickets and locusts) has not yet yielded unequivocal answers as to how the insect's compass system works; and the prospect that the underlying mechanisms will soon unfold in front of the electrophysiologist's eyes is daunting. Hence, on the basis of the pieces of evidence we have about the hardware of the insect's polarization channel, hypotheses and modelling approaches are needed to knit these loose threads of neurophysiological evidence into a theoretical fabric that is capable of accounting for the behavioural data.

Second, however, modelling approaches can be carried out at various levels and for various purposes. For example, one can run computer simulations to model the responses of particular types of macro-analyzers as the animal scans the celestial E-vector patterns: either to investigate theoretically what array of micro-analyzers and what connectivity patterns of macro-analyzers would yield the most reliable compass responses (see above; G. D. Bernard and R. Wehner, in preparation; Wehner, 1996), or to demonstrate how macro-analyzers that have been found to possess particular tuning angles and fields of view would behave in simulation experiments (Petzold and Labhart, 1994; J. Petzold, in preparation). Another approach is to use an opto-electronic hardware device mimicking the physiological properties of the cricket's polarization-opponent interneurons to perform measurements in the natural sky (Labhart, 1996; Labhart and Duemmler, work in progress). Finally, one can put the hardware version of any computer model into action by designing a robot – an autonomous agent – and testing its behaviour in the field. Most recently, we have entertained this approach in cooperation with colleagues from the New Artificial Intelligence camp (Lambrinos et al., in press) by building a robot that ran on wheels across the very same North African desert plains on which the biological agents – the *Cataglyphis* ants – were tested for their navigational performances. The robot (Fig. 22) was equipped with three pairs of wide-field polarization-opponent analyzers that shared the same zenith-centred field of view but differed in their tuning angles by 60° from each other. Provided with this sensory outfit, the robot could be programmed to perform according to either Hypotheses I or II. As it performed reasonably well in either situation, the robotics approach can now be used to test how particular sensory characteristics and model assumptions affect the precision and robustness with which the robot steers its compass courses.

If, in the end, I were asked to formulate the most general conclusion to be drawn from our behavioural analyses, neurophysiological studies and computer simulations on the mechanisms underlying the insect's skylight compass, I would make the following educated guess. The insect dissects the high-level task into a number of digestible low-level bits by employing a monochromatic polarization channel and a polarization-insensitive spectral channel (parallel coding) and uses special-purpose subroutines to deal with these particular bits. Take, for example, the polarization channel which provides the insect with a most detailed compass scale. It comprises

Figure 22. Mobile robot navigating by polarized skylight. The robot is equipped with six polarized light sensors (photo-diodes combined with linear polarizers) arranged in three pairs. The two members of each pair form a "polarization-opponent unit" mimicking the conditions found in the POL areas of insect eyes. The orientation of the polarization axes of the three pairs of polarizers can be variably adjusted with respect to the long axis of the robot, e.g., at 0°, 60° and 120°. All polarization sensitive sensors share the same zenith-centred field of view (acceptance angle 60°). Diodes and filters are mounted in cylindrical tubes which prevent direct sunlight from reaching the diodes. In addition, a set of eight equally spaced light-intensity sensors (photoresistors) is mounted within a metallic cylinder on top of the robot. The fields of view of these sensors are oriented horizontally and hence provide the robot with information about azimuthal light-intensity gradients. For details see Lambrinos et al. (1997).

a rather *coarse-grain* detector system that deals with the *global* aspects of the celestial E-vector patterns. The insect is not interested, so to speak, in assessing the orientation of individual E-vectors *per se*, but rather in using E-vectors as labels for spatial positions in the sky – and it is this spatial context that matters. The coarse-grain, global approach renders the system more robust against atmospheric perturbations that might lead to local changes in the E-vector pattern. Apart from this spatial aspect of wide-field integration, there is also a temporal aspect: as the time-period of an insect's foraging excursion is rather short, dramatic changes in the E-vector pattern as caused by haze or clouds are not likely to occur. Owing to these spatial and temporal limitations, or boundary conditions, the insect's special-purpose subroutine works sufficiently well in dealing with E-vectors in the sky and thus coping with a navigational problem which otherwise, if treated *ab initio* – might have lain beyond an insect's ken.

Acknowledgements
I am very grateful to my former graduate student Martin Müller, who during his Ph.D. work on the mechanisms of path integration became one of the most skillful *Cataglyphis* behaviourists ever having cooperated with me in North Africa. Out of this cooperation came a number of experiments reported in this chapter. Financial support by the Swiss National Science Foundation grant no. 31-28662.90 is gratefully acknowledged.

References

Arago, D.F.J. (1811) Mémoire sur une modification remarquable qu'éprouvent les rayons lumineux dans leurs passage à travers certains corps diaphanes, et sur quelques autres nouveaux phénomènes d'optique. *Mém. Cl. Sci. Math. Phys.* 1:93–134.

Batschelet, E. (1965) *Statistical methods for the analysis of problems in animal orientation and certain biological rhythms.* Am. Inst. Biol. Sciences, Washington, D.C.

Batschelet, E. (1981) *Circular Statistics in Biology.* Academic Press, London, New York.

Benhamou, S., Sauve, J.P. and Bovet, P. (1990) Spatial memory in large scale movements: efficiency and limitation of the egocentric coding process. *J. Theor. Biol.* 145:1–12.

Bernard, G.D. and Wehner, R. (1977) Functional similarities between polarization vision and color vision. *Vision Res.* 17:1019–1028.

Boehm, G. (1940) Über maculare (Haidinger'sche) Polarisationsbüschel und über einen polarisationsoptischen Fehler des Auges. *Acta Ophthalmol.* 18:109–142.

Brines, M.L. and Gould, J.L. (1979). Bees have rules. *Science* 102:571–573.

Brines, M.L. and Gould, J.L. (1982) Skylight polarization patterns and animal orientation. *J. Exp. Biol.* 96:69–91.

Clarke, D. and Graininger, J.F. (1971) *Polarized Light and Optical Measurement.* Pergamon Press, Oxford, New York.

Duelli, P. (1975) A fovea for E-vector orientation in the eye of *Cataglyphis bicolor* (Formicidae, Hymenoptera). *J. Comp. Physiol.* 102:43–56.

Duelli, P. and Wehner, R. (1973) The spectral sensitivity of polarized light orientation in *Cataglyphis bicolor* (Formicidae, Hymenoptera). *J. Comp. Physiol.* 86:37–53.

Edrich, W. and Helversen, O. v. (1987) Polarized light orientation in honey bees: is time a component in sampling? *Biol. Cybern.* 56:89–96.

Edrich, W., Neumeyer, C. and Helversen, O. v. (1979) "Anti-sun" orientation of bees with regard to a field of ultraviolet light. *J. Comp. Physiol.* 134:151–157.

Fent, K. (1985) *Himmelsorientierung bei der Wüstenameise* Cataglyphis bicolor: *Bedeutung von Komplexaugen und Ocellen.* Ph.D. Thesis, Zürich.

Fent, K. (1986) Polarized skylight orientation in the desert ant *Cataglyphis. J. Comp. Physiol.* A 158:145–150.

Fent, K. and Wehner, R. (1985) Ocelli: a celestial compass in the desert and *Cataglyphis. Science* 228:192–194.

Frisch, K.v. (1949) Die Polarisation des Himmelslichts als orientierender Faktor bei den Tänzen der Bienen. *Experientia* 5:142–148.

Frisch, K.v. (1967) *The Dance Language and Orientation of Bees.* Harvard University Press, Cambridge, MA.

Haidinger, W. (1844) Über das direkte Erkennen des polarisierten Lichts und der Lage der Polarisationsebene. *Ann. Phys., Leipzig* 63:29–39.

Hartline, P., Kass, L. and Loop, M.S. (1978) Merging of modalities in the optic tectum: infrared and visual integration in rattlesnakes. *Science* 199:1225–1229.

Helversen, O.v. and Edrich, W. (1974) Der Polarisationsempfänger im Bienenauge: ein Ultravioletrezeptor. *J. Comp. Physiol.* 94:33–47.

Homberg, U., Müller, M. and Vitzthum, H. (1996) The central complex: evidence for a role in polarized-light orientation. *Proc. Int. Congr. Entomol.* 20:204.

Horvàth, G. and Varjù, D. (1997) Polarization pattern of freshwater habitats recorded by video polarimetry in red, green and blue spectral ranges and its relevance for water detection by aquatic insects. *J. Exp. Biol.* 200:1155–1163.

Kirschfeld, K. (1972) Die notwendige Anzahl von Rezeptoren zur Bestimmung der Richtung des elektrischen Vektors linear polarisierten Lichtes. *Z. Naturforsch.* 27c:578–579.

Knudsen, E.I. and Konishi, M. (1978) A neural map of auditory space in the owl. *Science* 200:795–797.

Konishi, M. (1986) Centrally synthesized maps of sensory space. *Trends Neurosci.* 9:163–168.

Konishi, M., Takahashi, T.T., Wagner, H., Sullivan, W.E. and Carr, C.E. (1988) Neurophysiological and anatomical substrates of sound localization in the owl. *In*: G.M. Edelmann, W.E. Gall and W.M. Gowman (eds): *Auditory Function.* J. Wiley and Sons, New York, pp 721–745.

Labhart, T. (1986) The electrophysiology of photoreceptors in different eye regions of the desert ant, *Cataglyphis bicolor. J. Comp. Physiol.* A 158:1–7.

Labhart, T. (1988) Polarization-opponent interneurons in the insect visual system. *Nature* 331:435–437.

Labhart, T. (1996) An opto-electronic model of a polarization-sensitive insect interneuron. *Proc. Neurobiol. Conf. Göttingen* 24:356.

Lambrinos, D., Maris, M., Kobayashi, H., Labhart, T., Pfeifer, R. and Wehner, R. (1997) An autonomous agent navigating with a polarized light compass. *Adapt. Behav.* 6:175–206.

Lanfranconi, B. (1982) *Kompassorientierung nach dem rotierenden Himmelsmuster bei der Wüstenameise* Cataglyphis bicolor. Ph.D. Thesis, Zürich.

Malus, E. (1809) Sur une propriété de la lumière réfléchie par les corps diaphanes. *Bull. Sci. Soc. Philom.* 1:266–269.

Menzel, R. (1979) Spectral sensitivity and color vision in invertebrates. *In.* H. Autrum (ed): *Handbook of Sensory Physiology, Vol. VII/6A.* Springer-Verlag, Berlin, Heidelberg, New York, pp 503–580.

Mote, M. and Wehner, R. (1980) Functional characteristics of photoreceptors in the compound eye and ocellus of the desert ant, *Cataglyphis bicolor. J. Comp. Physiol.* 137:63–71.

Murphey, R.K. (1983) Maps in the insect nervous system, their implications for synaptic connectivity and target location in the real world. *In*: F. Huber and H. Markl (eds): *Neuroethology and Behavioral Physiology.* Springer-Verlag, Berlin, Heidelberg, New York, pp 176–188.

Oldfield, B.P. (1988) Tonotopic organization of the insect auditory pathway. *Trends Neurosci.* 11:267–270.

Petzold, J. and Labhart, T. (1994) Modelling polarization-opponent interneurons of insects: responses to the polarization patterns in the sky. *Proc. Neurobiol. Conf. Göttingen* 22:466.

Petzold, J., Helbling, H. and Labhart, T. (1995) Anatomy and physiology of four new types of polarization sensitive interneuron in the cricket, *Gryllus campestris. Proc. Neurobiol. Conf. Göttingen* 23:415.

Räber, F. (1979) *Retinatopographie und Sehfeldtopologie des Komplexauges von* Cataglyphis bicolor *(Formicidae, Hymenoptera) und einiger verwandter Formiciden-Arten.* Ph.D. Thesis, Zürich.

Ramskou, T. (1969) *Solstenen. Primitiv Navigation I Norden for Kompasset.* Rhodos, Kobenhavn.

Römer, H. and Rheinländer, J. (1989) Hearing in insects and its adaptation to environmental constraints. *In*: H.C. Lüttgau und R. Necker (eds): *Biological Signal Processing.* VCH Verlagsgesellschaft, Weinheim, pp 146–162.

Rossel, S. (1993) Navigation by bees using polarized skylight. *Comp. Biochem. Physiol.* A 104:695–708.

Rossel, S. and Wehner, R. (1982) The bee's map of the e-vector pattern in the sky. *Proc. Natl. Acad. Sci. USA* 79:4451–4455.

Rossel, S. and Wehner, R. (1984a) How bees analyse the polarization patterns in the sky. Experiments and model. *J. Comp. Physiol.* A 154:607–615.

Rossel, S. and Wehner, R. (1984b) Celestial orientation in bees: the use of spectral cues. *J. Comp. Physiol* A 155:605–613.

Santschi, F. (1911) Observations et remarques critiques sur le mécanisme de l'orientation chez les fourmis. *Rév. Suisse Zool.* 19:305–338.

Santschi, F. (1923) L'orientation sidérale des fourmis, et quelques considérations sur leurs différentes possibilités d'orientation. *Mém. Soc. Vaudoise Sci. Nat.* 4:137–175.

Seyfarth, E.A. and Barth, F. (1972) Compound slit sense organs on the spider leg: mechano-receptors involved in kinesthetic orientation. *J. Comp. Physiol.* 78:176–191.

Sparks, D.L. (1988) Neural cartography: sensory and motor maps in the superior colliculus. *Brain Behav. Evol.* 31:49–56.

Stockhammer, K. (1959) Die Orientierung nach Schwingungsrichtung linear polarisierten Lichtes und ihre sinnesphysiologischen Grundlagen. *Erg. Biol.* 21:34–56.

Strutt, J.W. (1871) On the light from the sky, its polarization and colour. *Phil. Mag.* 41: 107–120, 274–279.

Suga, N. (1990) Cortical computational maps for auditory imaging. *Neural Networks* 3:3–21.

Waterman, T.H. (1981) Polarization sensitivity. *In*: H. Autrum (ed.): *Handbook of Sensory Physiology, Vol. VII/6B.* Springer-Verlag, Berlin, Heidelberg, New York, pp 281–469.

Wehner, R. (1975) Space constancy of the visual world in insects. *Fortschr. Zool.* 23:148–160.

Wehner, R. (1982) Himmelsnavigation bei Insekten. Neurophysiologie und Verhalten. *Neujahrsbl. Naturforsch. Ges. Zürich* 184:1–132.

Wehner, R. (1983) Celestial and terrestrial navigation: human strategies – insect strategies. *In*: F. Huber and H. Markl (eds): *Neuroethology and Behavioural Physiology.* Springer-Verlag, Berlin, Heidelberg, New York, pp 366–381.

Wehner, R. (1991) Visuelle Navigation: Kleinstgehirn-Strategie. *Verh. Dtsch. Zool. Ges.* 84: 89–104.

Wehner, R. (1994a) The polarization-vision project: championing organismic biology. *Fortschr. Zool.* 39:103–143.

Wehner, R. (1994b) Himmelsbild und Kompassauge – Neurobiologie eines Navigationssystems. *Verh. Dtsch. Zool. Ges.* 87:9–37.

Wehner, R. (1996) Polarisationsmusteranalyse bei Insekten. *Nova Acta Leopoldina NF* 72:159–183.

Wehner, R. and Rossel, S. (1985) The bee's celestial compass – a case study in behavioural neurobiology. *Fortschr. Zool.* 31:11–53.

Wehner, R. and Strasser, S. (1985) The POL area of the honey bee's eye: behavioural evidence. *Physiol. Entomol.* 10:337–349.

Zeki, S. (1993) The representation of colours in the cerebral cortex. *Nature* 284:412–418.

Zollikofer, C.P.E., Wehner, R. and Fukushi, T. (1995) Optical scaling in conspecific *Cataglyphis* ants. *J. Exp. Biol.* 198:1637–1646.

Orientation and Communication in Arthropods
ed. by M. Lehrer

Magnetic orientation and the magnetic sense in arthropods

M. M. Walker

Experimental Biology Research Group, School of Biological Sciences, University of Auckland, Private Bag 92019, Auckland, New Zealand

Summary. The physical properties of the earth's magnetic field are summarized with the aim of emphasizing their significance as cues that can be exploited in orientational tasks. Past work has revealed magnetic orientation in vertebrates as well as invertebrates, including arthropods. The key finding to date has been that, as opposed to many vertebrates, the magnetic compass of arthropods responds to the polarity, rather than to the inclination of the earth's magnetic field. As in the case of vertebrates, the debate over how arthropods detect magnetic fields has yet to be resolved. Currently, evidence has been reported in support of a detection system based on magnetite crystals together with a variety of detection systems based on events occurring at the molecular level. Interactions between the magnetic and other compasses in orientation experiments suggest the existence of an area in the brain where spatial orientation information from magnetic and other stimuli converges. The slow advance of our knowledge on magnetic orientation in arthropods, as opposed to the much better understanding of magnetic orientation in vertebrates, arises from difficulties in identifying the appropriate behavioural contexts in which arthropods respond to magnetic fields in both laboratory and field situations. Arthropods thus present challenges not only in demonstrating magnetic orientation, but also in elucidating the sensory mechanisms involved in the perception of magnetic fields.

Introduction

The magnetic field of the earth (also called the geomagnetic field) is an ordered and relatively stable component of the environment that has been present throughout the history of life on earth and has only changed significantly during reversals of the magnetic field direction (Skiles, 1985). The physical properties of the earth's magnetic field provide orientation information to animals that can detect magnetic stimuli.

In contrast with the use of visual, olfactory, and acoustical cues, the use of magnetic cues for orientation is restricted to animals. Humans are not known to perceive any of the parameters of the earth's magnetic field, which might be one of the reasons why human investigators, despite extensive work over many decades, have gained only limited understanding of the use of the magnetic cues for orientation. Our own inability to detect magnetic fields may also be one of the reasons why magnetic orientation has always been, and still is, one of the most fascinating topics for workers concerned with sensory performances of animals.

In their recent review of magnetic orientation in animals, Wiltschko and Wiltschko (1995a) focused on whole-organism responses to magnetic fields and stressed the importance of discovering appropriate behavioural contexts for demonstrating magnetic orientation responses. In vertebrates, the necessary conditions for initial demonstration of magnetic orientation were discovered much more readily than in invertebrates, with the result that our understanding of magnetic orientation in invertebrates is much less advanced than is our understanding of magnetic orientation in vertebrates. Invertebrates are, however, being studied extensively, and recent results of studies on arthropods provide the opportunity to update the knowledge in the field following the review by Wiltschko and Wiltschko (1995a).

This chapter describes the physical properties of the earth's magnetic field and summarizes the information contained in it that can be used by animals in orientational tasks. It reviews what has already been learned about the use of this information by arthropods, particularly insects, to determine their position in space and to select their direction of locomotion. Furthermore, the behavioural contexts in which magnetic orientation by arthropods has been successfully demonstrated are discussed, with the aim of proposing new ways of overcoming difficulties towards a better understanding of magnetic orientation in arthropods. Finally, issues in detection and processing of magnetic field information and their relation to observed magnetic orientation responses leads to the suggestion that combined study of both magnetic orientation and the magnetic sense itself will lead to rapid advance in our knowledge on how arthropods use magnetic information for orientation.

The source of magnetism and the properties of the earth's magnetic field

Basic information on the physical properties of the earth's magnetic field necessary for understanding the arguments put forward in the present chapter is provided below. A detailed description and discussion of the earth's magnetic field is given by Skiles (1985).

The earth's magnetic field

By far the greatest contribution (usually $>95\%$) to the magnetic field measurable at any point on the earth is due to electric currents flowing in the molten iron core of the earth (the dynamo effect). This magnetic field is termed the main, or dipole, field. In essence, the earth behaves like a large bar magnet and we think of the magnetic lines of force as emerging out of the earth at one (the southern) magnetic pole, curving around the earth, gradually turning through 180° to re-enter the earth at the other (northern) magnetic pole (Fig. 1A). The total magnetic field observable at the surface of the earth is the sum of the strength of the dipole field and that of further magnetic fields produced by non-dipole components of the main field, magnetized crustal rocks, and the solar wind.

Declination

The axis of the dipole field (magnetic north and south) is not aligned with the axis around which the earth rotates (geographic north and south). The angle between the magnetic and the geographic axes is defined as the magnetic declination. The use of this parameter for magnetic orientation would require not only knowledge of the direction of the magnetic field vector, but also knowledge of the orientation of the geographic axis. Use of declination for orientation therefore appears doubtful, as knowledge of either magnetic or geographic north and south is sufficient for directional orientation.

Inclination and polarity

The rotation of the magnetic field lines as they pass between the two magnetic poles gives rise to the inclination (or dip) of the magnetic field. The inclination of the earth's magnetic field is defined as the angle it subtends with the horizontal plane at the surface of the earth. The inclination ranges between 90° (vertical to the surface of the earth) at the magnetic poles, and 0° (horizontal) at the magnetic equator (Fig. 1 A). The magnetic vectors point upwards (away from the earth's surface) in the southern hemisphere, but downwards (towards the earth's surface) in the northern hemisphere. Each magnetic vector can thus be divided into two components, a vertical one (upwards or downwards), and a horizontal one (from south to north) (Fig. 1 B). The direction of the vertical component determines whether the inclination is positive (northern hemisphere) or negative (southern hemisphere). The direction of the horizontal component gives the polarity of the field.

Intensity

The intensity (or strength) of the magnetic field is highest at the magnetic poles and lowest at the magnetic equator (Fig. 2). The unit of magnetic intensity is defined as the force a magnet exerts on a unit dipole at unit distance. The SI unit for magnetic field intensity is the Tesla ($= 10^4$ Gauss). Magnetic intensities commonly observed at the surface of the earth range from about 24 microTesla ($\mu T = 10^{-6}$ Tesla) at the magnetic equator to 60–70 μT at the magnetic poles.

A

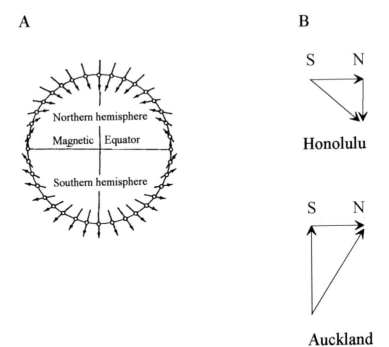

B

S   N

Honolulu

S   N

Auckland

C

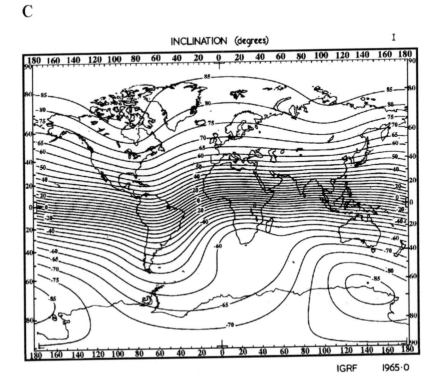

INCLINATION (degrees)

IGRF   1965·0

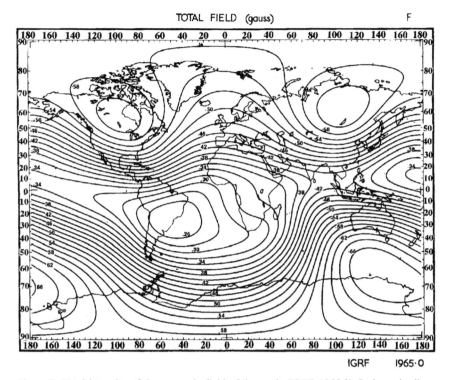

Figure 2. Total intensity of the magnetic field of the earth (IGRF 1965.0). Isointensity lines show the pattern of change in magnetic intensity between the magnetic equatorial region (24–35 µT) and the magnetic poles. Modified from Skiles (1985; with permission).

Figure 1. (A) A simplified model (Waterman, 1989) illustrating how the lines of magnetic force intersect the surface of the earth. The lines of force (arrows) intersect the surface of the earth at angles from 0° at the magnetic equator to 90° at the magnetic poles. The lengths of the arrows indicate relative magnetic intensity (used with permission from W. H. Freeman). (B) The vector diagrams illustrate resolution of the magnetic field vector at the surface of the earth into horizontal and vertical components for one southern (Auckland: 36°51′S, 174°46′E) and one northern (Honolulu: 21°18′N, 157°54′W) hemisphere location. At Auckland, inclination is large and negative (−62°) and the vertical component is large relative to the horizontal component. At Honolulu, inclination is positive (35°) and the horizontal and vertical components are close to equal. In experimental situations, the total field vector, the horizontal component and the vertical component can all be controlled using appropriately aligned electromagnetic coils that surround the experimental space. (C) Inclination or dip in the magnetic field of the earth (IGRF 1965.0). Isoinclination lines show the pattern of change in magnetic inclination between the magnetic equator (0°) and the magnetic poles (75°30′N, 101°W, on Bathurst Island, and 65°30′S, 143°18′E, on the Adelie coast in Antarctica). Positive (downward) inclination is shown in the northern hemisphere, and negative (upward) inclination in the southern hemisphere. (C) Modified from Skiles (1985; with permission).

The potential orientational information provided by the earth's magnetic field

The magnetic field contains several types of information that are of potential value to the animal's orientation performance, namely positional, directional, and temporal information.

Positional information provided by the main field

The systematic changes in the total intensity of the geomagnetic field illustrated in Figure 2 may provide the animal with information about its position in space. Information about position north and south of the magnetic equator may also be extracted from the systematic increase in inclination of the magnetic vector between the magnetic equator and poles (see Fig. 1). Over many regions on the earth, lines of equal inclination angles (see Fig. 1C), and lines of equal field intensity (see Fig. 2) do not run exactly parallel to each other, and thus they create a (non-orthogonal) bi-coordinate system that might enable the animal, at least theoretically, to determine its position in its environment with considerable accuracy (see Lohmann and Lohmann, 1996b).

Positional information provided by magnetic anomalies

The magnetic field observed at the surface of the earth is influenced by local and regional variations in crustal rocks and topography termed the residual field or magnetic anomalies (Fig. 3). The strengths of such fields are usually too small to deflect a compass needle. Measurement of the magnetic anomalies requires a magnetometer that detects very small changes in the total intensity of the field. It has been suggested, however, that pigeons use magnetic anomalies in determining position (Gould, 1982; Wiltschko and Wiltschko, 1995a). In arthropods, the use of this cue has been demonstrated only once (Walker and Bitterman, 1989a) (see further below).

Directional information

With the exception of periods in the history of the earth during which the magnetic field reversed its polarity (see Skiles, 1985), it has probably always been possible for an animal to determine direction relative to a goal using cues derived from the geomagnetic field. The polarity of the dipole field provides information on where north and south are and the inclination indicates the directions of the magnetic equator and the nearest magnetic

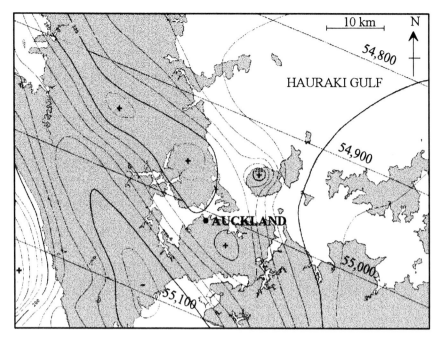

Figure 3. Magnetic map of the region around Auckland, New Zealand. Grey and white areas represent land and sea, respectively. Diagonal lines are contours of equal total intensity (in nanoTesla, nT). Other contours show the residual field variations (in nT) due to magnetic anomalies. Modified from Hunt and Syms (1977; Institute of Geological and Nuclear Sciences; with permission).

pole. A compass that responds to magnetic polarity can use the earth's magnetic field as a reference direction for selecting a particular alignment in space, or a particular direction of locomotion relative to this vector's direction. A compass that responds to magnetic inclination does not distinguish between north and south, but can recognise the direction toward the magnetic equator or magnetic pole in either hemisphere.

Temporal information

Apart from providing directional and positional information, the geomagnetic field may also offer *temporal* information if an animal is sufficiently sensitive to changes in the strength of the magnetic field. Small variations in the geomagnetic field are caused by ion currents in the ionosphere that are produced by the solar wind, a stream of charged particles (primarily protons) emanating from the sun. The ion currents are strongest in the middle of the day and produce a small deflection on a plot of magnetic intensity against time of day. The moon acts to affect the size of the diurnal signal produced by the sun. The effects are strongest at about new moon and full moon, and weakest around the middle of the lunar month. Although the temporal variations in the geomagnetic field are very small and are, in addition, embedded in substantial noise (Skiles, 1985), it has been suggested that insects are able use the diurnal signal as a *Zeitgeber* (for references, see Wiltschko and Wiltschko, 1995a).

Relevant concepts in orientation

The significance of the spatial information that can be extracted from the earth's magnetic field can best be appreciated when the various mechanisms known to participate in spatial orientation are considered.

Stages in spatial orientation

Several models for the use of spatial information for orientation have been proposed. Based on studies of birds, Griffin (1952) and Schmidt-Koenig (1965) developed a widely accepted classification of orientation mechanisms observed in animals: piloting using familiar landmarks (Type I orientation), compass orientation (Type II orientation) and true navigation (Type III orientation). Piloting requires direct or indirect sensory contact with the goal. In compass orientation, on the other hand, the animal uses an external stimulus as a directional reference to set and maintain a course towards a goal that is out of sensory contact. Research on true navigation has been guided by the "map and compass" hypothesis proposed by Kramer (1953). According to this hypothesis, navigation takes place in two steps. In the "map step", the animal first determines its current position relative to a goal such as its nest. In the "compass step", it sets a compass course (Type II orientation) for the goal using directional stimuli such as the sun, stars, polarised light and the geomagnetic field. Thus, positional as well as directional information are needed for successful navigation. As we have seen above, the magnetic field provides both of these types of information.

Path integration and the use of landmarks

Extensive work on homing of bees and ants has shown that these insects determine the home direction by integrating the directions and distances travelled during the outward journey (Mueller and Wehner, 1988; Kirchner and Braun 1994; Wehner et al., 1996). When foraging ants or bees are captured at the feeding site and are released at a different site, they select, for homing, the vector that they would have used at the site where they have been captured. The use of this home vector can be excluded by capturing the forager at the home site as soon as it has returned from a foraging trip.

Such animals no longer possess a homing vector, and therefore, when released at a new site, they can only rely on landmarks to return home. Ants treated in this way were able to return home when they were released in an area that was familiar to them, but they were lost when released in an unfamiliar area (Wehner et al., 1996). Similarly, honeybees were unable to locate a familiar food source after displacement unless they could see familiar landmarks (Dyer, 1991). Another example involves Euglossine bees that have been shown to forage up to 25 km from their nests visiting specific pollen and nectar sources that are widely dispersed (Janzen, 1971). During their outward journeys, foragers were found to visit the various food sources in the order in which they had initially found them. That is, they used a familiar route, rather than the shortest route among the food sources (Janzen, 1971). The orientation performance observed in these examples can be considered as map-like in that the animals are able to navigate effectively, even after displacement. However, they can do so only within a familiar area and not outside it.

A simple model (Cartwright and Collett, 1987; see also T.S. Collett and J. Zeil, this volume) for "map-like" behaviour by bees and wasps combines memorised retinotopic "snapshots" of the distribution of visual landmarks with compass vectors for the direction and distance to the nest. According to this model, to home after displacement, the animal would select the snapshot that gives the closest fit to its current retinal image and then use the direction vector to the nest associated with that particular snapshot to set a course for home. The course can be updated as the animal approaches the nest. Summation of the home vectors between the current position and a variety of known destinations would allow the animal to establish a novel route within its familiar area after displacement (Cartwright and Collett, 1987; see also Collett, 1996; Menzel et al., 1996).

Although the model would not permit navigation outside a familiar area, it can readily explain map-like behaviour within the area bounded by the most distant landmarks known to the animal. Results of recent experiments suggest that honeybees use the magnetic compass to adopt a preferred viewing direction when learning the spatial relations among landmarks, and to distinguish between the learned landmark constellation and one that differed from it only in being rotated by 90° about the vertical axis (Collett and Baron, 1994; Frier et al., 1996). The magnetic compass could thus provide directional information to become associated with the snapshots and so contribute to landmark-based orientation within a familiar area.

Experimental evidence for the perception of magnetic cues by arthropods

The main body of evidence for use of magnetic cues in orientational tasks by arthropods has been derived from experimental results in which the animal's performance under natural magnetic field conditions was compared with its response to artificially produced magnetic fields. In the following, some of the most important results are reviewed.

Preferences for particular directions relative to the magnetic vector

Many examples demonstrating the perception of the earth's magnetic field by arthropods involve the animal's tendency to align its body according to the cardinal directions of the magnetic vector. For example, resting termites (Roonwal, 1958), as well as resting flies (Becker, 1965; Wehner and Labhart, 1970) and honeybees (Altman, 1981) select body alignments that are preferentially parallel, anti-parallel or perpendicular to magnetic field lines. When the natural geomagnetic field is altered using a strong magnet, bees align their body position according to the cardinal directions of this artificial field. In flies, when the earth's magnetic field is compensated, the tendency to align in a particular direction disappears (Becker, 1965).

Further examples involve building activity of social insects. Some termites were found to build galleries oriented in the cardinal directions of the magnetic field (Becker, 1975). Introduction of artificial magnetic fields terminates building activity in several termite species (see references in Wiltschko and Wiltschko, 1995a). Further, when honeybees are moved from their hive to an empty circular hive, they build combs that are aligned in the same direction as in the parent hive (Lindauer and Martin, 1972, De Jong, 1982). In the presence of an artificial magnetic field, they build abnormal combs (Lindauer and Martin, 1972). Because no visual directional cues are available during building in darkness, and because artificial magnetic fields interfere so obviously with the building activity, these results suggest that the building activity is organised using the magnetic cue.

Further evidence comes from experiments on the dance language of honeybees (see von Frisch, 1967). When dancing on vertical combs, which is the natural situation, honeybees use the gravity vector as a directional reference. The direction of the sun is transposed to the upwards direction, with the angle between the dance direction and upwards indicating the direction of the food source relative to the sun. The dances contain, however, systematic errors, termed residual misdirection (*Restmissweisung*) (von Frisch, 1967). The magnitude of the misdirection varies as the sun azimuth changes during the course of the day (Fig. 4). The misdirection disappears, however, when the dances are oriented along the magnetic field lines, or (after a latency of 40–60 min) when the external magnetic field is reduced below 5% of its normal strength using Helmholtz coils (Lindauer and Martin, 1968). This result suggests that misdirection arises from a tendency of the bee to align its direction of locomotion with the direction of the magnetic vector, similar to the tendency of aligning resting positions and comb orientation with this direction (see above).

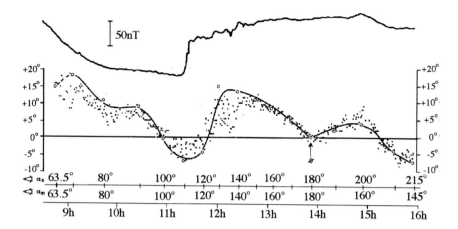

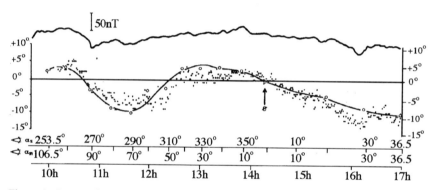

Figure 4. Curves of misdirection of the waggle runs of honeybees dancing on a vertical comb over the course of a day. Misdirection of the waggle runs (points) of the dances were recorded on the northern and southern faces of a comb aligned east-west. The solid line through the points in each misdirection plot is the curve generated by the mathematical description of misdirection by Martin and Lindauer (1977). The solid line above each plot is the variation in total intensity of the ambient magnetic field over the course of the day of recording. The vertical scale bar indicates the change in intensity (in nT) in both plots. Misdirection, i.e., the deviation in degrees of the observed dance angle from the expected dance angle, is shown on the ordinate in each plot. α_s and α_m on the abscissa in each plot are the expected dance angles with respect to the gravity and magnetic vectors, respectively. The plots of misdirection pass through the zero misdirection line up to several times per day. The zero crossings at Φ mark the points on the plots where dances are expected to run antiparallel (180°) to the magnetic inclination. Note that the dance data and the curves fitted to them also cross the zero misdirection line at magnetic azimuths of 80° (upper plot) and approximately 95° (lower plot) away from Φ. Data from Martin and Lindauer (1977; with permission).

Misdirection also frequently disappears when dance directions are at or near right angles to the external field vector (Martin and Lindauer, 1977). A mathematical model of misdirection developed by Martin and Lindauer (1977) incorporates the rate of change in total field intensity, as well as the sine of the angle between the dance direction and the projection of the magnetic field vector onto the dance surface. Curves generated by the model gave good fits to actual misdirection data (Martin and Lindauer, 1977), including the disappearance of misdirection in dances parallel or perpendicular to the external field vector (Fig. 4). The preference for alignments in the cardinal magnetic directions may thus reflect a fundamental property of the magnetic field detection system of arthropods. R. Campan (this volume) reviews, among other things, several experimental results that demonstrate the basic tendency of many arthropods to orient to the magnetic field (magnetotaxis), and discusses the possible function of this tendency in calibrating celestial compass mechanisms during the ontogeny of oriented behaviour.

When honeybees are forced to dance on a horizontal surface under light conditions that provide no directional information, the waggle runs of the dance eventually become preferentially oriented to the cardinal magnetic directions (Fig. 5). Under these conditions, in contrast to dances on vertical combs or in light that provides a directional reference, the directions of the waggle runs are not related to the directions of the food sources the dancers have visited. Amplification or reduction of the intensity of the magnetic field increases or reduces, respectively, the preference for the cardinal magnetic directions (Martin and Lindauer, 1977) (Fig. 5).

Conditioning experiments

In conditioning experiments, honeybees were found to readily learn to discriminate between the presence and the absence of a localized intensity anomaly superimposed on the uniform background field of the earth (Walker and Bitterman, 1985; Kirschvink and Kobayashi-Kirschvink, 1991). However, they do so only if the conditioned response requires movement (Walker et al., 1989). The bees' threshold sensitivity to changes in intensity (26 nanoTesla) ($nT = 10^{-9}$ Tesla) associated with the anomalies (Walker and Bitterman, 1989a) (Fig. 6) is probably sufficient for use in the map-step of navigation (see above), as well as for detecting the diurnal signal in the geomagnetic field. The sensitivity to changes in field intensities demonstrated in these experiments is similar to that found in fish (Walker, 1984).

In contrast to the success of conditioning to changes in intensity, it has been extremely difficult to condition animals to the direction of a magnetic vector. For example, honeybees trained to exit either the magnetic

Compensated field

N

n=10541

Normal field

N

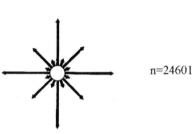

n=24601

Field amplified to 350 µT

N

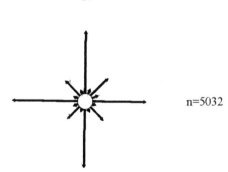

n=5032

Figure 5. Preferential alignment of dances by honeybees in the magnetic cardinal directions. Dances took place on a horizontal comb in the absence of directional visual stimuli. *Top panel*: dances recorded in a field of less than 5% of the normal intensity. *Middle panel*: dances recorded in the local geomagnetic field. *Bottom panel*: dances recorded in a field of 7.5 times the normal intensity. The numbers beside each plot indicate the total number of dances observed in each field, while the lengths of the arrows within each plot indicate the percentages of dances in the different directions. Data from Martin and Lindauer (1977; with permission).

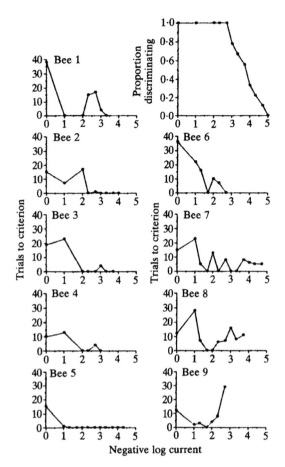

Figure 6. Discrimination thresholds for magnetic intensity anomalies in honeybees. Bees were trained individually to choose between two identical targets, one of which carried food (50% sucrose solution) while the other carried plain water. The availability of the sucrose solution was signalled by the presence of a localised magnetic anomaly projected over the target. The anomaly and sucrose solution were offered at the left target on half of the bee's visits, and at the right one on the other half of the visits (in balanced, quasi-random order). The bees were trained until they reached a criterion accuracy of six correct choices in a row, or seven correct responses out of eight visits. The performance of each of nine bees is plotted in terms of the number of trials required to reach this criterion for each of a series of anomalies of decreasing intensity indexed by the current (in amps, A) passed through the coils used to produce the anomalies. With a current of 1 A (corresponding to the value zero on the logarithmic abscissa), the intensity of the anomaly at the point of reinforcement with sucrose solution was about 30 times the background intensity, and with a current of 10^{-5} A (corresponding to the value 5, abscissa), at which the best animal failed, the intensity of the anomaly was 0.03% of background. The training of an animal ended if, at any current level beyond the first, it failed to meet the criterion of discrimination in 32 visits, and the plot for each animal terminates at the lowest intensity discriminated. Plotted in the upper right hand corner is the proportion of animals meeting the criterion at each current level. From Walker and Bitterman (1989a).

north or south arm of a T-maze (Kirschvink and Kobayashi-Kirschvink, 1991) frequently made long runs of same-direction choices, but the runs of choices were equally likely to be wrong as to be right. There was also no temporal pattern in the appearance of the runs of same-direction choices, which would have been the case in a successful conditioning experiment (Kirschvink and Kobayashi-Kirschvink, 1991). It is possible that the necessary conditions for training animals to a particular direction of the magnetic vector were not met in this experiment. We shall return to this point below.

Orientation in time

Many arthropod species would benefit from the ability to detect the diurnal signal in the earth's magnetic field to time their activities appropriately. The experimental evidence is, however, conflicting. Lindauer (1977) found that a 24-h feeding rhythm established in honeybees held in constant conditions was abolished when the normal magnetic field was replaced by an aperiodic field of ten times the normal intensity. Neumann (1988), however, found that artificially generated fields of up to twice the earth strength had no influence on rhythmic behaviour of honeybees.

For several reasons it is premature to conclude that loss of rhythmicity in a strong magnetic field is evidence for the use of the magnetic field as a *Zeitgeber*. First, in the absence of time training, individual bees exhibit classical free running rhythms that take no account of the diurnal signal in the geomagnetic field (Moore and Rankin, 1985). Second, strong fields (> 10X earth strength) also affect the duration (Tomlinson et al., 1981) and misdirection of bee dances (Lindauer and Martin, 1968), as well as comb building (Lindauer and Martin, 1972). Although they are consistent with the ability to detect magnetic fields, such disparate responses also suggest that strong fields have non-specific effects on the bee's behaviour – that is, the mechanism by which such effects occur may not necessarily depend on the animal having a magnetic sense.

Experimental evidence for the use of a magnetic compass

Direct evidence for compass orientation in arthropods was provided in experiments on migratory moths, *Noctua pronuba* and *Agrotis exclamationis* (Baker, 1987; Baker and Mather, 1982), on honeybees (Schmitt and Esch, 1993), and on wood ant foragers (Camlitepe and Stradling, 1995), using circular orientation arenas. In all of these experiments, the directions chosen by the animals in both normal and artificially altered magnetic fields were consistent with the expected directions of locomotion under the assumption that magnetic directional cues were used in the task.

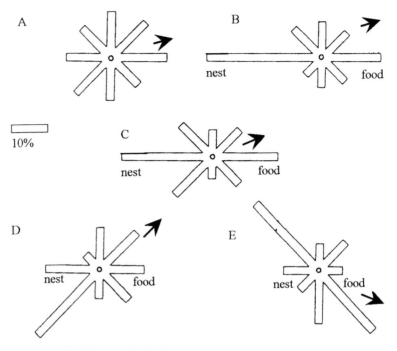

Figure 7. Magnetic orientation by the harvester termite *Trinervitermes germinatus* in normal and altered magnetic fields. The lengths of the bars represent the proportions of groups of up to 200 termites choosing among eight radially arranged pheromone trails in an orientation arena. The open circle in the centre of each plot indicates the point where the animals were released into the orientation arena. Arrows indicate the direction of magnetic north within the orientation arena. Control treatments show orientation in the normal local field (A, B) and with a pair of magnets aligned parallel to the horizontal component of the local field (C). In (A), odour trails were present in the arena but the termites were not carrying food. In (B), the termites were carrying food back towards the nest and were well oriented in the arena. In (C), the termites oriented towards the nest after amplification of the earth's field by magnets. In the experimental treatments (D) and (E), the direction of the magnetic field in the arena was shifted by 33° by aligning the pair of magnets with the oblique trails. The termites now chose the oblique trails and were well oriented. Data from Rickli and Leuthold (1988; with permission).

Magnetic compass orientation has also been revealed through associations of magnetic stimuli with other stimuli or activities, which may not themselves be directional. For example, forager termites placed in a laboratory arena and offered a choice among radially arranged odour trails moved in the homeward direction as indicated by the magnetic compass, but would only do so if they were carrying food (Rickli and Leuthold, 1988) (Fig. 7). Similarly, sandhoppers (*Talitrus saltator*) oriented magnetically in an orientation arena, but did so only if they jumped, and not if they crawled (Arendse and Kruyswijk, 1981).

A rather different use of the magnetic compass was reported by Leucht and Martin (1990) who showed that honeybee foragers use the magnetic compass to assist polarized light compass orientation. When foragers dancing on a horizontal surface had their view of the sky restricted to a 20° or 40° sector around the zenith, their dances were unimodally oriented towards the food source. However, when the strength of the magnetic field was reduced to less than 4% of the normal value, the dances became bimodal about the axis of the direction to the food source. Leucht and Martin (1990) concluded that the bees had used the magnetic field as an independent reference to resolve the 180° ambiguity in the directional information provided by the E-vector of polarised light under the experimental conditions (see also R. Wehner, this volume). Another example of the participation of the magnetic compass in an orientational task has already been described above: Collett and Baron (1994) suggested that bees use directional magnetic cues in the process of acquiring snapshots of landmarks.

Magnetic compass orientation can also replace orientation to another stimulus when that stimulus no longer provides directional information. For example, the flour beetle *Tenebrio molitor* normally orients away from light (probably to avoid desiccation). Arendse and Vrins (1975) showed that the beetles, when placed in a circular orientation arena under isotropic illumination, orient in directions that coincide with the direction away from light in their home containers. Their orientation tracked a shift of 120° in the direction of magnetic north and disappeared when the magnetic field was compensated. Similarly, magnetic compass orientation by marine isopods (*Idotea baltica basteri*) can be modified by allowing the animals to form an association between the floor slope in an aquarium and a magnetic field direction that is different from the direction of the slope at their home beach (Ugolini and Pezzani, 1995).

The relevant information for directional orientation: Inclination or polarity?

Although it was recognised quite early that the magnetic compass of vertebrates is based on magnetic field inclination (Wiltschko and Wiltschko, 1972), it was only recently demonstrated experimentally that the magnetic compass of arthropods is not. It seems instead to be based on field polarity.

The critical experiment for distinguishing between responses to magnetic field inclination and polarity is to invert the vertical component without changing the horizontal component of the external field (see Fig. 1B). Under these conditions, animals that respond to magnetic field inclination will reverse their orientation, whereas the orientation of animals that respond to magnetic field polarity will not change (Wiltschko and Wiltschko, 1972). Similarly, animals that use a magnetic inclination compass are expected to exhibit a bimodal orientation when the horizontal component is present, but the vertical component is cancelled, whereas the orientation of animals that respond to magnetic field polarity is expected to be unimodal even under these conditions (Wiltschko and Wiltschko, 1972).

Spiny lobsters (*Panulirus argus*) that had been fitted with eyecaps to prevent response to visual stimuli reversed their orientation direction when the horizontal component of the field was inverted, but not when the vertical component was inverted (Lohmann et al., 1995). The flour beetle (*Tenebrio molitor*) showed unimodal orientation to magnetic field direction even when the vertical component of the magnetic field was cancelled (Arendse and Vrins, 1975). These results show that these animals use a magnetic polarity compass, rather than an inclination compass.

Clearly, however, to support or contradict this conclusion, similar experiments on further arthropod species are needed.

How are the magnetic cues detected?

The finding that animals can orient by using magnetic cues implies that they possess a sensory system to mediate magnetic orientation. Sensory studies need make no *a priori* assumptions about the use of the magnetic sense, but rather focus first on testing for response to the different dimensions of the magnetic stimulus. Searching for the site and the nature of the detectors involves further, independent steps.

Out of several possible sensory mechanisms that have been proposed to mediate magnetic orientation (see review by Wiltschko and Wiltschko, 1995a), we here consider only two that have been discussed most frequently.

Detection based on interactions with visual pigment molecules

One hypothesis, based on experimental results obtained mainly from vertebrates, suggests that magnetic orientation is mediated by interactions between molecules of particular visual pigments and magnetic vectors. This conclusion is drawn from several findings showing that the accuracy of magnetic orientation performance depends on the wavelength of the light stimulus under which the performance is measured, and is supported by several theoretical analyses involving the behaviour of pigment molecules that have reached, by the absorption of light photons, an excited level that would render them sensitive to the direction of the magnetic field (for more details, see Wiltschko and Wiltschko, 1995a).

In arthropods, however, so far there has been only one experimental result that supports this idea. Phillips and Sayeed (1993) reported a 90° shift in magnetic compass orientation of *Drosophila melanogaster* caused by changing the wavelength of light in an orientation arena. Although this result is consistent with the hypothesis mentioned above, the data do not exclude the possibility that changing the wavelength of light exerts a non-specific effect on the animal's performance (Phillips and Sayeed, 1993). In addition, the hypothesis is contradicted by many reports on magnetic orientation in complete darkness by arthropods (Lindauer and Martin, 1972; Becker, 1975; Arendse and Vrins, 1975; Arendse, 1978; De Jong, 1982; Kirschvink and Kobayashi-Kirschvink, 1991; Schmitt and Esch, 1993). Even in vertebrates, magnetic orientation has been shown to occur despite the absence of visual cues (Lohmann and Lohmann, 1996a,b).

Detection based on ferromagnetic particles: The magnetite hypothesis

The second hypothesis is based on the discovery of magnetic particles (magnetite) in a large variety of species, both vertebrates and invertebrates (see Tab. 9.3. in Wiltschko and Wiltschko, 1995a). The role of magnetite in magnetic orientation was first demonstrated in bacteria (Blakemore, 1975). In arthropods, the first discovery of magnetite that might be used to detect magnetic fields was made by Gould et al. (1978) who found magnetite in the anterior dorsal abdomen of honeybees. Several attempts to locate magnetite *in situ* have, however, failed (Kuterbach et al., 1982; Schiff, 1991; Hsu and Li, 1994; Nesson, 1995; Kirschvink and Walker, 1995; Nichol and Locke, 1995).

The discovery of magnetite was followed by attempts to test for dependence of responses to magnetic fields on magnetite. In experiments by Gould et al. (1980), magnetic orientation was not affected by attempts to demagnetize the magnetite particles. However, discrimination of magnetic anomalies in conditioning experiments was abolished when magnetic wires were mounted over the magnetite particles, but not if the wires were mounted on other parts of the body (Walker and Bitterman, 1989a,b). The field produced by the wires decays proportionally to the cube of distance, and is less than 100 µT at a distance of 2 mm from the wires. Because the fields available to the bees in the experiments ranged from at least 250 µT to 1300 µT, the wires mounted on the anterior dorsal abdomen were probably located within 2 mm of magnetite-based magnetoreceptors in the bees.

Indirect evidence in support of the magnetite hypothesis could be provided by showing that the alignment of the horizontal dances of honeybees reported by Martin and Lindauer (1977) follows the pattern predicted by the Langevin function of classical paramagnetism (Reitz et al., 1979). This idea was put forward by Kirschvink (1981) who demonstrated statistically that the concentration of the dances in the magnetic cardinal directions depends on the strength of the magnetic field. The Langevin function predicts that the accuracy of the dance alignment should increase rapidly with increasing field strength until the field reaches earth-strength, and then increase further *asymptotically* in fields above earth-strength. However, the data available on misdirection in both the horizontal and vertical dances (Martin and Lindauer, 1977) do not include results for fields ranging between 5% and 200% of earth-strength. Data for fields in this range are required to decide whether or not misdirection indeed follows the Langevin function.

Further hypotheses are based on the idea that both of the mechanisms outlined above are involved in magnetoreception. Magnetic compass orientation may be mediated by detectors in the visual system, whereas position determination relies on magnetite-based receptors (Phillips, 1986; Wiltschko et al., 1994; Beason et al., 1995; Wiltschko and Wiltschko, 1995b).

Discussion

Although our understanding of magnetic orientation in arthropods lags behind that of magnetic orientation in vertebrates, much has been learned and can rapidly be learned through a combination of behavioural, anatomical and electrophysiological approaches. Key experimental results on magnetic orientation by arthropods reported in recent years have been the tentative finding that the magnetic compass of arthropods is based on the perception of magnetic field polarity, rather than inclination, and that magnetic information may be used to assist in determination of position. Better understanding of the conditions under which magnetic orientation can be examined in arthropods, as well as advances towards the discovery of the sensory system involved, are expected to stimulate future research in this field.

Magnetic compass orientation

A compass that responds to magnetic field polarity, as seems to be the case in arthropods, is fundamentally different from the magnetic inclination compass that is widespread among vertebrates. For migratory vertebrates, magnetic inclination indicates the equatorwards and polewards directions in both hemispheres. In birds, turtles, and presumably other migratory vertebrates as well, decreasing daylength as winter approaches should lead to orientation towards the equator, whereas increasing daylength as summer approaches shold lead to polewards orientation. However, information on inclination is ambiguous near the magnetic equator and the poles. For migratory arthropods, orientation by magnetic polarity will not lead to this type of ambiguity, but compass responses to magnetic polarity that are appropriate in one hemisphere will be inappropriate in the other. A polarity compass thus appears less flexible than an inclination compass.

This difference between the magnetic compasses of arthropods and vertebrates may reflect differences in the ecological demands involving long-distance orientation between arthropods and vertebrates during their separate evolutionary histories.

Despite the differences mentioned above, the magnetic compass of arthropods has much in common with that of vertebrates. Both can control direction of movement on their own or in concert with other compasses, and can also substitute for other compass mechanisms. The ability to associate directional responses across compasses has often, in both vertebrates and arthropods, facilitated the demonstration of the use of a magnetic compass.

Magnetic maps?

The existence and nature of magnetic maps in both vertebrates and invertebrates is likely to remain an open question for some time. Current experimental data (Janzen, 1971; Dyer, 1991; Kirchner and Braun, 1994) suggest that migratory and homing arthropod species do not require a complex magnetic map for orientation. Most of the data available can be explained by simple systems such as travel on familiar routes (Camlitepe and Stradling, 1995; Rickli and Leuthold, 1988; Janzen, 1971), path integration (Mueller and Wehner, 1988; Kirchner and Braun, 1994; Wehner et al., 1996), and the use of landmarks (Dyer, 1991; Collett, 1996; Menzel et al., 1996; T. S. Collett and J. Zeil, this volume) within a familiar area. There are no data indicating that arthropods can navigate outside a familiar area.

Based on results obtained in experiments on honeybees, two types of response to magnetic field information may contribute to map-like behaviour. First, as demonstrated by Collett and Baron (1994) and Frier et al. (1996), bees can integrate the magnetic field vector with other spatial information when learning and locating the positions of food sources. Indeed, if landmarks are to serve as directional cues, they must first be associated with a particular direction. Once the learning process has been successful, the use of the magnetic field as a directional cue can be abandoned. Therefore, the finding that visual marks are necessary for navigation after displacement within a familiar area (Dyer, 1991; Wehner et al., 1996) does not contradict the use of magnetic cues. Second, the threshold sensitivity of honeybees to changes in magnetic field intensity (Walker and Bitterman, 1989a) is consistent with the idea that bees can use local magnetic field information to determine their current position relative to a reference position, such as the nest or the feeding place. Thus, both types of information needed for finding the goal, namely knowledge of (i) the direction of the goal, and of (ii) the animal's current position in space, can be extracted from magnetic cues.

Why is the magnetoreceptor so difficult to find?

The experimental results demonstrating the use of magnetic cues imply, of course, the existence of a sensory system that detects these cues. The greatest frustration of workers on magnetic orientation (and presumably also of readers of papers concerned with this topic) is the fact that neither in vertebrates nor in arthropods has a sense organ for magnetic field detection yet been identified. Why is it so difficult to find the receptors involved in magnetic orientation?

The reason mainly lies in the fact that the magnetosensory system need not be large or elaborate, and therefore the sense organ involved is not expected to be as conspicuous as other sense organs, such as eyes or ears. Magnetic fields are relatively simple stimuli that vary only slowly in space and time and pass unaltered through tissues. A large number of sensory cells and nerves, large accessory sensory structures (such as lenses), and complex central processing elements are therefore not necessary for magnetic field detection.

It has also been argued that two separate sensory mechanisms exist, one for detecting magnetic intensity, the other for determining direction (inclination or polarity). If so, then after one of them has been identified, the search for the other is likely to continue.

Convergence of orientational information in the brain

The existence of a multimodal area in the brain for processing spatial information is suggested by orientation experiments as well as by models of spatial orientation. Convergence of orientational information in the brain would facilitate the interactions among compasses observed in the directional choices made by animals (Wiltschko and Wiltschko, 1995a; R. Wehner, this volume), and by the integration of magnetic information with other types of spatial information (Collett and Baron, 1994; Frier et al., 1996; see also R. Campan, this volume). Under the assumption that different types of spatial information converge in a particular region of the brain, cells in the multimodal area would be expected to receive directional information from both the visual and the magnetic sense organs, and to adjust their activity according to weighting factors given to the different sources of directional information under different conditions. For example, reliable directional information should be available from the sun under a clear sky. Under a cloudy sky, however, greater weight might be given to polarised light or magnetic field information.

Outlook: Behavioural contexts required for investigating the use of magnetic cues in arthropods

Arthropods present various problems involving the identification of the conditions necessary for encouraging them to display magnetic orientation. These problems can be quite specific and appear to be related to the animals' capacity of using cues others than magnetic ones to produce a directional response. Problems such as these can be overcome not only by knowledge of the biology of the species concerned, but also by selecting appropriate experimental procedures.

Motivation

The search for behavioural contexts suitable for eliciting responses to magnetic fields in both laboratory and field conditions should include explicit consideration of the animal's motivation to respond to the magnetic field. For example, understanding the importance of relative humidity in the behaviour of flour beetles (Arendse, 1978) and food-carrying in the behaviour of termites (Rickli and Leuthold, 1988) was critical for successful investigation of magnetic orientation in these insects. Similarly, more recent experiments (Schmitt and Esch, 1993; Camlitepe and Stradling, 1995; Lohmann et al., 1995) show that arthropods readily respond to magnetic field direction when they are motivated to orient to a particular goal. Therefore, future work on magnetic orientation in arthropods should exploit situations in which magnetic orientation is relevant to the task that is presented to the animal by the experimenter.

Choice of appropriate experimental conditions

Migratory vertebrates respond readily to magnetic field direction in laboratory orientation arenas in the absence of other biologically relevant stimuli. They seem to be capable of relying exclusively on magnetic cues. For example, demonstration of magnetic compass orientation by homing pigeons only requires that the sun be invisible (e.g., Keeton, 1969). Arthropods, however, rarely seem to rely exclusively on magnetic cues. In fact, magnetic compass orientation in arthropods has only occasionally been demonstrated under conditions that exclude the animal's use of any other orientational cue.

Two observations suggest appropriate methods for motivating animals to respond to magnetic cues rather than to others. First, when an animal orients towards a goal that promises a reward (for example, home, food, or a shelter), any cue present at the goal or *en route* to the goal (e.g., colour, shape, and odour of the goal itself, or compass cues others than magnetic on the way to the goal) can be used, and thus magnetic cues are not dominant. Orientation arena experiments measure the animal's attempt to initiate movement in the direction of the goal (e.g., Wiltschko and Wiltschko, 1972, Lohmann and Lohmann, 1996a,b). Presumably the animal selects the direction of locomotion either when initiating movement, or just before. Successful locomotion in the direction of the goal is likely to be rewarding for the animal. When, however, movement in a particular magnetic direction permits the animal to avoid or to escape from an *aversive* stimulus or situation, the animal should pay attention to the magnetic cue and ignore other stimuli. For example, Towne and Kirchner (1989) demonstrated detection of airborne sound in bees by applying electric shock whenever the bee approached the sound source. Similarly, Kirschvink and Kobayashi-Kirschvink (1991) trained bees to escape from a T-maze using magnetic field direction, but did not achieve adequate control over the behaviour of the animals. Still, the results were promising and their approach could be worth pursuing.

Another promising experimental approach is pairing the magnetic directional cue with another directional cue that is easily learned (e.g., Arendse and Vrins, 1975; Arendse, 1978). This approach is expected to be particularly useful in discrimination training experiments where the stimulus of interest is difficult to condition directly, as is indeed the case in conditioning to magnetic cues. Once conditioning to the combined stimulus has been successful, the easily conditioned stimulus is reduced in salience and eventually removed. If the animal is capable of detecting the second stimulus, for example the direction of the magnetic vector, then the conditioned response is expected to be transferred to the magnetic vector when the salient stimulus is removed.

Conclusion

Current knowledge conveys a sense of considerable opportunity for rapid advance in the understanding of both magnetic orientation and the magnetic sense in arthropods. A key advantage of working with arthropods is that the simplicity of their behaviour permits relatively direct access to underlying mechanisms once the appropriate conditions for revealing magnetic orientation and conditioning are identified. Recent studies of magnetic orientation in arthropods provide clues to the necessary conditions for future orientation and conditioning experiments. The arthropods thus present an opportunity for concerted study of magnetic orientation and magnetosensory capacity that is at present unmatched in any other animal taxon.

Acknowledgements
I thank Professor Wolfgang Wiltschko and Dr. Roswitha Wiltschko and members of the Experimental Biology Research Group for helpful discussions. Jennifer Ashbey assisted with the redrafting of the figures. The patience and tolerance of the editor of this volume, Dr. Miriam Lehrer, are most gratefully acknowledged.

References

Altman, G. (1981) Untersuchung zur Magnetotaxis der Honigbiene, *Apis mellifica L. Anz. Schaedlingskunde Pflanzenschutz Umweltschutz* 54:177–179.

Arendse, M.C. (1978) Magnetic field detection is distinct from light detection in the invertebrates *Tenebrio* and *Talitrus. Nature* 274:358–362.

Arendse, M.C. and Kruyswijk, C.J. (1981) Orientation of *Talitrus saltator* to magnetic fields. *Neth. J. Sea. Res.* 15:23–32.

Arendse, M.C. and Vrins, J.C.M. (1975) Magnetic orientation and its relation to photic orientation in *Tenebrio molitor* L. (Coleoptera, Tenebrionidae). *Neth. J. Zool.* 25:407–437.

Baker, R.R. (1987) Integrated use of moon and magnetic compass by the heart-and-dart moth, *Agrotis exclamationis. Anim. Behav.* 35:94–101.

Baker, R.R. and Mather, J.G. (1982) Magnetic compass sense in the large yellow underwing moth, *Noctua pronuba* L. *Anim. Behav.* 30:543–548.

Beason, R.C., Dussourd, N. and Deutschlander, M.E. (1995) Behavioural evidence for the use of magnetic material in magnetoreception by a migratory bird. *J. Exp. Biol.* 198:141–146.

Becker, G. (1965) Zur Magnetfeld-Orientierung bei Dipteren. *Z. Vergl. Physiol.* 51:135–150.

Becker, G. (1975) Einfluss von magnetischen, elektrischen und Schwerefeldern auf den Galeriebau von Termiten. *Umschau* 75:183–185.

Blakemore, R.P. (1975) Magnetotactic bacteria. *Science* 19:377–379.

Camlitepe, Y. and Stradling, D.J. (1995) Wood ants orient to magnetic fields. *Proc. R. Soc. Lond.* B 261:37–41.

Cartwright, B.A. and Collett, T.S. (1987) Landmark maps for honeybees. *Biol. Cybern.* 58:85–93.

Collett, T.S. (1996) Insect navigation *en route* to the goal: multiple strategies for the use of landmarks. *J. Exp. Biol.* 199:227–235.

Collett, T.S. and Baron, J. (1994) Biological compasses and the coordinate frame of landmark memories in honeybees. *Nature* 368:137–140.

De Jong, D. (1982) Orientation of comb building by honeybees. *J. Comp. Physiol.* A 147:495–501.

Dyer, F.C. (1991) Bees acquire route-based memories but not cognitive maps in a familiar landscape. *Anim. Behav.* 41:239–246.

Frier, H.J., Edwards, E., Smith, C., Neale, S. and Collett, T.S. (1996) Magnetic compass cues and visual pattern learning in honeybees. *J. Exp. Biol.* 199:1353–1361.

Frisch, K., von (1967) Dance Language and Orientation of Bees. Harvard University Press, Cambridge MA.

Gould, J.L. (1982) The map sense of pigeons. *Nature* 296:205–211.

Gould, J.L., Kirschvink, J.L. and Deffeyes, K.S. (1978) Bees have magnetic remanence. *Science* 201:1026–1028.

Gould, J.L., Kirschvink, J.L., Deffeyes, K.S. and Brines, M.L. (1980) Orientation of demagnetized bees. *J. Exp. Biol.* 86:1–8.

Griffin, D.R. (1952) Bird navigation. *Biol. Rev. Camb. Philos. Soc.* 27:359–400.

Hsu, C.Y. and Li, C.W. (1994) Magnetoreception in honeybees. *(Apis mellifera). Science* 265:95–97.

Hunt, T.M. and Syms, M.C. (1977) SHEET 3, Auckland Magnetic Map of New Zealand, 1:250,000, Total Force Anomalies, first edition. Dept. of Scientific and Industrial Research, Wellington.

Janzen, D.H. (1971) Euglossine bees as long-distance pollinators of tropical plants. *Science* 171:203–205.

Keeton, W.T. (1969) Orientation by pigeons: is the sun necessary? *Science* 165:922–928.

Kirchner, W.H. and Braun, U. (1994) Dancing honeybees indicate the location of food sources using path integration rather than cognitive maps. *Anim. Behav.* 48:1437–1441.

Kirschvink, J.L. (1981) The horizontal magnetic dance of the honeybee is compatible with a single-domain ferromagnetic magnetoreceptor. *Biosystems* 14:193–203.

Kirschvink, J.L. and Gould, J.L. (1981) Biogenic magnetite as a basis for magnetic field detection in animals. *Biosystems* 13:181–201.

Kirschvink, J.L. and Kobayashi-Kirschvink, A. (1991) Is geomagnetic sensitivity real? Replication of the Walker-Bitterman magnetic conditioning experiment in honeybees. *Am. Zool.* 31: 169–185.

Kirschvink, J.L. and Walker, M.M. (1995) Honeybees and magnetoreception. *Science* 269: 1889.

Kramer, G. (1953) Wird die Sonnenhöhe bei der Heimfindeorientieurng verwendet? *J. Ornithol* 94: 201–219.

Kuterbach, D.A., Walcott, B., Reeder, R.J. and Frankel, R.B. (1982) Iron-containing cells in the honey-bee (*Apis mellifera*). *Science* 218: 695–697.

Leucht, T. and Martin, H. (1990) Interactions between e-vector orientation and weak, steady magnetic fields in the honeybee, *Apis mellifica*. *Naturwissens.* 77: 130–133.

Lindauer, M. (1977) Recent advances in the orientation and learning of honeybees. *In: Proc. XV Int. Congr. Entomol.* Washington DC, pp 450–460.

Lindauer, M. and Martin, H. (1968) Die Schwereorientierung der Bienen unter dem Einfluss des Erdmagnetfeldes. *Z. Vergl. Physiol.* 60: 219–243.

Lindauer, M. and Martin, H. (1972) Magnetic effects on dancing bees. *In*: S.R. Galler, K. Schmidt-Koenig, G.J. Jacobs and R.E. Belleville (eds): *Animal Orientation and Navigation.* US Government Printing Office, Washington DC, pp 559–567.

Lohmann, K.J. and Lohmann, C.M.F. (1996a) Detection of magnetic field intensity by sea turtles. *Nature* 380: 59–61.

Lohmann, K.J. and Lohmann, C.M.F. (1996b) Orientation and open-sea navigation in sea turtles. *J. Exp. Biol.* 199: 73–81.

Lohmann, K.J., Pentcheff, N.D., Nevitt, G.A., Stetten, G.D., Zimmer-Faust, R.K., Jarrard, H.E. and Boles, L.C. (1995) Magnetic orientation of spiny lobsters in the ocean: Experiments with undersea coil systems. *J. Exp. Biol.* 198: 2041–2048.

Martin, H. and Lindauer, M. (1977) Der Einfluss des Erdmagnetfeldes auf die Schwereorientierung der Honigbiene (*Apis mellifica*). *J. Comp. Physiol.* A 122: 145–187.

Menzel, R., Geiger, K., Chittka, L., Joerges, J., Kunze, J. and Müller, U. (1996) The knowledge base of bee navigation. *J. Exp. Biol.* 199: 141–146.

Moore, D. and Rankin, M.A. (1985) Circadian locomotor rhythms in individual honeybees. *Physiol. Entomol.* 10: 191–197.

Mueller, M. and Wehner, R. (1988) Path integration in desert ants. *Proc. Natl. Acad. Sci. USA* 85: 5287–5290.

Nesson, M. (1995) Honeybees and magnetoreception. *Science* 269: 1889–1890.

Nichol, H. and Locke, M. (1995) Honeybees and magnetoreception. *Science* 269: 1888–1889.

Neumann, M.F. (1988) Is there any influence of magnetic or astrophysical fields on the circadian rhythm of honeybees? *Behav. Ecol. Sociobiol.* 23: 389–393.

Phillips, J.B. (1986) Two magnetoreception pathways in a migratory salamander. *Science* 233: 765–767.

Phillips, J.B. and Sayeed, O. (1993) Wavelength-dependent effects of light on magnetic compass orientation in *Drosophila melanogaster. J. Comp. Physiol.* A 172: 303–308.

Reitz, J.R., Milford, F.J. and Christy, R.W. (1979) *Foundations of Electromagnetic Theory.* Third ed. Addison-Wesley, Reading Mass.

Rickli, M. and Leuthold, R.H. (1988) Homing in harvester termites: evidence of magnetic orientation. *Ethology* 77: 209–216.

Roonwal, M.L. (1958) Recent work on termite research in India (1947–57). *Trans Bose Res. Inst.* 22: 77–100.

Schiff, H. (1991) Modulation of spike frequencies by varying the ambient magnetic field and magnetite candidates in bees (*Apis mellifera*). *Comp. Biochem. Physiol.* 100 A: 975–985.

Schmidt-Koenig, K. (1965) Current problems in bird orientation. *Adv. Study. Behav.* 1: 217–278.

Schmitt, D.E. and Esch, H.E. (1993) Magnetic orientation of honeybees in the laboratory. *Naturwissens.* 80: 41–43.

Skiles, D.D. (1985) The geomagnetic field: its nature, history and biological relevance. *In*: J.L. Kirschvink, D.S. Jones and B.J. MacFadden (eds): *Magnetite Biomineralization and Magnetoreception in Organisms.* Plenum Press, New York, pp 43–102.

Tomlinson, J., McGinty, S. and Kish, J. (1981) Magnets curtail honey bee dancing. *Anim. Behav.* 29: 307–308.

Towne, W.F. and Kirchner, W.H. (1989) Hearing in honeybees: detection of air-particle oscillations. *Science* 244: 686–688.

Ugolini, A. and Pezzani, A. (1995) Magnetic compass and learning of the Y-axis (sea-land) direction in the marine isopod *Idotea baltica basteri*. *Anim. Behav* 50:295–300.

Walker, M.M. (1984) Learned magnetic field discrimination in yellowfin tuna, *Thunnus albacares*. *J. Comp. Physiol.* A 155:673–679.

Walker, M.M. and Bitterman, M.E. (1985) Conditioned responding to magnetic fields by honeybees. *J. Comp. Physiol.* A 157:67–71.

Walker, M.M. and Bitterman, M.E. (1989a) Honeybees can be trained to respond to very small changes in geomagnetic field intensity. *J. Exp. Biol.* 145:489–494.

Walker, M.M. and Bitterman, M.E. (1989b) Attached magnets impair magnetic field discrimination by honeybees. *J. Exp. Biol.* 141:447–451.

Walker, M.M., Baird, D.L. and Bitterman, M.E. (1989) Failure of stationary but not of flying honeybees (*Apis mellifera*) to respond to magnetic field stimuli. *J. Comp. Psychol.* 103: 62–69.

Wehner, R. and Labhart, T. (1970) Perception of geomagnetic field in the fly *Drosophila melanogaster*. *Experientia* 26:976–986.

Wehner, R., Michel, B. and Antonsen, P. (1996) Visual navigation in insects: coupling of egocentric and geocentric information. *J. Exp. Biol.* 199:129–140.

Wiltschko, R. and Wiltschko, W. (1995a) *Magnetic Orientation in Animals.* Springer-Verlag, Berlin, Heidelberg.

Wiltschko, W. and Wiltschko, R. (1972) Magnetic compass of European robins. *Science* 176:62–64.

Wiltschko, W. and Wiltschko, R. (1995b) Migratory orientation of European robins is affected by the wavelength of light as well as by a magnetic pulse. *J. Comp. Physiol.* A 177:363–369.

Wiltschko, W., Munro, U. Beason, R.C., Ford, H. and Wiltschko, R. (1994) A magnetic pulse leads to a temporary deflection in the orientation of migratory birds. *Experientia* 50: 697–700.

Orientation and Communication in Arthropods
ed. by M. Lehrer
© 1997 Birkhäuser Verlag Basel/Switzerland

Chemo- and mechanosensory orientation by crustaceans in laminar and turbulent flows: From odor trails to vortex streets

M. J. Weissburg

School of Biology, Cherry-Emerson Building, Georgia Institute of Technology, 310 First Ave., Atlanta, GA 30332-0230, USA

Summary. Crustaceans use odor and fluid mechanical cues to extract information from their environment. These cues enable animals to find resources, orient to water currents, or escape predators. Because the properties of the fluid environment affect the transmission and structure of relevant signals, a better understanding of sensory and behavioral mechanisms will be aided by considering, at the same time, the hydrodynamic context of chemo- and mechanosensory behaviors. Crustaceans occupy aquatic habitats where flows range from almost completely laminar to nearly fully turbulent. The considerable scope of hydrodynamic properties is mirrored by equally extreme variations in the complexity of the signals entrained in these flows. Ambient noise and stochastic variation increase in increasingly energetic, turbulent conditions. The sensory and behavioral mechanisms of animals that orient in turbulent environments suggest that they have, in the course of evolution, been shaped by the flow properties. Here, sensory systems are geared to extract rapidly fluctuating signals against a noisy background. They sometimes have elaborate noise filtering mechanisms that enable the detection of rather coarse types of signal features to improve the signal-to-noise ratio. In contrast, the simpler and more predictable structure of signals carried in laminar flows may allow more accurate orientation and discrimination to occur, and free animals from the burden of supporting complex noise-filtering circuitry. Future comparative investigations of sensory physiology and behavior of animals in relation to their flow environment promise to increase our understanding of orientation by means of chemo- and mechanoperception.

Introduction

All life exists in a fluid medium. Although air and water differ in the magnitude of their defining properties, such as density and viscosity, from a purely physical perspective both obey the same laws governing the deformation and movement of fluids. As scientists from diverse disciplines have come to appreciate, properties of fluid flow, be it air or water, influence many aspects of the design of bodies and appendages, as well as the mechanics of locomotion. Numerous studies indicate that, in insects that move through air, even the design and operation of sensory systems that code fluid-borne information (i.e., chemo- and mechanosensation) are affected by features of flow (see e.g., K.-E. Kaissling, this volume). These findings are not surprising, because it is the underlying fluid dynamic forces that structure the signals that insects use to orient to their environment.

In the case of crustaceans, our knowledge on the correlation between the properties of the hydrodynamic environment and the animal's sensory capacities is less advanced than in the case of insects. This may be due to purely capricious historical reasons, but also to the fact that methods for measuring and controlling water-borne signals have only recently become sufficiently sophisticated to promote a rigorous analysis of how functional properties of sensory systems relate to the ways in which relevant stimuli are transported by flow. Recent studies on crustaceans have been concerned not only with the physiology and behavior of these animals, but, in addition, with the fluid dynamics and signal structure in the aquatic environments in which they live. The new findings may help forge a better understanding of how the animal's sensory systems function in the complex world created by moving water. This chapter reviews the results of recent and older studies in an attempt to uncover general principles involved in orientation, and to suggest promising future experimental approaches.

The two sensory modalities that are most affected by properties of flow are those of chemo- and mechanoperception. This review will focus on the behavior and underlying sensory mechanisms that permit orientation to chemical and near-field fluid mechanical stimuli in crustaceans. One of the goals of this effort is to stress how properties of flow provide insights into these two systems. To this end it is necessary to review some of the relevant fluid dynamical principles that govern the transport of both momentum and molecules through fluids. A thorough discussion of fluid dynamics is beyond the scope of this review, but the material should suffice to imbue a general sense of appreciation of the power of fluid dynamics in understanding chemo- and mechanoperception. Following this primer, I will discuss the chemosensory and mechanosensory capabilities of aquatic crustaceans, ultimately with respect to the hydraulic environment and to the differences between laminar and turbulent flow conditions.

Flow characterization and signal structure

Hydrodynamic considerations

Attempts to relate flow properties such as the degree of turbulence to the magnitude of simple mathematical entities are best regarded as rules of thumb, rather than as strict laws. Even so, a few relatively straightforward descriptors of flow can roughly define a given fluid mechanical environment.

The most universally useful definition of fluid flow is the Reynolds number (*Re*):

$$Re = 1U/v \tag{1}$$

where 1 is a length factor, U is the fluid velocity, and v is kinematic viscosity (Schlichting, 1979). These parameters have various physical manifestations depending on the exact flow scenario. The original derivation of Equation (1) was obtained by examining pipe flows, where U corresponded to the speed of water moving through the pipe whereas 1 was the pipe diameter (see Vogel, 1994). This formulation also has been used to estimate the turbulence of fluid jets emerging from biological sources, such as bivalve siphons (Monismith et al., 1990; Weissburg and Zimmer-Faust, 1993). For calculating the Re around an object, such as an appendage or animal immersed in a fluid, 1 is the maximum length of that object parallel to the flow, whereas U is the velocity of the object relative to that of the fluid, or simply the fluid velocity if the object is stationary.

The Re represents the ratio of the inertial to viscous forces in a fluid, or alternatively, a ratio of the time scales of inertial transport in eddy structures to dissipation of momentum in the eddy via viscous interactions. At a low Re (<1), viscous interactions between molecules of the fluid dominate the fluid flow. Eddies are either non-existent or short lived, because these molecular interactions diffuse the energy in a rotating eddy rapidly relative to its rotation time. Consequently, flow streamlines remain parallel to each other and the flow is laminar, orderly and well-behaved. In laminar flows, the local fluid motion is always in one direction, that is, parcels of fluid always move "down-current". In low Re environments, objects placed in the flow do not create a disturbance, because the fluid tends to creep around the object without causing streamlines to cross. At a high Re, on the other hand, inertial forces dominate the fluid, generating eddies and imposing large shearing forces within the fluid or between the fluid and a solid boundary. Flow is chaotic and turbulent; streamlines are not always parallel to one another, so that local flow may not always be in the direction of the bulk flow. Thus, in turbulent flows, descriptions of flow direction and velocity are based on time averages, and instantaneous velocity exhibits substantial variation due to momentum transfer occurring within the fluid. In turbulent conditions, objects placed in the flow will further enhance turbulence as vortices develop in the object's wake (Vogel, 1994).

Unfortunately, there is no universal value for Re that adequately demarcates the transition from laminar to turbulent flow, because the relationship of Re to turbulence intensity is heavily dependent on the flow scenario. Nonetheless, it is clear that flows conveying mechanical and chemical signals to crustaceans span several orders of Re magnitude, ranging from completely laminar to almost fully turbulent flow. The Re for the flow around copepod appendages varies from 0.1 to 5 (Koehl and Strickler, 1981; Andrews, 1983; Yen et al., 1991), which coincides with a quasi-laminar flow. Gill currents used by lobsters in social interactions have a Reynolds number approximately equal to 1500 at the point where the jet exits the excurrent siphon, and exhibit a high degree of turbulent structures in the resulting plume (Atema, 1988).

The interaction between fluids and solids immersed in them has a most fundamental effect on the nature of the resulting flow properties. Due to molecular interactions, a no-slip condition operates at any fluid-solid boundary; the velocity of the fluid in contact with the solid surface is zero, whereas far away from the interface the fluid reaches its maximal free-stream velocity, U (sometimes U_∞). A velocity gradient is thus established, where fluid velocity increases logarithmically with distance from the fluid-solid boundary, thus creating the eponymously called boundary-layer. The boundary layer begins at the zero velocity region of the fluid-solid inter-face, and is somewhat arbitrarily defined as extending to the height (z) at which the fluid velocity reaches 99% of the free-stream speed (Fig. 1, left panel). To further complicate matters, the thickness of the boundary layer is not constant. It increases at a rate proportional to the square root of the distance from the leading edge (i.e., the upstream edge) of the solid surface (Schlichting, 1979). The degree of turbulence at a specific point along a boundary layer can be estimated by calculating a local Re using Equation (1), where the length parameter 1 now represents the distance from the leading edge.

Several aspects regarding the implications of boundary layers warrant further discussion. First, using 1 in its normal guise as the length of the object parallel to the flow results in a calculation of the local Re of the corresponding boundary layer at the most downstream position of the object. It is thus a measure of the average turbulence level experienced by that body. Second, it should be apparent that the boundary layer developing over an object of sufficient length will eventually undergo a transition from laminar to turbulent flow, which occurs at a local Re of roughly 3×10^5 (Schlichting, 1979). A laminar boundary layer, as the name implies, consists entirely of well ordered laminar flow. In contrast, a turbulent boundary layer consists largely, if not exclusively, of turbulent flow, where local fluid motion sometimes may be in directions other than down-current. Finally, flow is relative to a body's frame of reference, and two or more boundary layers may exist simultaneously, each with its own relevant physical scale. Consider a parasitic copepod on the surface of a fish. The free-stream velocity around the fish is given by the fish's own swimming velocity (relative to any water current), and will generally result in turbulent flow. The "free-stream" velocity around the copepod is determined by the velocity gradient (boundary layer) around the fish; the copepod may thus find itself in a laminar boundary layer nestled within a turbulent flow. Vogel's (1994) cogent and fluid discussion of boundary layers may be consulted for more details, as well as for further interesting biological examples.

For the many benthic crustaceans, cues such as metabolites released from the excurrent siphons of bivalves, or vortices shed by a moving fish, will be transmitted within the boundary layer that develops over the sea-floor. Because the seafloor is essentially an infinitely long solid surface, the resulting (benthic) boundary layer developing over the substrate will generally exhibit some turbulence. A turbulent (as opposed to laminar)

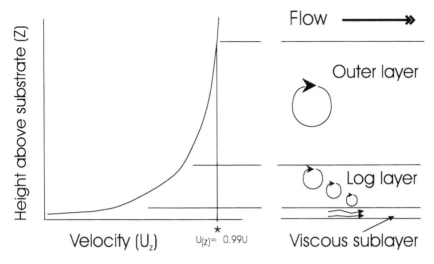

Figure 1. Dependency of flow velocity U(z) on height above the substrate (z), and structure within a turbulent benthic boundary layer. The boundary layer is defined as extending to the height where U(z) reaches 99% of the free-stream velocity, marked on the x-axis. The flow profile of a turbulent boundary layer is qualitatively depicted in the right panel, which shows the generation of large eddies in the outer layer, momentum transfer and subsequent generation of smaller eddies in the log layer (the Kolmogorov cascade), and laminar flow in the viscous sublayer. Note that whereas this example shows the boundary layer developing over a substrate, a turbulent boundary layer can develop over any surface immersed in flow.

boundary layer is composed of three somewhat distinct flow environments (Fig. 1). Nearest the substrate, where the no-slip condition exerts its influence, there is a thin region of laminar flow dominated by viscosity, which is referred to as the viscous or laminar sub-layer. Because of its orderly flow, the viscous sub-layer may represent a particularly effective environment for the transmission of chemical cues. This region is rarely more than several millimetres thick even in the most benign of flows, and can shrink to a thickness of less than 100 μm in stronger flows. At increasing distances above the substrate, inertial forces become more important. The region immediately above the viscous sub-layer is termed the log- or transition layer. Here both intertial and viscous interactions shape the flow, as the kinetic energy of the flow is entrained in small highly turbulent eddy structures. As a result, this region experiences the most extreme fluctuations of fluid velocity. The log-layer together with the viscous sub-layer encompasses about 30% of the boundary layer thickness. Thus, unless a benthic crustacean is small, it may not always experience the flow within the viscous sub-layer, but may often be (or have sensors) within the log-layer. The flow region beyond the log-layer is termed the outer layer, where the fluid approaches its maximal free-stream velocity. Flow velocity here is fairly constant and only the largest eddy structures are found in this region.

For benthic boundary layers we can define a roughness Reynolds number, Re*, which is an alternative to the local Re described above:

$$Re* = u*D/v \tag{2}$$

Here, the length constant is D, the hydraulic roughness length, which is a measure of the propensity of the bed substrate to disturb the flow. In some cases D is equivalent to the average diameter of the substrate. For many natural benthic environments, however, the hydraulic roughness length may be a complicated function of the surface topography, and may be influenced by the height of roughness elements such as ripples, sediment mounds, and encrusting benthic fauna. The velocity parameter is no longer U, the free-stream speed, but, instead, u*, the shear velocity. It measures the average strength of correlated velocity fluctuations (turbulence) occurring within the boundary layer, and is proportional to the degree of eddy diffusivity within the boundary layer. As a rule of thumb, u* is approximately 10% of the average free-stream velocity (Denny, 1988).

The roughness Reynolds number measures the degree of penetrance of turbulent eddies into the boundary layer. Turbulence begins to affect the outer reaches of the boundary layer at Re* values between 3.5 and 6, and extends all the way to the substrate at values between 75 and 100 (Schlichting, 1979). In these turbulent flows, the viscous sub-layer may effectively vanish. The thickness of the viscous sub-layer may be estimated by setting Equation (2) to 6, and solving for the value of D (Weissburg and Zimmer-Faust, 1993). In essence, this procedure renders the height of roughness elements that would begin to feel the effect of turbulent eddies, and thus it estimates the outer limits of the quasi-laminar viscous flow region.

Estimates of boundary layer turbulence structure are rare for near-shore environments occupied by the majority of benthic crustaceans. However, available data suggest that many flow environments will be composed of at least smooth-turbulent flows (Re* ca. 3.5), and flows that are transitional between smooth- and rough-turbulent regimes (3.5 < Re* < 100) will be common. Palmer and Gust (1985) measured shear velocities as high as 0.8 cm/s in a saltmarsh tidal creek in South Carolina. Although D was not measured, the silty sediment showed roughness elements (faecal pellets, worm tracks) approximately 0.5 mm tall, and substituting this for D yields a Re* that suggests smooth-turbulent or transitional flows. Shear velocities of 1–6 cm/s have been recorded in similar types of tidal channels and sandflats (Palmer 1988; J. E. Commins and R. K. Zimmer-Faust, unpublished data). Under these conditions, flows will be smooth-turbulent or transitional unless sediments are extremely fine (D < 100 μm) and the bed is without relief. Gross and Nowell (1983) measured shear velocity and hydraulic roughness length in Puget Sound, WA, USA, and found typical values of Re* of about 40.

Hart et al. (1996) recently measured hydrodynamics of stream flows in Pennsylvania, USA, using hot-film anemometry to examine small scale flow parameters on the surfaces of stones. They concluded that boundary layers were thin (< 10 μm), and flow was more turbulent than suggested by calculations of local Re. In most cases, boundary layer profiles failed to meet the theoretical expectations of logarithmic increases in velocity with distance from the substrate, making local estimates of shear velocity inappropriate in determining average turbulent intensity. Local shear velocities were approximately 10% of local free-stream velocity. The lack of correspondence with boundary layer theory suggests that these highly complicated flow patterns reflect the fact that turbulence does not develop locally, as in simple flows over plates or smooth surfaces, but as a result of flow separation and other instabilities produced upstream of the measured region (Hart et al., 1996). Based on Fourier analysis of temporal velocity profiles, (Hart et al., 1996) concluded that ambient hydrodynamic noise is concentrated in frequencies below 16 Hz. Thus, at least in this particular case, ambient noise can be intense enough to affect the perception of low frequency emanations from biotic sources (see below).

Properties of chemical signals

The processes that determine the characteristics of turbulent plumes are poorly understood. Odor plumes (or, more properly, odor jets) released into ambient flow are initially subjected to shearing forces such as would be felt on the surface of a solid object in flow (List, 1982). These shearing forces cause intrusion of water into the plume, and entrain filaments of the plume into the ambient flow. Odorants become quickly packaged in eddy structures, creating odor patches that may separate from the plume due to shear instability, and that begin to grow as they move downstream. The disruption of the plume proceeds as the eddies that initially entrain the odor become stretched and sheared into smaller eddies (termed the Kolmogorov cascade), particularly within the log-layer. The initial injection of turbulent kinetic energy specifies the length scale, η, of the smallest eddies produced in a given flow environment (Tennekes and Lumley, 1972):

$$\eta = (kZv^3/u^{3*})^{1/4} \tag{3}$$

where Z is height above the substrate, u^* is shear velocity, and k is Von Karman's constant. Turbulent energy is insufficient to sustain eddies smaller than η before they are dissipated as heat by viscous interactions within the fluid. At scales smaller than η, odorants can be redistributed only through molecular or shear diffusion. As predicted by Equation (3), greater turbulent energy (indexed by u^*) sustains a smaller minimum eddy size. Under moderately turbulent conditions ($Re^* = 3.6$), eddy sizes as small as

200 µm have been measured in benthic boundary layers in flume studies (Moore et al., 1992).

Plumes transported by turbulent flows do not produce smooth odor concentration gradients at spatial and temporal scales that can render distance and directional cues to macroscopic benthic crustaceans. Nonetheless, turbulent flows impose a certain degree of spatial dependency on the plume, because the characteristics of the distribution of odorants within the plume change with the distance from the source. Investigations using high resolution chemical microelectrodes show that turbulent odor plumes may contain information useful for chemosensory guided search (Moore and Atema, 1991; Moore et al., 1994b). Downstream from the odor source, time-averaged odor concentrations become increasingly dilute. Odors are scattered in intermittent patches that are variable in both their duration and spatial distribution. Odor pulses typically last for several seconds, with long intervening periods of no detectable odor concentration (Moore et al., 1994b). Patch boundaries become smeared due to eddy diffusion and possibly by shear-smoothing, so that odor pulses display shallow onset slopes and low peak concentrations. Near the odor source, on the other hand, peak concentrations are quite high, and fluctuations are more frequent and abrupt as compared to conditions downstream. Similar patterns occur as one proceeds transversely across the plume: odor patches are most intense, distinct, and numerous in the plume's core (Moore and Atema, 1991). Thus, the spatial and temporal structure of the odor plume may render (at least rough) information about the source's location even in turbulent flows.

The profiles of turbulent odor plumes change with the hydrodynamic setting (Fig. 2). In general, greater turbulence increases the variance in odor plume concentration, reduces the concentration in odor patches, and decreases both onset slope and the ratio of pulse concentration to onset slope (Moore et al., 1994b). In essence, plume features become more fuzzy and indistinct. J. E. Commins and R. K. Zimmer-Faust (unpublished data) determined that, in natural turbulent flows (u^* ca. 5), average concentrations are diluted 1000-fold only 5 cm away from the source, whereas peak concentrations are reduced by a factor of 250. In contrast, peak odorant concentrations in slow flows ($u^* = 0.4$) are diluted by less than 10-fold even 1 m downstream.

In addition to affecting concentration and concentration variance of odorants, turbulence magnitude also may determine how the plume structure changes as it is advected downstream. Interestingly, greater turbulence increases the correlation between downstream distance and certain features of the odor plume, particularly characteristics of individual odor patches (Moore et al., 1994b). For odor plumes emanating from bivalve mimics in smooth flows (Re* ca. 1; $Re_{jet} = 170$), patch concentrations change little except very close (< 25 cm) to the excurrent siphon, and onset slopes are not predictably different with respect to downstream distance within the first 1 m. However, when turbulence of the ambient flow increases (Re*

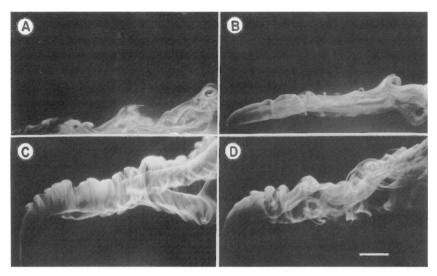

Figure 2. Odor plumes created by a model clam, and visualized using fluoresceine dye. In (A–C), Re* = 0.4, 1.6, 6.3, U = 1.0, 3.8, 14.4 cm/s, respectively (for explanation of Re*, see text), and the substrate is fine sand. D illustrates the effect of large roughness elements in generatig turbulence. Here, Re* = 2.7, U = 1.0 cm/s, and the substrate is small gravel (compare D with A, where Re* is different whereas velocities are similar, and D with B, where Re* is similar despite differences in velocity). Scale bar = 2 cm. From Weissburg and Zimmer-Faust (1993).

ca. 3.6), onset slope as well as concentration become significantly correlated with distance from the odor source for at least the first 1 m downstream. In addition, when the free-stream velocity increases, the main axis of the plume becomes advected downstream more quickly than in slower flows. This causes the plume's center of mass to be lower in the water column, so that free-stream velocity may determine which appendages or body surfaces are presented with chemical cues. By changing the release rate of odorants delivered through a tube parallel to the flow in an unquantified boundary layer, Dittmer et al. (1995), generated jets with an Re_{jet} = from approximately 2.7 to 143. At the lowest release rate, only peak pulse concentration was correlated with distance to the odor source. As the jet became increasingly turbulent, other parameters, such as the ratio of pulse height to pulse slope also changed predictably with downstream distance.

Flows transporting odor cues may also operate in low Re realms dominated or heavily influenced by visous forces. Copepods generate feeding currents to capture prey, and which they also use to inspect their olfactory environment (Koehl and Strickler, 1981; Yen et al., 1991; Yen and Strickler, 1996). The resulting flow fields are highly laminar, and are exquisitely structured such that the parallel stream lines flow past antennal sensilla to converge near the mouthparts. In this vicinity the flows range in Re from

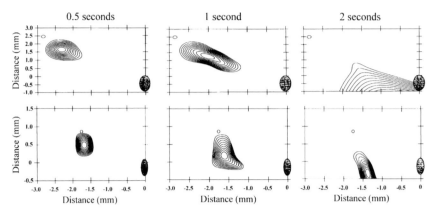

Figure 3. Visualizations (using electrochemical tracers) of odor patches deformed by copepod feeding currents. Top row: *Pleuromamma xiphias*. Bottom row: *Eucheata rimana*. In reach row, the odor profiles are mapped at three different times following stimulus introduction. Contours represent odor concentration isolines at decrements of 4% from the initial concentration of 50 mM (center). Coordinates denote distance, in mm, from the rostrum of the copepod. Oval at the right is the location of the head of the copepod, whereas the electrochemical sensor used to monitor the patch deformation is located at coordinates $x = -0.75$, $y = 0$. Note that the base of the copepod antennule would be at $x = 0$, $y = 0$, and extend for approximately 1000 microns parallel to the x-axis. Variations in structure of the deformed odor patch represent differences in the feeding current produced by each animal. (J. Yen and P.A. Moore, unpublished data.)

0.02 to 5. Velocity is highest where the feeding currents are created by the mouthpart movements, and declines exponentially with radial distance from the body axis. The copepod's self-generated flow results in an intense shear field that deforms microscale odorant patches around prey (i.e., algal cells) into a long, thin plume than can reach the sensors well in advance of the prey (Andrews, 1983; Moore et al., 1994a; Fig. 3). As noted by Strickler (1985), these flow fields place the copepod in the center of a watery web not unlike that of a spider, where the streamlines potentially furnish precise spatial information of odorants transported along these paths to particular setae along the antennal array. The energy of the self-generated flow field is greater than that of the ambient oceanic turbulence at a similar spatial scale, and thus it overrides the effect of ambient small scale turbulence (Yen et al., 1991). Thus, unlike larger animals at the mercy of turbulent inertial flows, small crustaceans may have considerable control over the stimulus environment even at some distance away from the sensory appendage.

Properties of fluid mechanical signals

Near-field fluid mechanical signals have received even less attention than have odor plumes. Again, flow visualization techniques (such as particle

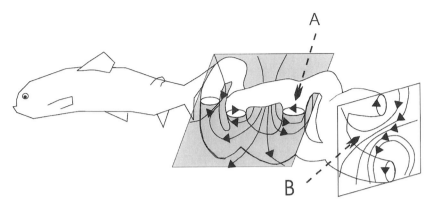

Figure 4. Vortex street shed by a swimming fish. The large arrows denote regions A and B of the flow field that are depicted in Figure 7, whereas the small arrowheads show flow direction. Redrawn after Blickhan et al. (1992), with authors' permission.

tracking and visualization using optical density gradients) have only recently become sophisticated enough to map the complex fields produced by moving animals. Bleckmann et al. (1991) used laser-Doppler velocimetry to measure the frequency of hydrodynamic disturbances generated by a variety of moving aquatic animals. Locomoting fish, frogs, and crustaceans were all found to produce power spectra characterized by strong low frequency components. Single strokes produce displacements only at frequencies < 10 Hz, whereas sustained swimming extends the frequency range to approximately 100 Hz. Power spectra for particle accelerations, which may be a more important parameter for near-field movement sensors, display frequency components from 25–200 Hz. In swimming fish, higher frequency components probably arise from vortices shed by movements of the caudal fin (Bleckmann et al., 1991; Blickhan et al., 1992). During swimming, the wakes are composed of a vortex street of rings in a ladder-like arrangement (Fig. 4), similar to that observed in flying birds (Spedding, 1987; Vogel, 1994). Each half-beat of the tail sheds a roughly circular vortex into the water, which remains coherent for some time and moves roughly perpendicular to the direction of motion (Breithaupt and Ayers, 1996).

Fluid disturbances due to appendage movements also have been measured in copepods. In *Euchaeta rimana*, repetitive movements of the antennules, mouthparts, swimming legs and urosomes all produce disturbances with frequencies between about 10 and 50 Hz (Yen and Strickler, 1996). Because many other planktonic crustaceans display similar beat frequencies for these appendages, it is likely that 10–50 Hz signals are typical for swimming crustacean zooplankton. During high speed escape bursts, copepods shed a series of small vortices of 1–2 cm dia., at frequencies between 30 and 100 Hz (Yen and Strickler, 1996). Although intense, small disturbances in these viscous flow environments are extremely transient. According to Kirk (1985), it takes only 600 ms for near-field vibrations of *Daphnia* appendage movements to decay to levels that are at the detection threshold of their predators, larvae of the phantom midge, *Chaoborus trivittatus*. Kirk (1985) found an even more rapid attenuation of signal strength with distance (d) than that specified by the dipole field equations that had been used to model small scale disturbances produced by other crustaceans (Lenz and Yen, 1993). In the model, decay was proportional to d^{-3}, rather than to d^{-14} as measured by Kirk (1985), but the reason for this discrepancy is not clear.

So far we have discussed the properties of fluid flow and their effects on propagation of chemical and mechanical stimuli in the aquatic animal's natural environment. Let us now turn to the animal and examine how it uses these stimuli for orientation and guidance.

Chemosensory orientation

Chemosensory orientation in benthic crustaceans

In crustaceans, predation is frequently guided by chemical cues, and so is the search for shelters or dwelling sites, as well as for mates (Gleeson, 1972; van Leeuwen and Maly, 1991; Yen et al., 1996). Due to their ability to track odor cues from potential prey, lobsters have long been model systems for the study of chemosensory orientation. Both the Florida spiny (*Panulirus argus*) and the American lobster (*Homarus americanus*) orient to chemical cues originating from prey flesh, which constitute a complex cocktail of amino acids and adenine nucleotides (Carr, 1988). Ablation experiments suggest the aesthetasc tufts on the lateral antennular filaments provide the primary input guiding long distance search, but that chemoreceptors on the walking legs may also mediate localization, particularly when the animal is close to the target (Devine and Atema, 1982; Moore et al., 1991b). Mate localization also seems to be mediated by sensors on the antennule (Gleeson, 1982; Bamber and Naylor, 1996). Other species may rely on different chemosensory appendages. Behavioral (Weissburg and Zimmer-Faust, 1994) and physiological data (M.J. Weissburg et al., unpublished data) suggest that blue crab orientation is mediated by chemosensory input from dactyls and possibly the walking legs.

In turbulent odor plumes generated in flow, typical orientation manoeuvres consist of slow upstream walking, often interspersed with crossstream tacking. The tacking behavior is variable in terms of its frequency, as well as the angle of the tack. Some animals proceed quite directly to the source, whereas others engage in extensive cross-stream excursions. The absence of stereotypic behavior and consistent turning angles suggests that, unlike in moths (Arbas et al., 1993), chemical information does not simply trigger a motor program that is subsequently guided by other cues such as flow direction. Rather, chemical cues seem intimately associated with loco-motory behavior of the searching animal, so that variation in stimulus pro-perties is directly reflected in variation of the animal's trajectory. Lobsters in turbulent plumes turn towards the side receiving the strongest stimulus, and more asymmetric stimuli result in sharper turns, suggesting that klino- and tropotactic mechanisms may be used to decode features of the plume (Reeder and Ache, 1980; Devine and Atema, 1982; Basil and Atema, 1994). Orientation in turbulent plumes has been termed chemotaxis by various authors, but it is unlikely to be a taxis in the strict sense, that is, a behavior guided by detecting a concentration gradient. As indicated above, smooth concentration gradients do not exist in naturally turbulent waters at a spatial scale perceived by larger crustaceans.

Studies that explicitly link locomotory behavior to quantified flows can help resolve the mechanics of orientation, but, unfortunately, such studies are rare. When foraging for live or artificial bivalves in carefully control-led flows, blue crab search behavior is heavily dependent on the hydraulic context. Behaviors are similar to those described in lobsters, consisting of upstream-directed movements and non-stereotyped cross-stream tacking. However, crabs assume tack angles increasingly normal to the flow direc-tion at higher Re* values, as the plume becomes more widely distributed in space and time. Orientation performance becomes poorer with increasing Re* values, with a precipitous decrease in performance when the flow regime approaches smooth-turbulence (Re* = 3.5). This Re* value is asso-ciated with the thinning of the viscous sub-layer (see Fig. 1), so that this layer is no longer accessible to chemosensory appendages, indicating that quasi-laminar flows transmit chemical signals that are more easily tracked than signals in turbulent flows. The animal's failure to orient to olfactory signals in more turbulent flows (Re* > 3.5) may be due to either decreased odor concentrations or increased signal variability.

Although turbulence is deleterious to orientation efficiency, so is a complete absence of water movement. In spite of a concentration gradient generated via diffusion, blue crabs are unable to locate an odor source in still water. High speeds of locomotion and clearly defined search behaviors suggest that odorants are detected, but that localization is compromised by the absence of flow. Lobsters can follow a discrete trail laid down in still water. However, they are unable to determine its directionality and follow the trial regardless of whether it was laid down towards or away from them (McLeese, 1973).

Current data from various organisms suggest two very general models of orientation behavior, which differ in the assumptions concerning the relevant stimuli, i.e., chemical vs. non-chemical cues (Fig. 5). The first as-sumes that water movements *per se* do not represent the adequate stimulus, and that flow plays a role only in so far as it produces the necessary olfac-

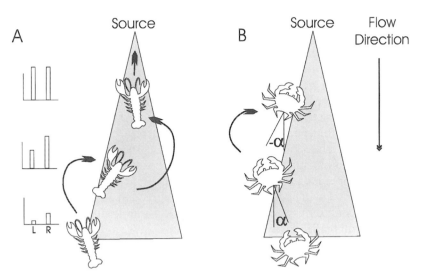

Figure 5. Two models (A and B) for chemosensory orientation towards an odor source. Shaded area depicts the region containing odorant stimuli emitted from the source. In A, a lobster uses simultaneous bilateral comparisons of odor intensity to turn into the plume, and sequential comparisons of odor intensity to progress towards the source within the plume. Bar diagrams show outputs of left and right sensors at each position. In this model, information on the source location is provided by the spatial and temporal properties of the intermittent olfactory stimulation, and not by the direction of flow. In B, a crab uses a binary comparison of odor presence-absence to determine that it has left the plume. In this model, the direction of flow is also crucial for orientation. The crab judges its exit angle (α) relative to flow direction and uses this information to re-enter the plume at the same angle at which it has exited.

tory signals that guide long distance search (e.g., Moore et al., 1991b). Relevant odor features, such as pulse onset slope or peak height, are a result of the nature of the plume in a turbulent flow. Lateral movements of the animal are guided by klinotactic mechanisms that keep it within the center of the plume, whereas successive temporal comparisons based on these same, or different odor features, establish the direction that leads towards the odor source. Odor sources in smooth concentration gradients cannot be localized, possibly because rapidly adapting chemosensors cannot continue to code chemical stimuli unless stimuli are intermittent. The behavior of lobsters in turbulent plumes is consistent with this model (Fig. 5A), recently summarized by Atema (1996).

Alternatively, flow may be important not only for generating intermittent signals, but, in addition, because flow direction can be used as a collimating stimulus to locate the source or the plume boundaries. Turning angles may reflect the joint action of bilaterally detected odor stimulus features, and the perceived current direction. Minimally, flow would provide information for initial movements once an odor cue has been detected (Weissburg and Zimmer-Faust, 1993). Orienting blue crabs always make their first steps upstream in the presence of odor, whereas movements in the absence of odor show no upstream bias. Subsequent movements in response to odors may be constrained by current direction, such that the animal always moves upstream using simple binary comparisons (odor presence/absence) to stay within the plume, or relocate it after initial detection (e.g., Zimmer-Faust et al., 1995). A more elaborate mechanism to relocate the plume would be to use flow direction to set the return tack angle by changing curse to re-enter the plume at the same angle at which the animal exited (Fig. 5B). According to this model, orientation is impossible in static water because of the animal's need of a cue that signals the upstream direction.

The importance of flow as a stimulus thus may be related to its predictability as a pointer for upstream direction and to the complexity of the resulting odor plume. In topographically simple habitats dominated by tidal flows (i.e., salt marshes, tidal flats, bathypelagic habitats), currents will be largely unidirectional, and will thus indicate the general heading towards the source of a plume. Unless water velocity is fairly rapid, these flows will have thick viscous sub-layers with stable and coherent odor profiles. In certain cases, flow may be depth-limited such than even in swift currents turbulence is low and plume boundaries are sharp (Zimmer-Faust et al., 1995). Such conditions may, again, be amenable to orientation based on flow direction, or on simple binary comparisons, or both. In contrast, habitats with a complex three-dimensional geometry are likely to produce currents with strong locally directional components that do not always point toward the source of odor emission. Odor plumes in these environments are expected to be quite complex and would give rise to more elaborate perceptual strategies. Clearly, comparative studies using animals from both simple and complex flow environments are necessary to reveal whether observed differences in behavior result from biological or from methodological factors.

Chemosensory orientation in planktonic crustaceans

Even planktonic crustaceans show an ability to orient to chemical cues, but far less is known about the behavioral and physiological mechanisms involved. Euphausids (krill), copepods, and crustacean larvae change their swimming behavior in response to chemical cues derived from prey (Poulet and Ouellet, 1982; Hamner et al., 1983; Gill and Poulet, 1988; Weissburg and Zimmer-Faust, 1991) and, when mature, from potential mates (van Leeuwen and Maly, 1991; Yen et al., 1996). The correlation between copepod abundances and levels of dissolved free amino acids in the field (Poulet et al., 1991) suggests that these behaviors allow animals to find areas of high food abundance in open waters. The observed changes in swimming behavior fit the classical definition of kinesis. In the presence of chemical stimuli animals generally show graded increases in speed and turning rates. Directional changes are not necessarily oriented towards the stimulus, but they nonetheless result in attraction to the source. Directed movements towards an odor source, such as the bilaterally mediated turns seen in lobsters and crabs, remain to be described in more detail. Krill may exhibit chemotaxis, because an animal leaving an odor patch quickly reverses direction and returns (Price, 1989), but other cues such as gravity and light could be used as collimating stimuli. Hamner and Hamner (1977) showed that pelagic shrimp follow sinking trails of amino acids, but here again the use of gravity as a directional cue is suspected.

However, within the context of a single individual reacting to discrete prey items at close range, copepods do display orientation abilities that cannot be explained by kinesis alone. As already noted, the low Re flows and high shear fields of copepod feeding currents draw a long thin 'trail' of odor into contact with sensory setae on the antennae. In response, *Eucalanus pileatus* redirects the feeding current (Koehl and Strickler, 1981), or changes its body orientation (Paffenhöfer and Knowels, 1978) to draw the prey particle into the central core of flow that passes by the oral region. Williamson and Vanderploeg (1988) describe a similar behavior for the freshwater copepod *Diaptomus pallidus* preying on rotifers. Ultrastructural and morphological investigations of copepod antennae have suggested the presence of aesthetasc-like sensilla (Elofsson, 1971; Strickler and Bal, 1973; Weatherby et al., 1994) that may mediate these behaviors. The chemosensory abilities of this appendage have been confirmed in recent physiological demonstrations of sensitivity to sugars and amino acids at concentrations as low as 10^{-10} to 10^{-9} M (M.J. Weissburg and J. Yen, unpublished data).

Olfactory perception at the neuronal level: Adaptation and disadaptation

Coding of chemical stimuli has been examined in the periphery, primarily on lobster aesthetasc chemosensory neurons. Many studies have focused on the dynamic properties of cells, because these may be especially critical in establishing the ability of an animal to respond to stimuli that vary both temporally and spatially. The results of many investigations suggest that these neurons are complex filters designed to extract rapid odor transients superimposed upon background signals of varying intensities.

Several studies indicate that lobster chemosensory neurons adapt extremely quickly and cease responding to a stimulus pulse within the first several hundred ms regardless of stimulus intensity. Features of a single pulse may be encoded in the first 200 ms of the phasic response (Gomez and Atema, 1996a), which is sufficient to encode the onset slope of even a very brief odor pulse. Responses to repetitive stimuli decline rapidly within the first few stimulus presentations, whereas the time for recovery to levels approaching that of pre-adapted cells takes several seconds (Voigt and Atema, 1990; Gomez and Atema, 1996b).

Rates of adaptation and disadaptation act jointly to determine the responsiveness of a cell, and seem important for several reasons. First, adaptation shifts the dynamic range of cells, so that they maintain their ability to encode any increase in stimulus concentration superimposed on a constant background, even when background concentrations become high (Borroni and Atema, 1988; Voigt and Atema, 1990). In essence, this mechanism improves the signal-to-noise ratio. Presumably, this is a useful feature for following odor plumes to their source, when cells become challenged with increasingly intense stimuli as the animal moves upstream.

Properties of adaptation/disadaptation also may help explain why animals stop their searching behavior close to the odor source, a question that has received much less attention than have the processes that govern long-distance orientation towards the stimulus. High background concentrations can increase the rate at which cells adapted to repetitive stimuli superimposed on these backgrounds (Voigt and Atema, 1990). Complete adaptation of aesthetasc chemosensory neurons in intense background concentrations may allow sensory elements on other appendages to guide the final phases of target localization. Although speculative, this idea is supported by several lines of evidence. For example, close to the odor source lobsters change search patterns (Moore et al., 1991b), possibly as other appendages become involved in mediating behavior. Animals lacking antennules can locate odor sources, indicating the existence of other chemosensory appendages (Reeder and Ache, 1980; Devine and Atema, 1982). These appendages sometimes have receptor populations that are tuned differently and are less sensitive than are those on aesthetasc sensilla (Johnson et al., 1984; Voigt and Atema, 1992; but see Derby and Atema, 1982), indicating that they can still respond at concentrations that would cause adaptation in antennular chemosensors. Further evidence suggests that rates of adaptation and disadaptation differ among appendages exposed to different temporal stimulus profiles. In fiddler crabs, chemosensory sensilla on legs are more habitually exposed to intense stimuli than are sensilla on the claw. Leg chemosensory neurons show slower rates of adaptation and disadapt more easily than do chemosensory neurons on the claw (Weissburg and Derby, 1995). Still, there is very little direct evidence showing how animals decide to cease locomotion, and behavioral investigations using selective deaferrentation of sensilla are necessary to establish the role of various appendages. Subsequent physiological studies could then serve to examine how neural properties of various chemosensory devices in specific environmental conditions establish their functional role.

Adaptation and disadaptation processes also may play a role in the ability of cells to discriminate among different odor pulse frequencies. Rapid adaptation rates result in cells that respond best to odor pulses with rapid onset slopes, and thus it sets the integration window of the cell. Disadaptation rates delimit the flicker fusion frequency; quicker disadaptation allows cells to resolve higher stimulus frequencies (Gomez et al., 1994; Moore, 1994). Depending on their adaptation-disadaptation properties, individual cells may operate as band pass filters within a variety of low frequency ranges (< 4 Hz) that match the dominant low frequency components in an odor plume (Gomez et al., 1994). Thus, an ensemble of cells may be able to code for changes in temporal stimulus features present in odor plumes, so that an animal perceives the odor environment through analysis of frequency components of plumes at varying distances from the source. Atema (1996) has recently suggested that the shape of an odor pulse (which includes both frequency and intensity components) is a characteristic feature of odor landscapes that may be used as a navigational cue.

Mechanosensory orientation

Similar to chemoperception, the perception of fluid mechanical disturbances is an important sensory modality in most crustaceans. The most commonly observed role of mechanical cues seems to be in mediating predator-prey interactions, which has been studied extensively in the crayfish. Stimulation of mechanoreceptors causes antennal and chelae sweeps that display a clear correlation between sweep angles and stimulus direction (Tautz, 1987; Schmitz, 1992; Breithaupt et al., 1995). Most commonly, appendages ipsilateral to the stimulus direction move backwards, whereas contralateral appendages sweep forwards, suggesting that stimulus direction is determined prior to appendage movement. These movements occur in response to water vibrations produced by dipole sources, water jets applied to the tailfan, and hydrodynamic stimuli produced by swimming fish. Here, antennal and chelae movements and directed body turns occur rapidly in a ballistic fashion, frequently resulting in contact between predator (crayfish) and prey (fish).

The results of numerous behavioral studies on the crayfish still do not provide a clear picture of the mechanisms that trigger these turns and sweeps. Although directionally sensitive sensor arrays have been found on various appendages (see below), their role in coding these complex fluid disturbances is still unknown. Stimulation of the tailfan appears essential for directed antennal and chelae movements in response to water jets applied from lateral and caudal directions (Schmitz, 1992). However, responses to vibratory stimuli and to water jets applied solely to the telson are less accurate than those observed in orientation to swimming fish (as judged by the frequency of ipsilateral-forward and contralateral-backward sweeps). Either the more complex hydrodynamic stimuli provide a richer cue to the tailfan system, or other sensory elements must be involved, or both. It is known that interneurons projecting to the CNS can have receptive fields that encompass not only the telson, but integrate, in addition, inputs from a number of other body regions or receptor organs (Taylor, 1968; Sigvardt et al., 1982; Wiese, 1988). Thus, for example, masking mechanosensory elements on the chelae diminishes the ability of interneurons to discriminate caudal from rostral stimuli (Wiese, 1988; Fig. 26.10).

Predator-prey interactions also can be mediated by fluid disturbances. Observations of several species of copepods indicate that predator capture responses are evoked as a result of detecting fluid mechanical signatures of their prey (Williamson and Vanderploeg, 1988; Yen and Fields, 1994). Akinetic prey are not detected. Vortex jets shed by rapidly moving prey, or possibly the signals produced by less intense oscillatory appendage movements during normal swimming, provide positional information used for orientation and capture. Copepods attempt to "capture" a water jet of size and intensity similar to those produced by natural prey (Yen and Strickler, 1996).

Copepod prey escape responses, similar to prey capture responses, are, again, mediated by mechanosensation. When copepods detect a predator's feeding current, they attempt to flee by jumping preferentially in the direction of lower shear, that is, away from the predator (Yen and Fields, 1992). In fact, copepod escape responses are evoked in several contexts in addition to predator-prey behavior. A long-standing frustration of oceanographers is that these small animals are adept at avoiding stationary objects (i.e., collectors) in flow, or moving nets and traps. Similar to their behavior in feeding currents, copepods react to large stationary objects in flow by jumping away. They appear to detect the objects at distances up to 7 mm (Strickler and Bal, 1973; Haury et al., 1980), which is impressive considering that copepods are generally less than 1 mm long.

A common factor in scenarios triggering copepod escape behavior is the presence of shear fields, induced by either boundary layer effects or non-uniform laminar sheared flows. As discussed by Yen and Fields (1992), copepods and other animals carried by flow require a spatially non-uniform velocity field (i.e., dv/ds, which is the definition of shear) to activate mechanosensory setae, as the relative velocity of an animal (with respect to the fluid) in a uniform flow is zero. However, sheared flow will expose different sensors to different relative velocities, thus containing both intensity and directional information. Theoretically, in sheared flow the paired antennae and their attendant mechanosensory sensilla (Yen and Fields, 1992) could be used to compare flow characteristics between left and right to establish directionality (Fig. 6). Thus, once the shear exceeds a certain level, the animal would execute its escape in the direction away from the shear field, i.e., towards the body side experiencing the least shear. Indeed, the position at which copepods first try to escape is related to the steepness of the local velocity gradient (Haury et al., 1980; Yen and Fields, 1992; Fields and Yen, 1996). However, measures of shear at escape points exhibit large variations, suggesting the participation of additional factors. It is

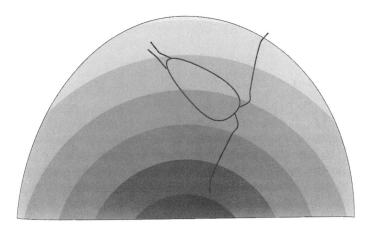

Figure 6. Copepod in shear flow (dv/ds) as might be produced in a copepod feeding current, near a stationary object, or in the vicinity of a gulping fish mouth. The intensity of shear increases from light to dark. In a flow field such as that shown here, mechanosensory setae on each antennule will each experience different relative flow speed due to the spatial gradient in velocity. Comparison of outputs of sensors between left and right antennules could, theoretically, indicate direction of increasing shear and mediate escape in the opposite direction.

possible, for example, that velocity differences (i.e., shear) interact with absolute velocity or acceleration to determine the degree of shear necessary to elicit avoidance or escape. Intense flows may cause sensors to operate close to saturation, implying that only a large change in velocities across the antennule can be coded. Note that absolute fluid velocity can be used to determine stimulus intensity only when the animal is stationary with respect to the flow, as in the case of benthic crustaceans.

Orientation to large-scale flows or currents is another context in which crustaceans utilize mechanical stimuli. Many decapods orient to the prevailing flow (Ebina and Wiese, 1984; Weissburg and Zimmer-Faust, 1993) with perceptual and physiological thresholds permitting detection of velocities less than 1 cm/s, perhaps as low as 0.1 cm/s (Laverack, 1962b; Ebina and Wiese, 1984; Weissburg and Zimmer-Faust, 1993). As already discussed for benthic crustaceans such as blue crabs, flow direction may provide important information for orientation to olfactory stimuli. The same holds true for bathypelagic crustaceans, whose approaches to deadfalls or bait come largely from the down-current direction (Thurston, 1979; Ingram and Hessler, 1983). However, orientation to flow may have other functions as well. Prawns orient to induced currents to reduce metabolic loads associated with respiration and feeding (Allanson et al., 1992), whereas lobsters may use the direction of wave surge to help locate their shelters (Herrnkind and McLean, 1971; Nevitt et al., 1995).

Mechanosensory structures

Peripheral mechanosensory elements that code fluid mechanical stimuli have been examined in a variety of crustacean species, revealing a bewildering variety of structures that are distributed on almost all body surfaces. This diversity has so far belied easy generalizations or statements on the correlation between structural properties and sensory function. As noted earlier, the role of various sensors in coding fluid disturbances is not completely clear. In some cases, setae may provide input necessary for controlling body or appendage movements, rather than for detecting biologically relevant extrinsic signals (e.g., Killian and Page, 1992). The most common mechanosensors are housed in erect hair-like sensilla that move in response to fluid displacements. It is believed that crustacean mechanosensory cells are of a single type relying on pivoting at the base of the hair, rather than bending along the hair's length, to transduce the signal (Crouau, 1996). They range in size from less than 100 to over 1000 µm. In crayfish, these sensors have displacement thresholds of $0.1-0.3$ µm (Wiese, 1976; Tautz and Sandeman, 1980), whereas thresholds for setae on copepod antennae are at least one order of magnitude lower (Lenz and Yen, 1993).

When examined using their electrical responses to oscillatory water movements (i.e., dipole sources), these receptors were found to respond best to the low frequency velocity and acceleration components characteristic of natural stimuli (Breithaupt and Tautz, 1990). For these sensors, the mass of water that surrounds the hair due to its boundary layer represents a (virtual mass) force. This force, as well as fluid induced drag are significant in determining displacement amplitude (Humphrey et al., 1993), and, therefore, neural response. Both forces increase with hair length, indicating that best frequencies should increase with decreasing hair length, and that shorter hairs are better acceleration detectors than are longer ones. Physiological data indicate that interneurons possess specificity for both high and low frequencies (Plummer et al., 1986), suggesting that peripheral elements fractionate the available frequency range. As yet, no one has attempted to correlate structural properties of crustacean sensilla with their frequency response or their function. Because the appropriate physical forces depend on fluid motion, one must be cautious in interpreting frequency-response curves generated by moving the hair mechanically. However, it may be technically difficult to stimulate and record from identified sensilla using oscillatory water movements.

Directional sensitivity of mechanoreceptors

The directional specificity of mechanosensors is a key feature in their ability to code fluid mechanical stimuli. Directional sensitivity can be achieved by several, not mutually exclusive mechanisms, such as a robust cuticular socket that restricts the directions of setal bending (Wiese, 1976; Ball and Cowan, 1977; Yen et al., 1992), asymmetrical pits that affect the water flow around the setae (Laverack, 1962a, 1962b), or a specific geometry of the mechanical coupling between the hair shaft and the neuron (Mellon, 1963; Wiese, 1976). Individual neurons within a single seta may exhibit opposite directional specificity, which is particularly common in antennal sensilla (e.g., Tazaki and Ohnishi, 1974; Wiese, 1976; Tazaki, 1977). Fractionation of the directional range may hold special advantages for coding, although such a system is not universal (e.g., Solon and Kass-Simon, 1981; Kouyama and Shimozawa, 1982).

For at least two reasons, the activity of a single seta, even one that is directionally sensitive, is rarely sufficient to provide a reliable directional signal. First, the activity of a single sensory neuron depends on both the stimulus intensity and direction, so that different combinations of these two signal characteristics may lead to similar neural responses. Second and more importantly, the complex nature of hydrodynamic stimuli produced by animals, or the decay of turbulent kinetic energy via the Kolmogorov cascade (Tennekes and Lumley, 1972), causes substantial spatial and temporal variation on a variety of scales. Thus, an ensemble of neurons may be required to filter out ambient noise and enhance the perception of relevant signals. The collection of procumbent hairs that monitor bending of the antennae is one such ensemble. This system will monitor the fluid forces impinging along the entire appendage to provide a more accurate representation of fluid motion than can be registered by a single sensory neuron. Antennal bending has been implicated in directing responses to tactile stimulation, but it also could mediate orientation to near-field fluid mechanical disturbances (Tautz et al., 1981). The group of directionally sensitive statocysts (Takahata and Hisada, 1982) could code the direction to fluid motion intensely enough to displace the animal, although whether or not this commonly occurs is unknown (Breithaupt et al., 1995).

Substantial evidence indicates that many features of crustacean mechanosensory systems are designed to provide the degree of filtering necessary for detecting relevant turbulent signals superimposed on a noisy background. Processing of mechanosensory information has been extensively studied in the crayfish, where it appears that lateral inhibition provides contrast enhancement to shape the response that is relayed to the CNS. Interneurons receive information from either headward or tailward sensitive sensory afferents. Some of these interneurons have bilateral receptive fields, where (relative to the soma) ipsilateral inputs are stimulatory, whereas contralateral inputs are inhibitory (Wilkens and Larimer, 1972; Wiese and Schulz, 1982; Tautz and Plummer, 1994). Thus, input that is equally weighted across the mid-line of the animal results in less activity than do stimuli that exert bilaterally asymmetrical effects. When stimuli are perfectly symmetrical, a balance point is achieved such that these interneurons are virtually silenced (Wiese and Schultz, 1982).

This system, often described as a mechanism for common mode rejection or contrast enhancement, has a number of important properties. First, the animal may actually use the balance point to orient itself. Activity in interneurons that receive only unilateral excitatory input can be used to determine whether quiescent neurons result from a lack of stimuli, or the animal is positioned so that stimuli are symmetrical across the mid-line (Wiese and Schultz, 1982). This would make a particularly effective mechanism to align to currents. Secondly, because the headward vs. tailward subsystems have separate inhibitory pathways, fluid moving in opposite directions in regions across the mid-line evokes a stronger interneuronal response, as compared to stimuli moving in the same direction across each side of the tailfan. Focal stimuli that elicit pronounced directionally sensitive responses in an interneuron can then further sharpen the directional tuning of other interneurons to which they are synaptically coupled.

This mechanism has been explored in the local directionally selective neuron (LDS), and its post-synaptic target, the spiking caudal photoreceptor (CPR), which is also light sensitive. Each is a pair of higher order interneurons that receives ispilateral excitatory and contralateral inhibitory stimulation. Because the LDS is non-spiking, it exerts a graded inhibitory effect to produce a highly directional response in the CPR (Tautz and Plummer, 1994). The CPR shows an extremely abrupt cut-off response to contralateral stimuli, thus providing a very accurate bilateral representation of stimulus strength. However, the CPR can further discriminate among water jets applied from different angles, although the angular resolution is at best $10°$ and is frequently even poorer than that (Tautz and Plummer, 1994).

The LDS-CPR system receives its input from a population of receptors that do not share similar directional tuning profiles, i.e., directional information is encoded by the summed excitatory and inhibitory effects of a spatially contiguous group of neurons, regardless of their individual directional sensitivity. Thus, information on the directional specificity of individual sensilla is not preserved, so that coding the precise shape of a disturbance is sacrificed for the sake of increased sensitivity to large asymmetrical disturbances, which are more likely to be biologically relevant. For example, a single vortex (such as that shed by a moving fish, see Fig. 4) that stimulates opposite sides of the tailfan with movement in opposing directions is expected to produce the most distinct response from neurons such as the CPR (Tautz and Plummer, 1994) (Fig. 7A). In addition, a vortex ring creates a strong unidirectional flow that could be detected as a well-defined transient signal as it crosses over the mid-line of the tailfan (Fig. 7B).

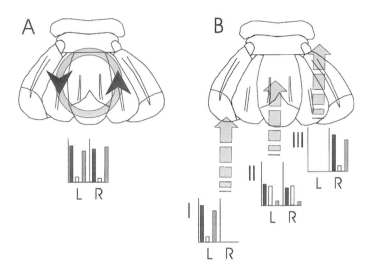

Figure 7. Two hypothetical scenarios for coding of mechanosensory information by the directionally sensitive interneurons in crayfish tailfan. A and B correspond to flow patterns produced in regions marked A and B in Figure 4. In the bar diagrams, black and white bars depict excitatory and inhibitory inputs, respectively, to left (L) and right (R) interneurons. In each case, the resulting output is depicted by the grey bar. In A, the vortex entrains flow that is moving in opposite directions to the left and the right side of the tailfan midline, resulting in symmetrical responses from interneurons on both sides. In B, the unidirectional flow in the center of a vortex is moving from left to right across the tailfan, resulting in the sequence of neural outputs depicted in I–III. When the vortex is on the left side (I), the left interneuron is maximally stimulated, whereas inhibition on the left interneuron is practically absent. The mirror-symmetrical situation is shown in III, when the vortex is on the right side of the tailfan. As the vortex crosses the mid-line (Situation II), excitatory and inhibitory inputs are balanced on the left and the right side, resulting in a transient loss of interneuronal output.

Conclusions

The response properties of chemo- and mechanosensory systems in aquatic crustaceans appear to have been finely honed by the features of their stimulus environment. To be effective, sensory systems must take advantage of the information present in the characteristics of natural signals by appropriate strategies to acquire and code this information. The reliance on properties of odor pulses rather than average concentrations, as well as low frequency tuning of hydrodynamic receptors, are manifestations of sensory systems well adapted to their environment. Sensory systems must, in addition, filter out extraneous information to maximize signal-to-noise ratio. Thus, chemosensory elements adapt to constant backgrounds to better resolve fluctuating signals, and mechanosensory systems use lateral inhibition to emphasize bilaterally asymmetrical fluid disturbances.

Table 1. Habitat characteristics for aquatic crustaceans, and representative organisms that live in them. Currently, only organisms occupying 3D/Low Re, and 2D/High Re habitats have been examined with respect to their mechano- and chemosensory systems, and the hydrodynamic setting of their habitat.

	Dimensionality	
Reynolds Number	3D-Pelagic	2D-Benthic
High	Krill, mysiids, shrimp	Crabs, lobsters
Low	Copepods, larvae	Juvenile crabs and lobsters, copepods

Investigating the coding of fluid-borne information by olfactory and mechanoreceptive systems is a fairly new pursuit, mainly because of the difficulties involved in manipulating and measuring flow and its entrained signals. Consequently, our knowledge on chemo- and mechanosensory systems in crustaceans currently comes from studies conducted in a relatively narrow range of flow environments. Flow regimes can be broadly characterized as turbulent or laminar, whereas animals may move in either two-dimensional (benthic) or three-dimensional (pelagic) realms. Of the four possible combinations that characterize aquatic habitats (Tab. 1), only two have received more than cursory scrutiny. Certainly, the other environments also pose several intriguing questions. Both mechanical and chemical information are important to large pelagic crustaceans for regulating feeding, schooling, and mating in a 3D-turblent environment (e.g., Hamner et al., 1983; Price, 1989; Wiese and Marschall, 1990). Small benthic forms that move within the boundary layer will receive signals transmitted in quasi-laminar (or perhaps viscous) flows. Even a single organism may experience several of these habitats during its lifetime. In its 3D habitat, a pelagic crustacean larva may detect food via chemosensation, or perhaps use chemical cues to locate suitable settlement sites. Upon metamorphosis to its benthic form, it occupies a 2D habitat and will become increasingly affected by turbulence as it matures. Because comparative investigations may sharpen our understanding of chemo- and mechanoperception in general, I would like to outline several areas where different properties of flow regime may have crucial implications on sensory properties.

Chemosensory orientation and neural properties in low-Re flows

In low-Re flows, viscous interactions allow persistence of odor plumes over relatively long time or spatial scales. Recent evidence (Yen et al., 1996) indicates that male copepods of certain species track a female using trails as old as 10 s, and follow that trail as far as 10 cm before reaching her. Rather than the filamentous plumes that occur at high Re, a copepod creates a stable trail-like odor signal with a width similar to the copepod's (ca. 100 μm). In low Re regimes, such trails maintain their spatial integrity to preserve the gradient of chemical stimulus intensity along the trail. True chemotaxis then becomes an effective strategy to track moving targets such as mates or prey. Indeed, conspecifics are located by orientation to such tracks (Yen et al., 1996), similar to trail-following terrestrial insects. At a scale relative to their body size, such a behavior would not be adequate in larger crustaceans, due to the evanescence to odor plumes in turbulent flows.

If integration properties defined by adaptation-disadaptation processes are designed to provide the appropriate temporal window, then animals tracking chemical cues in low Re flows should have low pass filters with extremely low cut-off frequencies. Because neurons in these environments do not experience the rapid odor transients produced in more turbulent regimes, adaptation should be sufficiently slow to allow the animal to monitor odor signals over longer time periods.

The expected correlation between temporal stimulus profiles, adaptation and flow regime also applies to the design of various appendages on a given organism. The Re describing flow around an appendage will depend on the appendage's length, shape, roughness etc., which will modify the thickness of the resultant boundary layer. Diffusion through the boundary layer partially determines stimulus profiles at the receptor, and evidence indicates that appendage morphology affects the microscale structure of chemical signals (Moore et al., 1991a). Particular adaptation-disadaptation time-courses may be an appropriate mechanism for responding to different flow regimes around particular appendages.

Predictability, filtering, and spatial processing

The spatial and temporal variance of odor and fluid mechanical cues in turbulent environments represents a significant challenge for sensory systems that extract information from these signals. For distance orientation, it is unlikely that olfactory or mechanosensory systems furnish immediate and precise spatial information, unless they operate within the laminar flow region of the boundary layer.

In most cases, chemosensory tracking in turbulent plumes requires elaborate sampling strategies, and animals make somewhat halting progress until they are quite close to the source. Because olfactory orientation requires repeated contact with the odor signals, the animal must continually re-assess its position relative to a target whose location under these conditions is indicated only roughly. Even mechanosensory systems seem to extract only extremely general information on the spatial relationships between the signal source and its receiver, and operate only at distances on the order of the receiver's body length. The initial orientation event may only move the animal into position to acquire more precise information from tactile sensory systems (e.g., Breithaupt et al., 1995). Because of the high level of ambient noise in turbulent flows, these systems have evolved a complicated array of sensors and interneurons for improving the signal-to-noise ratio. Indeed, the hydrodynamic noise created by water flow may produce different behaviors, as signal-to-noise ratios change in still vs. moving fluids (Bleckmann et al., 1991). Because most studies have been performed in static conditions, our current understanding of the capacities and limitations of mechanosensory orientation is incomplete.

One advantage of laminar flow regimes is the simplicity of structure, and therefore of the information conveyed by the fluid flow. When viscosity dominates, extraneous fluid motion that may create hydrodynamic noise or disrupt odor plumes is dampened. One immediate consequence may be that noise rejection is not a constraint on system design. An animal may possess sensors responsive to very small displacements without being overwhelmed by spurious signals. Such high sensitivity may be requisite in low-Re flows where signals attenuate rapidly, or may allow detection of more remote targets (when scaled to body size) than is possible in inertial realms.

Complicated noise-rejection circuitry also may be unnecessary in low-Re flows. In this more predictable fluid milieu, sensory arrays may be designed to pre-filter ambient fluid motion, rather than having the system rely on filtering by CNS elements. Sensors may have particular morphologies, or be socketed and positioned such that they do not move in the flow, or have response thresholds tuned so that they do not react to the ambient fluid movement. It is probably not coincidental that the longest and presumably most sensitive sensilla on a copepod antennae lay just at the periphery of the self-generated feeding current. Such adaptations also may be present in larger animals. However, because of the complicated flows produced by irregularly shaped bodies in turbulent flows, it may be difficult to correlate flow fields with properties of the attendant sensory structures.

In laminar, as opposed to turbulent flow, activation of a single mechano-sensory seta may well provide accurate directional information. In the absence of random fluid movement, there would be little need to use large spatial arrays that sacrifice directional selectivity in favour of signal enhancement. If directional information is coded by the setae and relayed unchanged to the CNS, then the activity of individual sensory neurons or small neuronal groups may be sufficient to determine directionality.

Similar arguments apply to chemosensory elements. In laminar flows, an odor filament carried by flow will contact sensilla whose position has a well-defined relationship to the direction to the signal source. Streamlines are parallel to one another, and each streamline will intersect the antennule at a specific point along its length. Thus, a copepod that creates its own feeding current collapses its sensory space onto an essentially two-dimensional grid with axes corresponding to distance from the antennule and position along the antennule. A similar 2D projection is expected to occur in animals in laminar boundary-layer flows. Thus, the location of a sensillum that is stimulated by an odor filament provides information on the direction of the odor source. Conceivably, a strict topographic projection of sensory afferents into the CNS that codes for the position of the receptor on the antennal array could inform the animal on the relative location of the source of the odor. Vertebrate auditory (Konishi, 1986) and chemosensory systems (Hayama and Caprio, 1989) are known to use topographic representations to encode stimulus direction, suggesting that even crustaceans might utilize a similar mechanism for localizing odor or mechanical cues in their aquatic world.

Acknowledgements
I am indebted to the many colleagues with whom I have discussed these or other ideas, and who have aided me in developing this perspective over the years of posters, talks and meetings. You know who you are. Drs. Chuck Derby, David Dusenbery and Jeannette Yen were especially helpful in providing comments on earlier drafts of this manuscript. Thanks also to the participants of the 1995 NATO sponsored MAST Turbulence Workshop for sharpening my hydrodynamic acumen. Drs. Thomas Breithaupt, Paul Moore, Jeanette Yen and Richard K. Zimmer-Faust graciously provided access to unpublished data, and Dr. R. Blickhan kindly gave permission to redraw his illustration of vortex streets shed by fish. Funding from National Institute of Health (#1R29CD273) is gratefully acknowledged; this work would not have been possible without their continued support.

Note added in proof
At the times of this writing, a special issue of *Philosophical Transactions of the Royal Society of London* is in press, which is devoted to mating in planktonic copepods. This issue contains a number of significant findings of relevance to issues discussed in this review, and a number of additional references for chemo- and mechanoperception that I was not able to present. Notably, several contributions further analyze chemosensory trail following mentioned in this review, and come to the conclusion that these behaviors are, at least in part, mediated by tropo- and/or klinotactic mechanisms. Additional papers discuss the structure of chemical and fluid mechanical trails in these low Re environments, extending and deepening our understanding of the signal environment experienced by small aquatic creatures.

References

Allanson, B.R., Skinner, D. and Imberger, J. (1992) Flow in prawn burrows. *Est. Coastal Shelf Sci*. 35:253–268.

Andrews, J.C. (1983) Deformation of the active space in the low Reynolds number feeding current of calanoid copepods. *Can. J. Fish. Aquat. Sci*. 40:1193–1302.

Arbas, E.A., Willis, M.A. and Kanzaki, R. (1993) Organization of goal oriented locomotion: pheromone modulated flight behavior of moths. *In*: R.D. Beer, R.E. Ritzmann and T. McKenna (eds): *Biological Neural Networks in Invertebrate Neuroethology and Robotics*. Academic, New York, pp 159–198.

Atema, J. (1988) Distribution of chemical stimuli. *In*: J. Atema, R.R. Fay, A.N. Popper, and W.N. Tavolga (eds): *Sensory Biology of Aquatic Animals*. Springer-Verlag, New York, pp 29–56.

Atema, J. (1996) Eddy chemotaxis and odor landscapes: exploration of nature with animal sensors. *Biol. Bull*. 191:129–138.

Ball, E.E. and Cowan, N. (1977) Ultrastructure of the antennal sensilla of *Acetes* (Crustacea, Decapoda, Natantia, Sergestidae). *Phil. Trans. R. Soc. Lond*. B 277:429–456.

Bamber, S.D. and Naylor, E. (1996) Mating behaviour of male *Carcinus maenus* in relation to a putative sex pheromone: behavioural changes in response to antennule restriction. *Mar. Biol*. 125:483–488.

Basil, J. and Atema, J. (1994) Lobster orientation in turbulent odor plumes: Simultaneous measurements of tracking behavior and temporal odor patterns. *Biol. Bull.* 187:272–273.

Bleckmann, H., Breithaupt, T., Blickhan, R. and Tautz, J. (1991) The time course and frequency content of hydrodynamic events caused by moving fish, frogs and crustaceans. *J. Comp. Physiol.* A 168:749–757.

Blickhan, R., Krick, C., Zehren, D., Nachtigall, W. and Breithaupt, T. (1992) Generation of a vortex chain in the wake of a subundulatory swimmer. *Naturwissens.* 79:220–221.

Borroni, P.F. and Atema, J. (1988) Adaptation in chemoreceptor cells I. Self-adapting backgrounds determine thresholds and cause parallel shift of dose-response function. *J. Comp. Physiol.* A 164:67–74.

Breithaupt, T. and Ayers, J. (1996) Visualization and quantitative analysis of biological flow fields using suspended particles. *In:* P.H. Lenz, D.K. Hartline, J.E. Purcell and D.L. Macmillan (eds) *Zooplankton: Sensory Ecology and Physiology.* Gordon Breach Publishers, Amsterdam, pp 117–129.

Breithaupt, T. and Tautz, J. (1990) The sensitivity of crayfish mechanoreceptors to hydrodynamic and acoustic stimuli. *In:* K. Wiese, W.-D. Krenz, J. Tautz, J. Riechert, and B. Mulloney (eds): *Frontiers in Crustacean Neurobiology.* Birkhäuser Verlag, Basel, pp 114–120.

Breithaupt, T., Schmitz, B. and Tautz, J. (1995) Hydrodynamic orientation in crayfish (*Procambarus clarkii*) to swimming fish prey. *J. Comp. Physiol.* A 177:481–491.

Carr, W.E.S. (1988) The molecular nature of chemical stimuli in the aquatic environment. *In:* J. Atema, R.R. Fay, A.N. Popper, and W.N. Tavolga (eds): *Sensory Biology of Aquatic Animals.* Springer-Verlag, New York, pp 3–27.

Crouau, Y. (1996) Association in a crustacean sensory organ of two usually exclusive mechanosensory cell. *Biol. Cell.* 85:191–195.

Denny, M.W. (1988) *Biology and mechanics of the wave-swept environment.* Princeton University Press, Princeton, NJ.

Derby, C.D. and Atema, J. (1982) The function of chemo- and mechanoreceptors in lobster (*Homarus americanus*) feeding behavior. *J. Exp. Biol.* 98:317–327.

Devine, D.V. and Atema, J. (1982) Function of chemoreceptor organs in spatial orientation of the lobster, *Homarus americanus*: differences and overlap. *Biol. Bull.* 163:144–153.

Dittmer, K., Grasso, F. and Atema, J. (1995) Effects of varying plume turbulence on temporal concentration signals available to orienting lobsters. *Biol. Bull.* 189:232–233.

Ebina, Y. and Wiese, K. (1984) A comparison of neuronal and behavioural thresholds in the displacement sensitive pathway of the crayfish *Procambarus. J. Exp. Biol.* 107:45–55.

Elofsson, R. (1971) The ultrastructure of a chemoreceptor organ in the head of copepod crustaceans. *Acta. Zool.* 52:299–315.

Fields, D.M., and Yen, J. (1996) The escape response of *Pleuromamma xiphias* in response to a quantifiable fluid mechanical disturbance. *Mar. Fresh. Behav. Physiol.* 25:323–339.

Gill, C.W. and Poulet, S.A. (1988) Responses of copepods to dissolved free amino acids. *Mar. Ecol. Prog. Ser.* 43:269–276.

Gleeson, R.A. (1982) Morphological and behavioral identification of the sensory structures mediating pheromone reception in the blue crab, *Callinectes sapidus. Biol. Bull.* 163:162–171.

Gomez, G. and Atema, J. (1996a) Temporal resolution in olfaction I: Stimulus integration time of lobster chemoreceptor cells. *J. Exp. Biol.* 199:1771–1779.

Gomez, G. and Atema, J. (1996b) Temporal resolution in olfaction II: Time course of recovery from adaptation in lobster chemoreceptor cells. *J. Neurophysiol.* 76:1340–1343.

Gomez, G., Voigt, R. and Atema, J. (1994) Frequency filter properties of lobster chemoreceptor cells determined with high resolution stimulus measurement. *J. Comp. Physiol.* A 174:803–811.

Gross, T.F. and Nowell, A.R.M. (1983) Mean flow and turbulence scaling in a tidal boundary layer. *Cont. Shelf Res.* 2:109–126.

Hamner, P. and Hamner, W.M. (1977) Chemosensory tracking of scent trails by the planktonic shrimp *Acetes sibogae australis. Science* 195:886–888.

Hamner, W.M., Hamner, P., Strand, S.W. and Gilmer, R.W. (1983) Behavior of Antarctic krill, *Euphausia superba*: chemoreception, feeding, schooling and molting. *Science* 220:433–435.

Hart, D.D., Clark, B.D. and Jasentuliyana, A. (1996) Fine-scale field measurement of benthic flow environments inhabited by stream invertebrates. *Limnol. Oceanog.* 41:297–308.

Haury, L.R., Kenyon, D.E. and Brooks, J.R. (1980) Experimental evaluation of the avoidance reaction of *Calanus finmarchicus. J. Plankt. Res.* 2:187–203.

Hayama, T. and Caprio, J.C. (1989) Lobule structure and somatotopic organization of the medullary facial lobe in the channel catfish *Ictalurus punctatus*. *J. Comp. Neurol.* 285:9–17.

Herrnkind, W.F. and McLean, R. (1971) Field studies of homing, mass emigration and orientation in the spiny lobster, *Panulirus argus*. *Ann. N.Y. Acad. Sci.* 188:359–377.

Humphrey, J.A.C., Devarakonda, R., Iglesias, I. and Barth, F.G. (1993) Dynamics of arthropod hairs. I. Mathematical modeling of the hair and air motions. *Phil. Trans. R. Soc. Lond.* B 340: 423–444.

Ingram, C.L. and Hessler, R.R. (1983) Distribution and behavior of scavenging amphipods from the central North Pacific. *Deep Sea Res.* 25:683–705.

Johnson, B.R., Voigt, R., Borroni, P.F. and Atema, J. (1984) Response properties of lobster chemoreceptors: tuning of primary taste neurons in the walking legs. *J. Comp. Phys.* 155: 593–604.

Killian, K.A. and Page, C.H. (1992) Mechanosensory afferents innervating the swimmerets of the lobster II. Afferents activated by hair deflection. *J. Comp. Physiol.* A 170:501–508.

Kirk, K.L. (1985) Water flows produced by *Daphnia* and *Diaptomus*: Implications for prey selection by mechansensory predators. *Limnol. and Oceanogr.* 30:670–686.

Koehl, M.A.R. and Strickler, J.R. (1981) Copepod feeding currents: food capture at low Reynolds number. *Limnol. Oceanogr.* 26:1062–1073.

Konishi, M. (1986) Centrally synthesized maps of sensory space. *TINS* 9:163–168.

Kouyama, N. and Shimozawa, T. (1982) The structure of a hair mechanoreceptor on the antennule of the crayfish (Crustacea). *Cell Tiss. Res.* 266:565–578.

Laverack, M. (1962a) Responses of cuticular sense organs of the lobster, *Homarus vulgaris* (Crustacea). I. Hair-peg organs as water current receptors. *Comp. Biochem. Physiol.* 5:319–335.

Laverack, M. (1962b) Responses of cuticular sense organs of the lobster, *Homarus vulgaris* (Crustacea). II. Hair-fan organs as pressure receptors. *Comp. Biochem. Physiol.* 6:137–145.

List, E.J. (1982) Turbulent jets and plumes. *Ann. Rev. Fluid Mech.* 14:189–212.

Lenz, P. and Yen, J. (1993) Distal setal mechanoreceptors of the first antennae of marine copepods. *Bull. Mar. Sci.* 53:170–179.

McLeese, D.W. (1973) Orientation of lobsters (*Homarus americanus*) to odor. *J. Fish Res. Board Can.* 30:838–840.

Mellon, DeF., Jr. (1963) Electrical responses from dually innervated tactile receptors on the thorax of the crayfish. *J. Exp. Biol.* 40:137–148.

Monismith, S.G., Koseff, J.R., Tompson, J.K., O'Riordan, C.A. and Nepf, H.M. (1990) A study of model bivalve siphonal currents. *Limnol. and Oceanogr.* 35:680–696.

Moore, P.A. (1994) A model of the role of adaptation and disadaptation in olfactory receptor neurons: implications for the coding of temporal and intensity patterns in odor signals. *Chem. Senses* 19:71–86.

Moore, P.A. and Atema, J. (1991) Spatial information in the three-dimensional fine structure of an aquatic odor plume. *Biol. Bull.* 181:408–418.

Moore, P.A., Atema, J. and Gerhardt, G.A. (1991a) Fluid dynamics and microscale chemical movement in the chemosensory appendages of the lobster *Homarus americanus*. *Chem. Senses* 16:663–674.

Moore, P.A., Scholz, N. and Atema, J. (1991b) Chemical orientation of lobsters, *Homarus americanus* in turbulent odor plumes. *J. Chem. Ecol.* 17:1293–1307.

Moore, P.A., Zimmer-Faust, R.K., BeMent, S.L., Weissburg, M.J., Parrish, J.M. and Gerhardt, G.A. (1992) Measurement of microscale patches in a turbulent aquatic odor plume using a semi-conductor based microprobe. *Biol. Bull.* 183:138–142.

Moore, P.A., Fields, D.M. and Yen, J. (1994a) The fine structure of chemical signals within the feeding current of calanoid copepods. *Eos* 75:163.

Moore, P.A., Weissburg, M.J., Parrish, J.M., Zimmer-Faust, R.K. and Gerhardt, G.A. (1994b) Spatial distribution of odors in simulated benthic boundary layer flows. *J. Chem. Ecol.* 20:255–279.

Murlis, J., Elkinton, J.S. and Cardé, R.T. (1992) Odor plumes and how insects use them. *Ann. Rev. Entomol.* 37:505–532.

Nevitt, G.A., Pentcheff, N.D., Lohmann, K.J. and Zimmer-Faust, R.K. (1995) Evidence for hydrodynamic orientation by spiny lobsters in a patch reef environment. *J. Exp. Biol.* 198:2049–2054.

Paffenhöfer, G.-A. and Knowles, S.C. (1978) Feeding of marine planktonic copepods on mixed phytoplankton. *Mar. Biol.* 48:143–152.

Palmer, M.A. (1988) Dispersal of marine meiofauna: a review and conceptual model explaining passive transport and active emergence with implications for recruitment. *Mar. Ecol. Pro. Ser.* 48:81–91.

Palmer, M.A. and Gust, G. (1985) Dispersal of meiofauna in a turbulent tidal creek. *J. Mar. Res.* 43:179–210.

Plummer, M.R., Tautz, J. and Wine, J.J. (1986) Frequency coding of waterborne vibrations by abdominal mechanosensory interneurons in *Procambarus clarkii*. *J. Comp. Physiol.* A 158:751–764.

Poulet, S.A. and Ouellet, G. (1982) The role of amino acids in the chemosensory swarming and feeding of marine copepods. *J. Plankt. Res.* 4:341–361.

Poulet, S.A., Williams, R., Conway, D.V.P. and Videau, C. (1991) Co-occurrence of copepods and dissolved free amino acids in shelf sea waters. *Mar. Biol.* 10:373–385.

Price, H.J. (1989) Swimming behavior of krill in response to algal patches: a mesocosm study. *Limnol. Oceanogr.* 34:649–659.

Reeder, P.B. and Ache, B.W. (1980) Chemotaxis in the Florida spiny lobster, *Panulirus argus*. *Anim. Behav.* 28:831–839.

Schlichting, H. (1979) *Boundary layer theory*. McGraw-Hill, New York, NY.

Schmitz, B. (1992) Directionality of antennal sweeps elicited by water jet stimulation of the tailfan in the crayfish *Procambarus clarkii*. *J. Comp. Physiol.* A 171:617–627.

Strickler, J.R. (1985) Feeding currents in calanoid copepods: two new hypotheses. *In*: M.S. Laverack (ed.): *Physiological Adaptations of Marine Animals. Symposium of Society for Experimental Biology* 39:459–485.

Strickler, J.R. and Bal, A.K. (1973) Setae of the first antennae of the copepod *Cyclops scutifer* (Sars): their structure and importance. *Proc. Natl. Acad. Sci.* 70:2656–2659.

Sigvardt, K.A., Hagiwara, G. and Wine, J.J. (1982) Mechanosensory integration in the crayfish abdominal nervous system: Structural and physiological differences between interneurons with single and multiple spike initiating sites. *J. Comp. Physiol.* A 148:143–157.

Solon, M. and Kass-Simon, G. (1981) Mechanosensory activity of hair organs on the chelae of *Homarus americanus*. *J. Comp. Physiol.* A 6:217–223.

Speeding, G.R. (1987) The wake of a kestrel (*Falco tinnunculus*) in flapping flight. *J. Exp. Biol.* 127:59–87.

Takahata, M. and Hisada, M. (1982) Statocyst interneurons in the crayfish *Procambarus clarkii* (Girard). II. Directional sensitivity and its mechanism. *J. Comp. Physiol.* 149:301–306.

Tautz, J. (1987) Water vibration elicits acitive antennal movements in the crayfish, *Oronectes limosus*. *Anim. Behav.* 35:748–754.

Tautz, J. and Plummer, M. (1994) Comparison of directional selectivity in identified spiking and non-spiking mechanosensory neurons in the crayfish *Oronectes limosus*. *Proc. Natl. Acad. Sci. USA* 91:5853–5857.

Tautz, J. and Sandeman, D.C. (1980) The detection of waterborne vibrations by the sensory hairs on the chelae of the crayfish. *J. Exp. Biol.* 88:351–356.

Tautz, J., Masters, W.M., Eicher, B. and Markl, H. (1981) A new type of water vibration receptor on the crayfish antennae. I. Sensory physiology. *J. Comp. Physiol.* 144:533–541.

Taylor, R.C. (1968) Water-vibration reception: A neurophysiological study in unrestrained crayfish. *Comp. Biochem. Physiol.* 27:795–805.

Tazaki, K. (1977) Nervous responses from mechanosensory hairs on the antennal flagellum in the lobster, *Homarus gammarus* (L.). *Mar. Behav. Physiol.* 5:1–18.

Tazaki, K. and Ohnishi, M. (1974) Responses from tactile receptors in the antenna of the spiny lobster *Panulirus japonicus*. *Comp. Biochem. Physiol.* 47A:1323–1327.

Tennekes, H. and Lumley, J.L. (1972) *A first course in turbulence*. MIT Press, Cambridge, MA.

Thurston, M.H. (1979) Scavenging abyssal amphipods from the North-East Atlantic Ocean. *Mar. Biol.* 51:55–68.

Van Leeuwen, H.C. and Maly, E.J. (1991) Changes in the swimming of male *Diaptomus leptopus* (Copepoda: Calanoida) in response to gravid females. *Limol. Oceanogr.* 36:1188–1195.

Vogel, S. (1994) *Life in moving fluids*, 2nd ed. Princeton University Press, Princeton, NJ.

Voigt, R. and Atema, J. (1990) Adaptation in chemoreceptor cells. III. Effects of cumulative adaptation. *J. Comp. Physiol.* A 166:865–874.

Voigt, R. and Atema, J. (1992) Tuning of chemoreceptor cells of the second antennae of the American lobster (*Homarus americanus*) with a comparison of four of its other chemoreceptor organs. *J. Comp. Physiol.* A 171:673–683.

Weatherby, T.M., Wong, K.K. and Lenz, P.H. (1994) Fine structure of the distal sensory setae on the first antennae of *Pleuromamma xiphias* Giesbrecht (Copepoda). *J. Crust. Biol.* 14: 670–685.

Weissburg, M.J. and Derby, C.D. (1995) Regulation of sex-specific feeding behavior in fiddler crabs: Physiological properties of chemoreceptor neurons in claws and legs of males and females. *J. Comp. Physiol.* A 176:513–526.

Weissburg, M.J. and Zimmer-Faust, R.K. (1991) Ontogeny versus phylogeny in determining patterns of chemoreception: Initial studies with fiddler crabs. *Biol. Bull.* 181:205–215.

Weissburg, M.J. and Zimmer-Faust, R.K. (1993) Life and death in moving fluids: Hydrodynamic effects on chemosensory-mediated predation. *Ecol.* 74:1428–1443.

Weissburg, M.J. and Zimmer-Faust, R.K. (1994) Odor plumes and how blue crabs use them to find prey. *J. Exp. Biol.* 197:349–375.

Wiese, K. (1976) Mechanoreceptors for near-field water displacement in crayfish. *J. Neurophysiol.* 39:816–833.

Wiese, K. (1988) The representation of hydrodynamic parameters in the CNS of the crayfish *Procambarus*. *In*: J. Atema, R.R. Fay, A.N. Popper, and W.N. Tavolga (eds): *Sensory Biology of Aquatic Animals*. Springer-Verlag, New York, pp 665–686.

Wiese, K. and Marschall, H.P. (1990). Sensitivity to vibration and turbulence of water in the context of schooling in Antarctic krill *Euphasia superba*. *In*: K. Wiese, W.-D. Krenz, J. Tautz, H. Riechert and B. Mulloney (eds): *Frontiers in Crustacean Neurobiology*. Birkhäuser Verlag, Basel, pp 121–130.

Wiese, K. and Schultz, R. (1982) Intrasegmental inhibition of the displacement sensitive pathway in the crayfish (*Procambarus clarkii*). *J. Comp. Physiol.* A 147:447–454.

Wilkens, L.A. and Larimer, J.L. (1972) The CNS photoreceptor fo crayfish: morphology and synaptic activity. *J. Comp. Physiol.* 80:389–407.

Williamson, C.E. and Vanderploeg, H.A. (1988) Predatory suspension-feeding in *Diaptomus*: prey defenses and the avoidance of cannibalism. *Bull. Mar. Sci.* 43:561–572.

Yen, J. and Fields, D.M. (1992) Escape responses of *Acartia hudsonica* (Copepoda) nauplii from the flow field of *Temora longicornis* (Copepoda). *Arch. Hydrobiol. Beih.* 36:123–134.

Yen, J. and Fields, D.M. (1994) Behavioral responses of *Eucheata rimana* to controlled fluid mechanical stimuli. *EOS, Trans, Am. Geophys. Union* 75:184.

Yen, J. and Strickler, J.R. (1996) Advertisement and concealment in the plankton: what makes a copepod hydrodynamically conspicuous. *Invert. Biol.* 115:191–205.

Yen, J., Lenz, P.H., Gassie, D.V. and Hartline, D.K. (1992) Mechanoreception in marine copepods: electrophysiological studies on the first antennae. *J. Plankt. Res.* 14:495–512.

Yen, J., Colin, S., Doall, M. and Strickler, J.R. (1996) Mate tracking in copepods: pheromones or species specific wakes? *EOS. Trans. Am. Geophys. Union.* 77:425–426.

Yen, J., Sanderson, B., Strickler, J.R. and Okubo, A. (1991) Feeding currents and energy dissipation by *Euchaeta rimana*, a subtropical pelagic copepod. *Limol. Oceanogr.* 36:362–369.

Zimmer-Faust, R.K., Finelli, C.M., Pentcheff, N.D. and Wethey, D.S. (1995) Odor plumes and animal navigation in turbulent flow: A field study. *Biol. Bull.* 188:111–116.

Orientation and Communication in Arthropods
ed. by M. Lehrer
© 1997 Birkhäuser Verlag Basel/Switzerland

Vibratory communication in spiders: Adaptation and compromise at many levels [*]

F. G. Barth

Biozentrum, Institut für Zoologie, Universität Wien, Althanstr. 14, A-1090 Wien, Austria

* *Dedicated to Professor Dr. Drs. h.c. H. Autrum on the occasion of his 90ᵗʰ birthday.*

Summary. Spiders are an exquisite choice of experimental animals for anyone interested in the vibrational sense and vibratory communication. For the vast majority of spiders, vibrations represent signals of overwhelming behavioral significance. The vibratory world spiders live in can only be adequately appreciated if we consider it in a broad biological context. Only then both the richness in adaptations and the diversity of selective pressures which must have been at work during evolution become apparent.

Taking the courtship behavior of *Cupiennius salei* (Ctenidae) and some of its close relatives as a representative example, this chapter illustrates some of these aspects at different levels of organization. *C. salei* is a large Central American wandering spider living on monocotyledonous plants. These plants (rather than a web) serve as transmission channels for its vibratory courtship signals. Roughly following a bottom-up approach, the vibration receptors, the vibratory signals, the neural responses to vibrations, species recognition and reproductive isolation, are related to each other. Apart from providing a sketch of the biological "design" of vibratory communication, the present chapter strives to emphasize the virtues of an organismic approach, blending reductionist experiments in the lab with observations in the field.

Introduction

Spiders are a very successful group of animals. With about 30 000 described species they are close in number to bony fishes and protists and crustaceans. In certain habitats there are armies of individuals, with corresponding ecological consequences. Fifty to 150 spiders may be found on one square meter. One ton of insects may be consumed on just 10 000 m² during one summer. England and Wales were estimated to be inhabited by about 2.2×10^9 spiders consuming 2×10^{14} insects per year (Bristowe, 1958; Kirchner, 1964). In addition to having conquered all sorts of terrestrial habitats, many spiders exploit the air by putting up their webs in order to catch flying insect prey.

Among the reasons for this evolutionary success, highly developed sensory systems take a prominent position. They provide the spiders with detailed information on what is going on around them (Barth, 1985a). In most spiders the mechanical senses are particularly important guides of behaviors like prey capture and courtship (in some spiders, however, this role is taken by their visual sense). Strain detection in the exoskeleton by slit sensilla, air movement detection by trichobothria and vibration detection are all highly developed not only with regard to the sensory periphery, but also to the central nervous system. This can be most readily seen from the spiders' behavioral performance.

This chapter is meant to familiarize the reader with the vibratory sense of spiders and its use in courtship communication. I will not so much concentrate on the "technical" view explaining details of biological "engineering" (Barth, 1985b, 1993, 1996). Instead, I will try to explain why it is rewarding to interpret vibratory communication within a broad biological framework where many factors act together to form a network of mutually dependent components, and where all adaptations are compromises coping with more than a single selective force.

Adaptation implies specialization. It narrows down the range of possible reactions, even in cases – probably the majority – where a system is not optimal, but just good enough to ensure its owner's reproductive success. To quantify "adaptation" in an ultimate sense is a task for evolutionary biology and not so much the goal of physiologists. For good reasons, physiologists tend to study systems with exaggerated traits. Such systems promise to reveal their "design" more readily than does the average result of nature's evolutionary tinkering. Even in such extreme cases, however, the "design" of a sensory system does not tell us much about fitness unless we explicitly study it. Even not so obviously streamlined systems may be the close-to-optimal result of simultaneous adaptation to many selective pressures and, in addition, they offer the advantage of preserving the possibility of further adaptation in a multitude of relationships.

Let us shortly consider such relationships relevant for vibratory communication in spiders before addressing any details. Sensory systems used in communication are related to signalling behavior. Signalling behavior is related to the signal properties and these in turn to the habitat where the signals are produced and through which they are transmitted. We must expect coupled evolution of all the above aspects. Regarding our spider case, the complexity of the communication system is emphasized by the following simple predictions.

(i) The vibratory signals sent by the courting male and female should have properties which limit signal loss by transmission to a tolerable extent under the relevant physical environmental conditions. (ii) The properties of the receiving sensory systems should cope with just these vibratory signals and be able to receive them under natural conditions, to identify them and to localize their origin. (iii) We must expect adaptive compromises for the following main reasons. (a) The vibratory sense of spiders has to deal with several types of vibrations, such as prey generated signals and courtship vibrations. (b) Vibratory courtship signals should be conspicuous for the conspecific, but cryptic for predator and prey. (c) The same signals may be used in different behavioral contexts, such as species recognition and male competition. (d) The place and time of signalling and receiving should favor effective and sufficiently undisturbed communication.

In the following, a sketch of the multifactorial reality will be drawn in which a sensory system has to work. The approach will be mainly bottom up, starting with the vibration receptors and ending with spider populations of different but closely related species. For most of the "proximate" details, the reader is referred to the extensive original literature and a recent review on the vibratory sense of spiders (Barth, 1997).

Vibration receptors

Slit sensilla

Among the several spider sensilla sensitive to substrate vibrations the metatarsal lyriform organ excels with regard to sensitivity and structural adaptation to its particular function. It is considered *the* spider vibration receptor. The metatarsal organ belongs to the slit sensilla which are kind of strain gauges built into the arachnid exoskeleton (Barth, 1985c) (Fig. 1). In the wandering spider *Cupiennius salei*, the hero of our story, there are more than 3000 such devices distributed in specific patterns over the body, in particular the extremities (Barth and Libera, 1970). The information provided by the ensemble of these mechano-electrical transducers permits the spider brain to draw a detailed picture of the mechanical events going on in its exoskeleton.

The adaptive value of the enormous variability of slit morphology can be understood as the result of its evolution under selective pressures inherent to the adequate stimulus. The key to an in-depth understanding of both the diversity and the adaptedness at the receptor level is an understanding of stimulus physics. Previous analyses demonstrated the effect of morphological features such as slit length and slit orientation (relative to the compressional lines of force/strain field, Fig. 1C) on the receptor's sensitivity. Likewise, the complex mechanics of parallel arrangements of up to c. 20 slits in lyriform organs becomes intelligible when considering its effects on absolute sensitivity, working range, and directional sensitivity (Barth et al., 1984). In selected cases, the specific position of slit sense organs on the walking leg turned out to be crucial for the differential measurement of strains due to muscle contraction force and hemolymph pressure, respectively (Fig. 1D). Slit sensilla thus impressively teach us how versatile a tiny hole or arrangement of holes in a piece of material can be to cope with the task of measuring minute displacements (strains, down to a few $\mu\varepsilon$) generated by various sources (Barth and Blickhan, 1984; Blickhan and Barth, 1985).

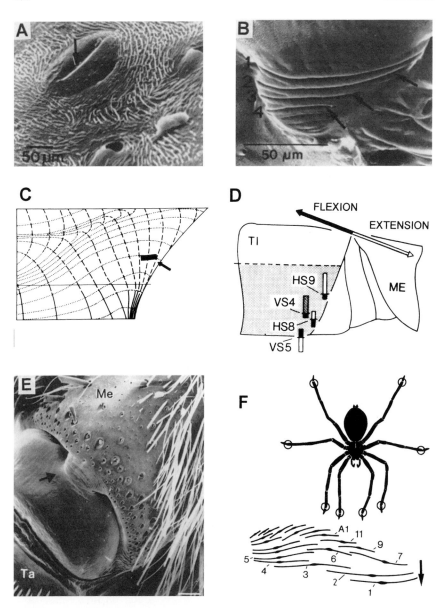

Figure 1. *Slit sensilla and metatarsal organ* (*Cupiennius salei*). (A) Single slit sensillum on opisthosoma; arrow points to dendrite attachment area. (B) Lyriform organ HS 8 on the posterior side of the distal tibia; numbers and arrows refer to particularly well studied slits (4 out of 8). (C) Lateral aspect of distal tibia with location of lyriform organ HS 8 (bar indicated by arrow); according to tension optical experiments using an araldite model under load equivalent to that generated by flexor muscle activity, the organ lies in an area where compression lines (– –) run roughly at a right angle to its long axis. Continuous lines ventrally indicate tension. (D) Strains at the site of different lyriform organs: During flexion by muscle forces, negative strains (see bars below horizontal lines indicating site of lyriform organ) implying slit compression are measured at the site of the organs laterally on the tibia (Ti; VS 4 anterior

Metatarsal lyriform organ

In contrast to the great majority of slit sensilla, the metatarsal lyriform organ does not mainly respond to proprioreceptive stimuli. It is a vibration sensitive exteroreceptor exploiting the potential contained in the *Bauplan* of a simple hole in the cuticle in a specific way.

The most obvious morphological adaptations of the metatarsal vibration receptor are its position on the walking legs and the orientation of its 21 constituent slits (Fig. 1E, F). The organ is found distally on the metatarsus and, unlike all other lyriform organs, dorsally on the leg segment. Its slits vary in length and are oriented at right angle to the long leg axis, which is another distinctive feature of this organ. The organ bridges a furrow in the metatarsal cuticle, behind a thickened area of heavily sclerotized cuticle bordering the metatarsal-tarsal-joint (Fig. 1E). The adaptive significance of these characteristics is as follows: Whenever the tarsus is moved up and down or sideways by vibrations of the substrate, the slits are stimulated by compression. The deep furrow at both lateral sides increases the organ's deformability and thus its mechanical sensitivity. In addition, the furrow focuses the compressional forces transmitted by the vibrating tarsus onto the organ.

When measuring the physiological threshold curves of individual slits of metatarsal organs (Barth and Geethabali, 1982), two features emerge which are particularly relevant in the present context: The shape of the threshold curve, and the values of absolute sensitivity.

Shape of threshold curve
The slits do not exhibit optimum curves, but rather behave like high-pass filters in the frequency range measured (0.1 Hz–3 kHz) which is believed to comprise all frequencies of biological significance. In terms of displacement, sensitivity is low up to about 10 to 40 Hz (depending on the slit examined), but increases by up to 40 dB/decade at higher frequencies, following constant acceleration. Assuming that the constancy of a stimulus parameter at threshold over a range of frequencies indicates the particular importance of this parameter in eliciting the threshold response, the particular shape of the curve suggests that the vibration receptor operates as a displacement receiver at low frequencies, and as an acceleration receiver at higher frequencies. This is exactly what one would use to measure vibra-

---

side; HS8 and HS9 posterior side) and positive strains at the site of the ventral organ VS5. During extension by hemolymph pressure, VS5 is the only one compressed. *Me* metatarsus. (E) The metatarsal organ (arrow) dorsally on the distal metatarsus; *Me* metatarsus; *Ta* tarsus; scale bar 100 µm. (F) Location of metatarsal organs on legs (circles) and arrangement of 21 slits in the organ; arrow points towards tarsus. (A–D), (F) from Barth (1985c), (E) from Barth (1986).

tions with technical transducers dealing with small acceleration values at low frequencies and with high acceleration values at high frequencies, even of low intensity.

Absolute threshold values

Sensitivity is 10^{-3} to 10^{-2} cm to about 10 to 40 Hz. It rapidly decreases at higher frequencies down to 10^{-6} to 10^{-7} cm at 1 kHz. The corresponding acceleration values are as low as 0.1 mm/s^2 in the very low frequency range (0.1–1.0 Hz) but they increase, depending on the individual slit, up to about 1 m/s^2 in the high frequency range.

Courtship vibrations on plants

The behavioral context in which the vibration receptors are of particular interest is the reciprocal vibratory communication between male and female during courtship. Our model system is a night-active neotropical wandering spider, *Cupiennius salei* (Ctenidae). This is a big Central American spider with a leg span of up to about 10 cm in both the female and the male. Like *C. getazi* and *C. coccineus*, the other representatives of the genus (Lachmuth et al., 1984) studied in the present context, *C. salei* lives on monocotyledonous plants like banana trees, agaves and bromeliads (Barth et al., 1988a). The spider's dwelling plant, rather than a web as in many other spiders, serves as the signal-transmitting medium and is thus the most relevant environmental factor in vibratory communication. Several important consequences of this fact will be discussed in the section on *Functions and adaptations*. First, we must identify the vibratory signals and their behavioral relevance (Fig. 2).

Male and female vibrations

The *male vibrations* (Schüch and Barth, 1985) come in trains (or series) of syllables (up to 50) (Fig. 2A). Syllables in turn consist of up to 12 pulses which result from oscillations of the opisthosoma and are transmitted to the plant via the legs; when oscillating, the opisthosoma does not touch the substrate (Dierkes and Barth, 1995). Silent intervals lasting for 10 s or more separate the trains of syllables. Syllable duration is about 100 ms, and the pause between two consecutive syllables lasts for about 250 ms. The frequency spectrum of the syllables shows a prominent peak around 75 to 100 Hz. There are additional vibrations due to pedipalpal scratching and drumming on the substrate. These are of short duration and contain high frequency components. Sometimes pedipalpal signals are missing altogether without any obvious effect on the success of the courtship. The opisthosomal syllables represent the only necessary and sufficient male courtship signal.

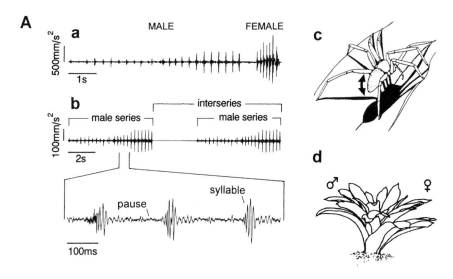

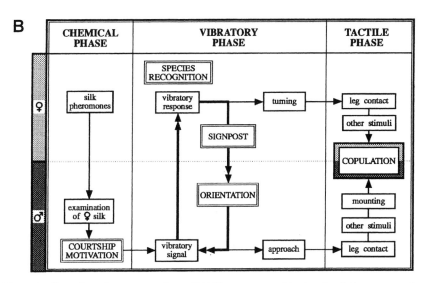

Figure 2. *Courtship vibrations on plants.* (A) Vibratory courtship signals of *Cupiennius salei* recorded by accelerometer on a bromeliad. (a) Male signal and female response; (b) temporal structure of male courtship vibration; (c) opisthosomal movements generating the male vibratory signal; (d) courtship situation on a plant. (B) The different phases of the courtship behavior of *Cupiennius salei*; heavy lines and double arrow heads in vibratory phase indicate that the respective behaviors may be repeated several times before the next stage of courtship. [A (a, b)] from Schmitt et al. (1993), [A (c, d)] and (B) from Barth (1994).

The *female response* (Fig. 2A) follows the male series within a short time interval (0.9 ± 0.5 s) (Rovner and Barth, 1981). It lasts from about 0.1 to 1.8 s, usually lacks the obviuos temporal structure of the male signal, and has its main frequency components around 20 to 40 Hz (Schüch and Barth, 1985; Baurecht and Barth, 1992).

Functions and adaptations

If we simply ask what might be good about these signals, we quickly realize that we stir up a rich mix of biological questions. A few selected answers shall illustrate this diversity.

Why vibratory signals to begin with?

Cupiennius represents a genus of night active spiders. Vibratory communication also works at night (this of course does not imply that all spiders using vibratory courtship signals are nightactive). Airborne signals are effectively produced only by large animals (for a dipole source $r_0 > \lambda/2\pi$) relative to the wavelength (λ) emitted. This is why so many arthropod singers use ultrasound and mechanisms of frequency multiplication, such as stridulation. Such limits do not exist for vibratory signals. Even very small animals may use very low frequencies. Vibratory signals, in contrast to most airborne acoustic, olfactory and visual signals, allow a kind of private communication, because signal transmission is rather limited in space (see also section on *Do signals fit their functions*), and signal presence may be kept very short. Thus, the danger of attracting potentially "listening" predators and of alarming competitors is reduced. It may add to this advantage that signal transmission is less diffuse along the leaves of a bromeliad or through the threads of a spider web than is the transmission of sound or chemical signals through the air. Considering the role that many a vibratory signal plays in the orientation of the receiver towards the source of the signal, the small amount of signal drift (compare airborne sound and chemicals in wind), and the relative unimportance of obstacles (compare visually mediated orientation) will be advantageous as well.

What functions do the courtship signals have?

The courtship of *Cupiennius* and the vibratory signals exchanged by the partners have several different functions (Fig. 2B). The male vibrations are species-specific and serve, in the first place, for the reproductive isolation of the species (Barth, 1993). This implies that it is mainly the female which must identify a partner as conspecific. The female's vibratory response informs the male about the presence of a conspecific potential mate (which would not have responded as reliably to a heterospecific male signal) (Barth and Schmitt, 1991). In addition, it serves the male to localize the female and to orient towards it. Further short-term functions of spider courtship vibrations involve the synchronization of the potential partners and the reduction of female aggressiveness, so that the male is not mistaken for prey. Sexual selection is likely to occur as well, but is hard to distinguish from signal recognition on the basis of the data available for *Cupiennius*. Long-term modulatory effects of vibratory signals on receiver motivation and behavior are also likely, but have never been studied so far.

In addition to all this, male "courtship" vibrations are used in male competition (see *Species recognition and reproductive isolation* and *Parental investment theory*).

Do the signals fit their functions?

Several parameters of the vibratory signals render them very suitable for fulfilling the functions specified above.

Low frequency. 1) With peaks of the frequency spectra at about 30 Hz (female) and 90 Hz (male), the courtship signals of *Cupiennius* are low frequency phenomena. To communicate with airborne sound at such low frequencies (wavelength several meters) would be as ineffective for a spider as it is for an insect, unless one allows for communication over very short distances and using nearfield air movement (to be picked up by hair-like structures) instead of farfield air pressure (e.g., Michelsen et al., 1987; Kirchner, 1994). Even *Cupiennius* may detect airborne nearfield vibrations in the contact phase of courtship with its trichobothria (Reißland and Görner, 1985; Barth et al., 1993, 1995). At an earlier stage of courtship, however, this would not work. A male wandering around at night in search of a female needs to cover longer distances (1 m and more) with its signals when probing a bromeliad or a banana plant for the presence of a receptive female[1]. 2) Low frequencies like those responsible for the main peaks in the spectra of the courtship vibrations are transmitted very well through the dwelling plants of *Cupiennius* (c. 0.3 dBcm^{-1}). In contrast, frequencies beyond a few hundred Hz are usually attenuated much more, one of the consequences being that vibrations due to pedipalpal drumming and scratching travel considerably shorter distances than do the courtship vibrations produced by opisthosomal bobbing (Baurecht and Barth, 1992). 3) We note that it is the female signal that contains particularly low frequencies. This makes sense considering the function of the female vibration in guiding the male to its partner (the *C. salei*-female remains stationary). The lower the frequency, the smaller signal attenuation and propagation speed. This decreased propagation speed implies larger time-of-arrival differences at the different legs of *Cupiennius* which are known

---

[1] The effect of signal spread on spider fitness is not known, like in all other cases of vibratory communication. We should be aware of our ignorance regarding questions like: What is the effect on fitness if the active space of the system is reduced by 10 or 20 or 50 cm? At the more proximate level of analysis, however, we can quantify the reach of the signals (see below).

to be used for orientation towards a nearby source of vibrations (Hergen-röder and Barth, 1983; Wirth, 1984).

Temporal pattern. The most obvious and distinctive feature of the male spider courtship vibration is its temporal pattern. 1) It sets these signals apart from both background noise and from the vibrations originating from prey animals which the spiders use to locate and roughly identify (as prey) their victim (Barth, 1985b; Barth et al., 1988b). 2) It is the temporal structure that most obviously differs among the various species of the genus *Cupien-nius* (Barth, 1993). Its significance for species recognition and reproduc-tive isolation was demonstrated experimentally by using synthetic "male" vibrations in order to quantify the female response frequency (Schüch and Barth, 1990) (Fig. 3). The durations of syllable and pause and their relation to each other (duty cycle) are strongly represented in the female releasing mechanism (and in the spike trains of the long slits of the metatarsal organ).

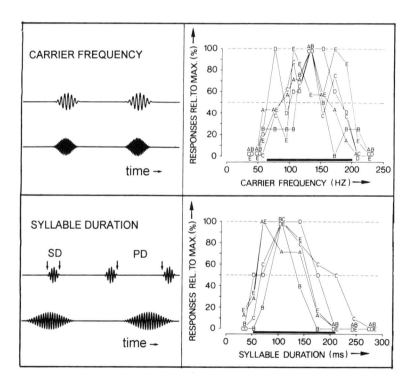

Figure 3. *Species recognition.* Responses of females to synthetic male vibrations (examples given in left panels) as a function of varied carrier frequency (above) and syllable duration (below); SD, syllable duration; PD, pause duration. Female responses (panels to the right) are given as percentages relative to maximum of response frequency. A–E symbolize values obtained from the five females tested. Bars on x-axes indicate effective ranges. From Schüch and Barth (1990).

3) Because the frequency contents of the male syllables hardly varies, propagation velocity in the plant is about the same for all syllables. As a consequence, the temporal pattern of consecutive syllables is well preserved during propagation. 4) The significance of temporal parameters has recently received further support by a study on the effect of ambient temperature on the vibratory courtship communication of *Cupiennius salei* (Shimizu and Barth, 1996). In order to compensate for temperature-dependent changes of male vibrations, the female can rely on both the adjustment of her receiving system and on temperature-invariant signal parameters. Among these, the duty cycle (i.e., the ratio between syllable duration and the duration of syllable plus pause), that remains unchanged between 13° and 34°C ambient temperature, is thought to be particularly important. 5) The occurrence of the female vibration in a narrow temporal window following the end of a series is likely to help the male recognize it as a response to his own signal. In addition, a kind of virtual synchronization of the female response will be advantageous in a noisy environment: The female often responds at the "supposed" onset time of a male syllable (had the male continued to signal after completing the series) (Shimizu and Barth, 1996). 6) Finally, the intermittent character of signalling (as opposed to continuous "calling") might make it difficult for predators to localize the source of vibrations. The actual advantage of the sender, however, has never been carefully studied and quantified.

Amplitude

The male vibration of *Cupiennius salei* reaches values up to about 1 m/s^2, the female vibration even up to about 1.6 m/s^2 (measured close to the sender). They are conducted through the plant as bending waves. Characteristic features of such bending waves are 1) that the main movement component is perpendicular to the surface, and 2) that propagation is dispersive (higher frequency components propagate faster than lower frequency components) and rather slow (from a few m/s to about 300 m/s between 30 Hz and 2 kHz on a dwelling plant of *C. salei*; Wirth, 1984; Barth, 1996).

Considering the low threshold acceleration (as small as 7 mm/s^2) of individual slits of the metatarsal vibration receptor in the range of the dominant frequency components of the female signal (Barth and Geethabali, 1982; Baurecht and Barth, 1993), the courtship vibrations of *Cupiennius* are large signals with a larger spread than one would naively expect. From the receiver's receptor thresholds (measured electrophysiologically), the values determined for signal attenuation in the transmission channel (i.e., the dwelling plants of *Cupiennius*) and the size of the sender's signal one can calculate the signal range. It is about 150 cm. The active range increases to about 200 cm due to a threshold decrease by up to c. 10 dB if one considers more naturalistic band limited noise (1/3 octave; $Q = 0.35$) instead of sinusoidal stimuli (Barth, 1985b).

How do these findings compare with real life? Taking the female vibratory response behavior as an indicator of the active space of the male signal, the present record mark measured with a male and a female (*C. coccineus*) on a large banana plant in Costa Rica is 380 cm (Barth, 1993). One of the tricks to achieve this good performance may be the convergence of the inputs from several vibration receptors in the CNS. From a neuroanatomical point of view, this possibility is quite feasible (Babu and Barth, 1989; Anton and Barth, 1993). There also exist interneurons showing such convergence (Speck-Hergenröder and Barth, 1987; Friedel and Barth, 1995), but no details on the network are available. Presumably the true distance record is still larger; so far, the size of the plants and the technical problems of monitoring freely moving courting spiders on them did not allow to test still longer distances.

Obviously there must be a lower limit of the male signal amplitude in order to get a female behavioral response. The amplitudes found must have evolved as a compromise response to a number of selective pressures: female vibration receptor threshold and central nervous mechanisms related to vibration sensitivity; distance to receiver over which communication should work (which in turn will depend on the population structure, plant community structure etc.); signal attenuation by the transmission channel (type and geometry of plant); background noise in the field.

The upper limit of vibration strength is likely to be determined by many different factors as well: energetic constraints on the sender; the risk to alarm predators and competitors; response properties of the receiver's vibration receptors or of its CNS, or of both.

Behavioral experiments showed that in order to elicit a female response, the amplitude (A) of the male signal not only has to be above a threshold of about 8 mm/s^2 (Schüch and Barth, 1990), but also below an upper limit which is indeed the limit of the natural range of signal strength (about 1000 mm/s^2). The enormous "working range" of the female reflects the variability of signal strength it is exposed to under natural conditions, and contrasts the considerably narrower filters in her releasing mechanism for signal parameters other than amplitude (A), such as syllable duration (SD) and pause duration (PD), and duty cycle (DC) (Fig. 3; the corresponding Q_{3dB} values[2] are: A: 0.3; SD: 1.4; PD: 1.1; DC: ≈ 2.1) (Schüch and Barth, 1990). This then looks like an adaptation which makes a lot of sense. But why does the female not respond to higher than normal vibration amplitudes? The answer to this question will be given in the next section and will remind us how much of filtering may be going on in the sensory periphery.

---

[2] This value is the dimensionless raito of the maximally effective value of the parameter considered and the 3dB bandwidth of the corresponding effectivity curve. The Q_{3dB} values thus serve as a measure of filter steepness.

Neural responses to vibrations

Receptor level

The electrophysiological measurement of receptor thresholds using sinusoidal stimuli of different frequencies permits only crude predictions of what the response to a natural stimulus with a temporal structure like that of the male courtship vibration will be. In addition, most of the courtship does not take place at threshold. When applying natural stimuli and synthetic male signals it turns out that the female vibration receptor is fine-tuned (adapted) to some parameters of the male signal (Baurecht and Barth, 1992, 1993).

Amplitude

When receiving signals larger than those occurring under natural conditions, the slits of the metatarsal organ no longer provide the female with precise information on the temporal structure of the signal (Fig. 4A). There is spike activity between the syllables. The resulting disappearance of the temporal pattern can be quantitatively expressed as a decrease in the synchronization between stimulus and response. Remarkably, the behavioral response probability of the female declines in the *same* range of amplitudes.

Parallel processing

Another feature of the metatarsal organ reflecting properties of the signals is the parallel processing of signal components. Whereas pedipalpal vibrations elicit responses from all slits examined, opisthosomal signals mainly elicit responses from long distal slits (see slits 1–3, Fig. 1F). It is these long slits which exhibit a logarithmic relationship of their response to stimulus intensity (acceleration) for as long as it increases within the natural amplitude range ($3-1000 \text{ mm/s}^2$) (Fig. 5A). Remarkably, the response decreases at still higher acceleration values. The coincidence with both the behavioral findings and the natural signal amplitude range is striking. The physiology of these slits is adapted to cover a large range of signal amplitudes (and just the right one) at the expense of the ability to discriminate signal amplitude differences. This may not be a disadvantage, because absolute signal amplitude is a relatively uncritical aspect of the male courtship signal ($Q_{3dB} \approx 0.3$) (Schüch and Barth, 1990). The response curve of the small slits (such as slit 11 in Fig. 5A) is *not* logarithmic but linear (between 10 and 10000 mm/s^2), underlining the significance of the argument (Baurecht and Barth, 1993).

Temporal pattern

A third filter property, not obvious from a reductionist threshold curve (based on sinusoidal stimuli) and indicating adaptedness of the female vibration receptor to a particular feature of the male vibration, brings us

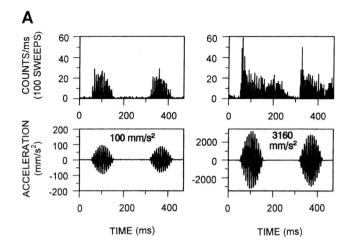

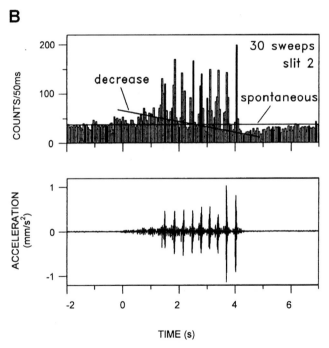

Figure 4. *Responses of the vibration receptor*. (A) Peri-stimulus time histograms (top panels) of the response of a long slit (No. 2) of the metatarsal organ to vibrations with different acceleration amplitudes (bottom panels). Best synchronization between syllables and response is at 100 mm/s^2; spontaneous activity is depressed during pause between syllables. At high acceleration amplitudes (e.g., 3160 mm/s^2), the receptor shows a prolonged activity between syllables. (B) Response of slit 2 of metatarsal organ to one male courtship series of nine syllables (below); peri-stimulus time histogram of response (above) showing the decrease of spike activity between syllables in the course of the stimulus series and post-stimulus depression following the last syllable. (A) from Baurecht and Barth (1993), (B) from Baurecht and Barth (1992).

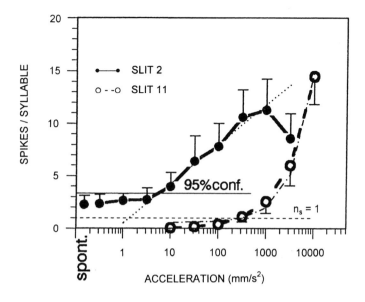

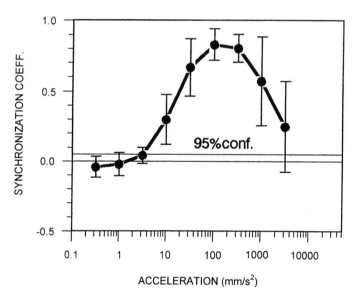

Figure 5. *Responses of the vibration receptor.* Response characteristics of slit 2 (long) and 11 (short) as a function of acceleration amplitude; n_s, numbers of spikes elicited during a syllable. 95% confidence interval of spontaneous activity indicates threshold of slit 2, whereas the mean value of one spike/syllable is regarded as the threshold of slit 11. Note the logarithmic response of slit 2 between 3 and 1000 mm/s^2, as opposed to the linear relation in the response of slit 11. The lower graph gives the mean synchronization coefficient and its standard deviation for five animals. From Baurecht and Barth (1993).

back to the temporal patterns. At least three syllables in a series are necessary to induce a female behavioral response to more than 50% of all male series; the response rate increases up to 12 syllables in the male series (Schüch and Barth, 1990). At the receptor level, the signal to noise ratio of the response is reduced with increasing syllable number by an increasingly perfect synchronization between stimulus and response which is due to a decrease of spike discharge between syllables (Baurecht and Barth, 1992). In addition, the spontaneous spike activity after the end of a series is reduced (poststimulus depression), as if to mark each package of syllables (Fig. 4B). In agreement with this finding, behavioral studies have shown that syllables are only effective if they come in series, i.e., in definite packages. A train of even 1000 consecutive syllables never elicited more than two female responses (Schüch and Barth, 1990).

Central nervous system

Parallel processing

There is evidence that the parallel processing of vibrations differing in frequency contents, already apparent in the metatarsal lyriform organ, is continued in the CNS: Opisthosomal vibrations (low frequency components) and pedipalpal vibrations (high frequency components) are represented by different types of interneurons. Interneurons representing syllables typically receive direct sensory input. They add to the frequency selectivity of the long slits in the vibration receptor (Fig. 6). Both the microstructure of the male courtship vibration (syllable structure) and its macrostructure (beginning and end of series) are copied by different types of plurisegmental interneurons (Friedel and Barth, 1995).

Threshold curves

The threshold curves of the receptor cells in the periphery and those of interneurons differ remarkably. The receptors of various spiders (leg nerve recordings in *Tegenaria* and *Zygiella*, Liesenfeld, 1961; single slit recordings in *Dolomedes triton* by Bleckmann and Barth, 1984 and in *Cupiennius salei* by Barth and Geethabali, 1982) all showed no tuning to a limited frequency range and very similar high pass characteristics (see above). In contrast, all threshold curves available for substrate vibration sensitive interneurons in the suboesophageal ganglionic mass (*C. salei*) do show a "best frequency range" (Speck-Hergenröder and Barth, 1987). The cells behave like bandpass filters with best frequencies in the low (80–100 Hz), medium (about 400 Hz), or high (about 800 Hz) frequency range, i.e., in the ranges of courtship and prey signals, respectively. We hypothesize that such filter properties participate in the recognition of different biologically relevant vibrations. The insensitivity of all of these neurons in the frequency range typical of background noise (≤ 10 Hz) supports this view

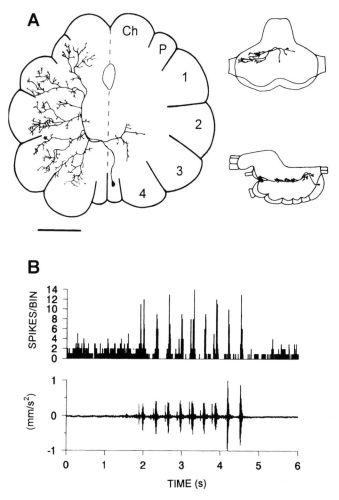

Figure 6. *Response of vibration sensitive interneuron* (bilateral, plurisegmental). (A) Dorsal, frontal, and sagittal view of cobalt-filled neuron. Ch, P, cheliceral and pedipalpal neuromer, 1–4 leg neuromeres; scale bar 500 μm. (B) Peri-stimulus time histogram of the responses (top panel) of 10 consecutive presentations of a male courtship series (bottom panel); stimulation of contralateral legs 2–4. From Friedel and Barth (1995).

and is likely to enhance the signal to noise ratio (Barth, 1985b; Barth et al., 1988b).

Species recognition and reproductive isolation

It is natural habitat, *Cupiennius* is exposed to all sorts of vibrations. These can be classified in three groups which the spiders are able to distinguish: background vibrations of abioitc origin (wind), prey vibrations, and courtship vibrations. The most distinctive features of these vibrations are a conspicuously narrow frequency spectrum with peaks below 10 Hz for the wind induced vibration, a broad-band frequency spectrum containing high frequencies (peaks between 400 and 900 Hz, walking cockroach) for moving prey insects and intermediate frequencies (*C. salei*: peaks male 75 to 100 Hz, peaks female 20 to 50 Hz) and a high temporal order (male signal) for courtship vibrations (Barth, 1985b; Barth et al., 1988b). Several arguments derived from electrophysiological and behavior experiments suggest that the spiders do indeed use these features to decide on no reaction, prey capture behavior, or courtship behavior (Barth, 1986).

In the field, *Cupiennius* will also be exposed to heterospecific courtship signals. The question of adaptation thus goes beyond the limits of one species. Courtship signals not only have to be distinguished from background noise and prey generated vibrations, but also from heterospecific courtship vibrations. At this point, even a physiologist must be prepared to address problems of taxonomy and ecology in order to know which species there are and which ones are likely to come across each other. When revising the genus *Cupiennius*, we boiled down the number of nominal species from 21 to seven, including a new species. Together with a second new species, *C. remedius* (Barth and Cordes, 1997) and a rediscovered species in South America (Brescovit and v. Eickstedt, 1995), nine known species currently make up the genus. Fieldwork in Central America has led to a reasonably detailed picture of the geographical distribution of the different species of *Cupiennius* and of the population structure and density of the three large species, *C. salei, C. getazi*, and *C. coccineus* (Barth et al., 1988a; Lachmuth et al., 1984; Schuster et al., 1994). The ranges of *C. salei* do not overlap with those of the other two species, whereas *C. coccineus* and *C. getazi* occur sympatrically.

A series of filters

Species recognition and reproductive isolation do not rely on vibratory signals alone. Although these are particularly important, the story is more complicated (Barth and Schmitt, 1991). Species recognition and discrimination is a multistage process with three principal phases named after the sensory system most obviously involved (Fig. 2B, Tab. 1). *(i) Chemical phase*: When moving about, females leave pheromone-laden draglines on the plant. Upon contact with female silk, the male gets aroused and initiates vibratory communication. (ii) *Vibratory phase*: The male vibration travels through the plant and reaches the female which responds with her own vibration. Reciprocal vibratory communication may go on for 1 h or more. During that time the partners approach each other; in *C. salei* the female remains stationary. (iii) *Tactile phase*: During the last of the pre-copulatory phases, the partners are in mechanical reach of each other. It forms an effective barrier in heterospecific pairings if their courtship has progressed that far.

Table 1. In *Cupiennius*, the probability $p^{(*)}$ for passing through all phases of courtship to copulation is the product of the probabilities for passing through the chemical phase (p_c, male response to female pheromones), the vibratory phase (p_v, female response to male vibrations) and the contact or tactile phase (p_t). Values are given for both conspecific and heterospecific pairings. From Barth and Schmitt (1991), modified.

Pairings Female × male	Probabilities			
	p_c ×	p_v ×	p_t =	$p^{(*)}$
C. cocc ♀ × *cocc* ♂	0.67	0.73	0.90	0.44
× *get* ♂	0.92	0.13	0.00	0.00
× *sal* ♂	0.50	0.33	0.00	0.00
C. get ♀ × *get* ♂	0.92	0.84	0.90	0.70
× *cocc* ♂	0.42	0.13	0.00	0.00
× *sal* ♂	0.20	0.00	–	0.00
C. sal ♀ × *sal* ♂	0.88	1.00	1.00	0.88
× *cocc* ♂	0.34	0.64	0.33	0.07
× *get* ♂	0.60	0.25	1.00	0.15

The effectiveness of courtship as a reproductive barrier among the species was quantified by measuring the contribution of its three components (i–iii) in the different pairings of *C. salei*, *C. getazi*, and *C. coccineus*. The probability of copulation equals the product of the probabilities of passing through each phase (Tab. 1) (Barth and Schmitt, 1991). The pheromone-laden female dragline is a rather anonymous signal for the males. Although contact with conspecific silk arouses the males significantly more often than that with heterospecific silk in most combinations, there is a large proportion (about 58%) of male responses to the female silk of the other species. Likewise, male taxis in a y-maze does not critically depend on the species specificity of the female silk. Upon contact with the conspecific female silk, the males began vibratory communication with a probability (p_c) of 0.67 (*C. coccineus*), 0.88 (*C. salei*), and 0.92 (*C. getazi*), respectively.

Judging from their own vibratory response, the females distinguish conspecific and heterospecific male vibrations very well. This is the main reason why we conclude that it is to a particularly large extent the task of the female to guarantee reproductive isolation. Whereas female *C. getazi* are the most selective (see p_v in Tab. 1), female *C. salei* are the least selective. Finally, the contact or tactile phase is never surmounted by heterospecific males if the females are *C. getazi* or *C. coccineus*, but it is a less effective barrier if the female is *C. salei* (see p_t, Tab. 1).

From taking all these probabilities together it follows that the behavioral barrier among the species alone is doing the job of species isolation. Whereas the overall probability of copulation ($p^{(*)}$, Tab. 1) is between 0.44 and 0.88 in conspecifics, it is 0.00 in four of the six heterospecific pairings, and only 0.04 and 0.15 in the remaining two. The postcopulatory barriers, however, are 100% in these cases. These findings imply that the differences in the morphologies of sexual structures, which are taken to be so important in distinguishing spider species, are not relevant in the given context and that their importance has to be reevaluated (Eberhard, 1985).

Sympatry and allopatry

What might sympatry and allopatry have to do with spider courtship vibrations? We expect that mating barriers are stronger between sympatric than between allopatric species which are geographically separated anyway. Taking courtship as a whole, reproductive isolation is indeed greater between the sympatric species, *C. coccineus* and *C. getazi*, than it is in allopatric pairings including *C. salei*. The picture becomes more complex when comparing the sexes. As expected, the females of *C. salei*, the allopatric species, are the least discriminative females. Regarding the males, *C. getazi* is the least selective species, whereas *C. salei* is as selective as *C. coccineus*. One may argue that this does not contradict theory: The female choice during the vibratory courtship phase is more important for the discrimination process than the male choice anyway.

We go on and ask why the females of *C. getazi* and *C. coccineus* are particularly unfriendly towards the males of the respective other sympatric species. And why do females of *C. salei* (allopatric species) respond more often to males of *C. coccineus* than to those of *C. getazi*, and why do females of *C. coccineus* respond to males of *C. salei* significantly more often than to males of *C. getazi*?

The mutual dislike of heterospecific but sympatric pairings may be due to character displacement by competition in the same habitat. Alternatively, it may simply indicate a small degree of phylogenetic relatedness. The preference of *C. salei* females for *C. coccineus* males (rather than for *C. getazi*) may have the same reason. The sequencing of DNA fragments of the mitochondrial 12S + 16SsrDNA genes (Huber et al., 1993; Felber, 1996) using the polymerase chain reaction ($\approx$ 684 nucleotides) and the results of a phylogenetic analysis of the data, not only support the view of a close phylogenetic relationship among *C. salei*, *C. coccineus*, and *C. getazi* and the monophyly of the genus *Cupiennius* (8 species studied), but also point to a particularly close relationship between *C. salei* and *C. coccineus*. Accordingly, our observations of courtship behavior may indeed reflect differences in phylogenetic relatedness. Character displacement still remains an alternative or additional explanation.

Time sharing

Cupiennius is a genus of wandering spiders. According to observations in the field and laboratory (Barth and Seyfarth, 1979; Seyfarth, 1980; Schmitt et al., 1990), the males are the truly wandering spiders. On average, males were 3.5 (*C. coccineus, C. getazi*) to 12.7 (*C. salei*) times more active than females. The most obvious explanation of this difference is sexually motivated searching behavior of the males which must find the female silken threads and the females themselves for reproducing. Differences in the daily activity patterns of the two sympatric species might be an additional mating barrier between them. In other words it may not only be important to know *how* they vibrate, but also *when* they vibrate.

More than 90% of all locomotor activity (males: 96%; females 91%; average of the three large species) occurs in the dark phase. Within 20 min after the onset of darkness all spiders show activity values larger than 50% of the maximum values. Light-off is an effective *Zeitgeber* which promptly activates the spiders. The absolute activity maximum occur long before the end of the dark phase. A comparison of the activity patterns of the two sympatric species shows interesting differences. (i) During the time period of absolute activity maxima of *C. getazi* males and females, there are relative minima in the activity of both males and females of *C. coccineus*. (ii) The activity period of *C. getazi* is shorter than that of *C. coccineus* males. *C. coccineus* spiders are most active when the activity of *C. getazi* is already decreasing (Fig. 7A).

Considering that the degree of overlap of activity periods determines – among other factors – the number of interspecific encounters in sympatric species, our data suggest that the activity patterns of *C. getazi* and *C. coccineus* do contribute to reproductive isolation.

Parental investment theory

Vibratory communication of spiders has to be seen in a still wider biological context if we strive to understand both the gross and more subtle differences in the behavior of the two sexes (Schmitt et al., 1994). Provided male investment in the next generation is small compared to that of the female, parental investment theory (Trivers, 1972) predicts that the males compete for females, that their threshold for sexual arousal is low, and that they are relatively indiscriminative with respect to mate choice. In contrast, females should be more selective and more able to reject heterospecific males and unsuitable conspecific partners (Thornhill, 1979; Alcock, 1989).

In *Cupiennius*, the male just inseminates the female without any further care for its offspring, whereas the female not only produces a large number of eggs but, in addition, takes care of its egg-sac for several weeks. In short: There are differences in parental investment. The consequences predicted by theory were investigated in some detail in *C. getazi* and were all confirmed (Schmitt et al., 1994). (i) The males of *C. getazi* are easily aroused. They begin vibratory courtship upon contact with conspecific female silk with a probability of 0.92. In contrast to females, they sometimes even start

A

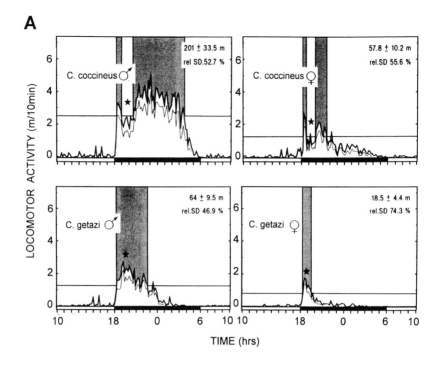

B

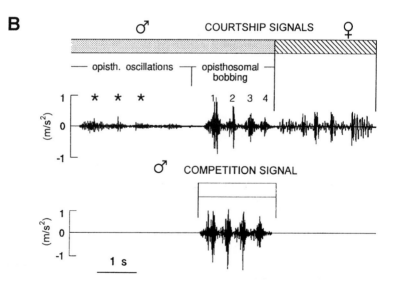

Figure 7. *Species recognition and reproductive isolation*. (A) Daily locomotor activity patterns of adult male and female spiders of *Cupiennius getazi* and *C. coccineus*. In all cases N = 10, and both the mean (thick line) and standard error (thin line; only lower limit shown) are given. The total amount of activity (m) is given by numbers in right upper corner; horizontal lines indicate 50% of maximum activity and shaded areas represent time period of maximum activity. Stars

courtship spontaneously, i.e., in the absence of any spider signals (Barth and Schmitt, 1991). (ii) Males approach displaying males and compete with them producing vibratory signals (Fig. 7B) (Schmitt et al., 1992). Male-male fights are ritualized (as opposed to female attacks on males). In the absence of a female, the outcome of fights is not correlated with age, leg length, body weight and rate of signalling. The heavier and the larger male does win, however, when a responding female is present. (iii) Males of *C. getazi*, similar to the other large species of *Cupiennius*, are less discriminative than are females. In behavioral experiments, females discriminated against vibrating heterospecific males, whereas males did not discriminate against vibrating heterospecific females (Barth and Schmitt, 1991).

C. getazi offers a splendid opportunity to compare the choosiness of the sexes and thus the distinctiveness of their releasers directly, using the same signal: Male vibrations not only induce vibratory responses of females, but also vibratory behavior in competing males. When assessing preferences of males and females to replayed and variously altered natural male signals by quantifying their propensity to respond with vibratory signals, males were found to indeed be less narrowly tuned than females to five out of the seven parameters tested (acceleration amplitude of syllables; number of syllables in a series; syllable duration; sequence duration; interseries duration) (Schmitt et al., 1992). This finding may also explain why female signals are less species-specific than are male signals.

Although these results agree with parental investment theory, they still allow two interpretations, which do not exclude each other: First, the vibratory innate releasing mechanism of the males is indeed less selective than that of the females; second, the males are simply more easily aroused sexually. Electrophysiological studies are needed to further clarify this issue.

Conclusions

From the results presented in this chapter and the gaps of knowledge illustrated by them it should be obvious that a full understanding of a communication system such as that used by the courting spiders asks for a multi-level analysis within a broad biological framework. We have to ask

---

mark time of maximum activity in *C. getazi* and time of relative minimum in *C. coccineus*; black area on x-axis indicates dark period (18.00–06.00). (B) Vibratory signals observed in *C. getazi*. Above: male vibratory courtship signal (final 5s) consisting of opisthosomal oscillations, pedipalpal drumming (stars) and opisthosomal bobbing, and female vibratory response. Below: competition signal of a male. (A) from Schmitt et al. (1990), (B) from Schmitt et al. (1992).

both proximate and ultimate questions and to combine laboratory and field work.

The amount of quantitative data and the diversity of research methods needed to at least roughly understand the tangle of relationships typical of "real life" may be discouraging, but we do depend on them. A close collaboration of evolutionary biologists and sensory biologists (which, unfortunately, is so rare) would be helpful indeed in bridging the gap between those mainly interested in describing physiological mechanisms and proximate causes at the level of the individual animal, and those judging the effectiveness of a signal system on the reproductive success at the population level. If our goal is to be able to fully appreciate the effectiveness of a communication system and its many and diverse aspects of adaptation and compromise, we must consider both levels and do our work as close to the animal's behavior in its natural environment as possible, striving for the best possible blend of reductionist science with natural history.

Acknowledgements
I am grateful to all the loyal associates who over the years have contributed important chapters to our story. The Austrian Science foundation (FWF) generously supported our work. The secretarial and technical help of M. Wieser and H. Halbritter in preparing this manuscript is much appreciated.

References

Alcock, J. (1989) *Animal Behavior*. Sinauer Associates, Sunderland, Massachusetts.

Anton, S. and Barth, F.G. (1993) Central nervous projection patterns of trichobothria and other cuticular sensilla in the wandering spider *Cupiennius salei* (Arachnida, Araneae). *Zoomorphology* 113:21–32.

Babu, K.S. and Barth, F.G. (1989) Central nervous projections of mechanoreceptors in the spider *Cupiennius salei* KEYS. *Cell Tissue Res.* 258:69–82.

Barth, F.G. (ed.) (1985a) *Neurobiology of Arachnids*. Springer-Verlag, Berlin-Heidelberg-New York-Tokyo.

Barth, F.G. (1985b) Neuroethology of the spider vibration sense. *In*: F.G. Barth (ed.): *Neurobiology of Arachnids*. Springer-Verlag, Berlin, pp 203–229.

Barth, F.G. (1985c) Slit sensilla and the measurement of cuticular strains. *In*. F.G. Barth (ed.): *Neurobiology of Arachnids*. Springer-Verlag, Berlin, pp 162–188.

Barth, F.G. (1986) Vibrationssinn und vibratorische Umwelt von Spinnen. *Naturwiss.* 73(9): 519–530.

Barth, F.G. (1993) Sensory guidance in spider pre-copulatory behavior. *Comp. Biochem. Physiol.* 104A:717–733.

Barth, F.G. (1994) Courtship and vibratory comunication in the spider *Cupiennius salei* Keys. (Ctenidae). Commentary on film C 2318 of the Austrian Institute for the Scientific Film. *Wiss. Film* 45/46:71–76.

Barth, F.G. (1997) The vibrational sense of spiders. *In*: R.R. Hoy, A.N. Popper and R.R. Fay (eds): *Comparative Hearing: Insects*. Springer Handbook of Auditory Research. Springer-Verlag New York; *in press*.

Barth, F.G., and Blickhan, R. (1984) Mechanoreception. *In*. J. Bereiter-Hahn, A.G. Matoltsy, and R. Richards (eds): *Biology of the Integument*. Springer-Verlag, Berlin, pp 554–582.

Barth, F.G. and Cordes, D. (1997) *Cupiennius remedius* new species (Araneae, Ctenidae), and a key for the genus. *J. Arachnol.*; *in press*.

Barth, F.G. and Geethabali (1982) Spider vibration receptors. Threshold curves of individual slits in the metatarsal lyriform organ. *J. Comp. Physiol. A* 148:175–185.

Barth, F.G. and Libera, W. (1970) Ein Atlas der Spaltsinnesorgane von *Cupiennius salei* Keys., Chelicerata (Araneae). *Z. Morph. Tiere* 68:343–369.

Barth, F.G. and Schmitt, A. (1991) Species recognition and species isolation in wandering spiders (*Cupiennius* spp., Ctenidae). *Behav. Ecol. Sociobiol.* 29:333–339.

Barth, F.G. and Seyfarth, E.-A. (1979) *Cupiennius salei* Keys (Araneae) in the highlands of central Guatemala. *J. Arachnol.* 7:255–263.

Barth, F.G., Ficker, E. and Federle, H.-U. (1984) Model studies on the mechanical significance of grouping in compound spider slit sensilla. *Zoomorphology* 104:204–215.

Barth, F.G., Seyfarth, E.-A., Bleckmann, H. and Schüch, W. (1988a) Spiders of the genus *Cupiennius* SIMON 1891 (Araneae, Ctenidae). I. Range distribution, dwelling plants, and climatic characteristics of the habitats. *Oecologia* 77:187–193.

Barth, F.G., Bleckmann, H., Bohnenberger, J. and Seyfarth, E.-A. (1988b) Spiders of the genus *Cupiennius* SIMON 1891 (Araneae, Ctenidae). II. On the vibratory environment of a wandering spider. *Oecologia* 77:194–201.

Barth, F.G., Wastl, U., Humphrey, J.A.C. and Devarakonda, R. (1993) Dynamics of arthropod filiform hairs. II. Mechanical properties of spider trichobothria (*Cupiennius salei* KEYS.). *Phil. Trans. R. Soc. Lond. B* 340:445–461.

Barth, F.G., Humphrey, J.A.C., Wastl, U. Halbritter, J. and Brittinger, W. (1995) Dynamics of arthropod filiform hairs. III. Flow patterns related to air movement detection in a spider (*Cupiennius salei* KEYS.). *Phil. Trans. R. Soc. Lond. B* 347:397–412.

Baurecht, D. and Barth, F.G. (1992) Vibratory communication in spiders. I. Representation of male courtship signals by female vibration receptor. *J. Comp. Physiol. A* 171:231–243.

Baurecht, D. and Barth, F.G. (1993) Vibratory communication in spiders. II. Representation of parameters contained in synthetic male courtship signals by female vibration receptor. *J. Comp. Physiol. A* 173:309–319.

Bleckmann, H. and Barth, F.G. (1984) Sensory ecology of a semiaquatic spider (*Dolomedes triton*). II. The release of predatory behavior by water surface waves. *Behav. Ecol. Sociobiol.* 14:303–312.

Blickhan, R. and Barth, F.G. (1985) Strains in the exoskeleton of spiders. *J. Comp. Physiol. A* 157:115–147.

Brescovit, A.D. and von Eickstedt, V.R.D. (1995) Occurrência de *Cupiennius* Simon na América do Sul e redescrição de *Cupiennius celerrimus* Simon (Araneae, Ctenidae). *Revta. bras. Zool.* 12(3):641–646.

Bristowe, W.S. (1958) *The World of Spiders*. Collins, London.

Dierkes, S. and Barth, F.G. (1995) Mechanism of signal production in the vibratory communication of the wandering spider *Cupiennius getazi* (Arachnida, Araneae). *J. Comp. Physiol. A* 176:31–44.

Eberhard, W.G. (1985) *Sexual Selection and Animal Genitalia*. Harvard Univ. Press, Cambridge.

Felber, R. (1996) *The phylogenetic relationships of spiders in the genus* Cupiennius *deduced from mitochondrial DNA Sequences*. Diploma Thesis, University of Vienna.

Friedel, T. and Barth, F.G. (1995) Responses of female interneurons to male courtship vibrations in a spider (*Cupiennius salei* Keys., Ctenidae). *J. Comp. Physiol. A* 177:159–171.

Hergenröder, R. and Barth, F.G. (1983) Vibratory signals and spider behavior: How do the sensory inputs from the eight legs interact in orientation? *J. Comp. Physiol. A* 152:361–371.

Huber, K.C., Haider, T.H.S., Müller, M.W. Huber, B.A., Schweyen, R.J. and Barth, F.G. (1993) DNA-sequence data indicates the polyphyly of the family Ctenidae (Araneae). *J. Arachnol.* 21:194–201.

Kirchner, W. (1964) Bisher Bekanntes über die forstliche Bedeutung der Spinnen. *Waldhygiene* 5(6/7):161–198.

Kirchner, W.H. (1994) Hearing in honeybees: the mechanical response of the bee's antenna to near field sound. *J. Comp. Physiol. A* 175:261–265.

Lachmuth, U., Grasshoff, M. and Barth, F.G. (1984) Taxonomische Revision der Gattung *Cupiennius* SIMON 1891 (Arachnida; Araneae). *Senckenbergiana biol.* 65:329–372.

Liesenfeld, F.J. (1961) Über Leistung und Sitz des Erschütterungssinnes von Netzspinnen. *Biol. Zbl.* 80:465–475.

Michelsen, A., Towne, W.F., Kirchner, W.H. and Kryger, P. (1987) The acoustic near field of a dancing honeybee. *J. Comp. Physiol. A* 161:633–643.

Reißland, A. and Görner, P. (1985) Trichobothria. *In*: F.G. Barth (ed.): *Neurobiology of Arachnids*. Springer-Verlag, Berlin, pp 138–161.

Rovner, J.S. and Barth, G.G. (1981) Vibratory communication through living plants by a tropical wandering spider. *Science* 214:464–466.

Schmitt, A., Schuster, M. and Barth, F.G. (1990) Daily locomotor activity patterns in three species of *Cupiennius* (Araneae: Ctenidae): The males are the wandering spiders. *J. Arachnol.* 18,3:249–255.

Schmitt, A., Schuster, M. and Barth, F.G. (1992) Male competition in a wandering spider (*Cupiennius getazi*, Ctenidae). *Ethology* 90:293–306.

Schmitt, A., Friedel, T. and Barth, F.G. (1993). Importance of pause between spider courtship vibrations and general problems using synthetic stimuli in behavioral studies. *J. Comp. Physiol.* A 172:707–714.

Schmitt, A., Schuster, M. and Barth, F.G. (1994) Vibratory communication in a wandering spider (*Cupiennius getazi*, Ctenidae): Female and male preferences of various features of the conspecific male's releaser. *Anim. Beh.* 48:1155–1171.

Schüch, W. and Barth, F.G. (1985) Temporal patterns in the vibratory courtship signals of the wandering spider *Cupiennius salei* KEYS. *Behav. Ecol. Sociobiol.* 16:263–271.

Schüch, W. and Barth, F.G. (1990) Vibratory communication in a spider: female responses to synthetic male vibrations. *J. Comp. Physiol.* A 166:817–826.

Schuster, M., Baurecht, D., Mitter, E., Schmitt, A. and Barth, F.G. (1994) Field observations on the population structure of three ctenid spiders (*Cupiennius* specc., Araneae). *J. Arachnol.* 22:32–38.

Seyfarth, E.-A. (1980) Daily patterns of locomotor activity in a wandering spider. *Physiol. Entomol.* 5:199–206.

Shimizu, I. and Barth, F.G. (1996) The effect of temperature on the temporal structure of the vibratory courtship signal of a spider (*Cupiennius salei* Keys.). *J. Comp. Physiol.* A 179:363–370.

Speck-Hergenröder, J. and Barth, F.G. (1987) Tuning of vibration sensitive neurons in the central nervous system of a wandering spider, *Cupiennius salei* Keys. *J. Comp. Physiol.* A 160:467–475.

Thornhill, R. (1979) Male and female sexual selection and the evolution of mating strategies in insects. *In*: M.S. Blum and N.A. Blum (eds): *Sexual Selection and Reproductive Competition in Insects*. Academic Press, London, pp 81–121.

Trivers, R.L. (1972) Parental investment and sexual selection. *In*. B. Campbell (ed.): *Sexual Selection and the Descent of Man*, 1871:1971. Aldine, Chicago, pp 136–179.

Wirth, E. (1984) *Die Bedeutung von Zeit- und Amplitudenunterschieden für die Orientierung nach vibratorischen Signalen bei Spinnen*. Diploma Thesis, J.W. Goethe-Universität, Frankfurt am Main.

Orientation and Communication in Arthropods
ed. by M. Lehrer
© 1997 Birkhäuser Verlag Basel/Switzerland

Acoustical communication in social insects

W. H. Kirchner

Universität Konstanz, Fakultät für Biologie, Postfach 5560 M657, D-78457 Konstanz, Germany

Summary. Airborne and substrate-borne sounds play a crucial role in intraspecific communication in many social insect species, and often also in interspecific information transfer. Acoustical signals are involved in a variety of social interactions and serve several functions, such as alarming or warning nestmates, recruiting nestmates to a profitable foraging site in the near or the far surrounding of the nest, or to a site within the nest where urgent activity is required. Sounds are also used for communication among members of the reproductive castes. Some sound signals enhance the effect of other, non-acoustical signals, and some have still unknown functions that remain to be unravelled. The present chapter reviews results of early and recent studies on the various mechanisms of sound production and sound perception in different social insects, on the physical parameters of the acoustical signals involved, and on the nature and the capacities of the respective sensory organs. It also describes experimental results demonstrating the various functions of acoustical communication in social insects, and discusses their evolutionary and ecological significance.

Introduction

The social way of life of termites, ants, wasps and bees adds a new level of complexity to their communication systems. Whereas solitary insects communicate mainly with potential mates or potential competitors, there is a variety of other contexts in which social insects communicate. Caste differentiation and division of labour require social coordination and information exchange within the colonies. Many species have evolved systems of alarm communication. And sophisticated systems of recruitment communication allow social insects to exploit resources fast and efficiently.

Because most social insects spend much of their time in the darkness of their nests, vision is of limited use for social information transfer. Instead, social insects use mainly pheromones and sound to communicate. Especially in ants (Hölldobler and Wilson, 1990), but also in bees (Free, 1987), pheromones serve as sources of information in various situations. However, it has become increasingly clear that, in all groups of social insects, sound and vibration are used widely for communication, particularly in situations in which such signals are superior to chemical signals for reasons that will become obvious below.

This chapter reviews results of studies on a variety of social insect species in an attempt to summarize the current state of our knowledge in this many-sided field of research. In the first and second sections, several mechanisms of signal production and signal perception, respectively, will be described. The third section will deal with the communicatory functions of sound and vibrations in various behavioural contexts. The last section discusses the ecological significance of acoustical and vibrational communication.

Mechanisms of sound production in social insects

In social insects, several different mechanisms of sound production occur. Accordingly, the physical properties of the signals differ among species.

Sound production by stridulation

Various ant species of the subfamilies Ponerinae, Nothomyrmicinae, Pseudo-myrmicinae and Myrmicinae possess specialized sound emitting structures that enable them to emit acoustical signals by stridulation (Markl, 1973; Hölldobler and Wilson, 1990). Amongst these subfamilies, stridulatory organs occur mostly in species that nest in the soil and tend to be absent in species that nest in plants, rotten logs and leaf-litter (Markl, 1973). The stridulatory organs consist of a sharp scraper that rubs against a file of transverse ridges on the surface of the preceding segment (Spangler, 1967; Markl, 1968). The anatomical locations of both structures are identical in almost all ants. The file is located on the surface of the fourth abdominal tergite, which is rubbed against a scraper on the third abdominal tergite by dorsoventral movement of the gaster. Only two exceptions are described in the literature (Hölldobler and Wilson, 1990). In *Nothomyrmecia macrops*, the scraper and the file are in the same segments as in the other ants, but on the ventral side of the abdomen. In the genus *Rhytidoponera*, there are dorsal and ventral files on the forth abdominal segment.

The stridulatory movements of ants induce either airborne sound, or substrate vibrations, or both. In the leaf-cutter ant, *Atta sexdens*, an audible airborne sound is emitted primarily by the gaster (Masters et al., 1983). In most other species, the major fraction of the energy is transmitted to the substratum and induces substrate vibrations. Stridulatory signals consist of repetitive short pulses of sound (Fig. 1). The pulse repetition rate is determined by the velocity of scraping; generally it is at least several hundred Hz. The carrier frequency is determined by mechanical properties of the ants' body, but it is generally above 5 kHz and ranges well into the ultrasonic frequencies.

Sound production by drumming

Other ant species and termites produce sound by banging or rapping body parts, such as the head or the abdomen, against the substrate. Amongst ants,

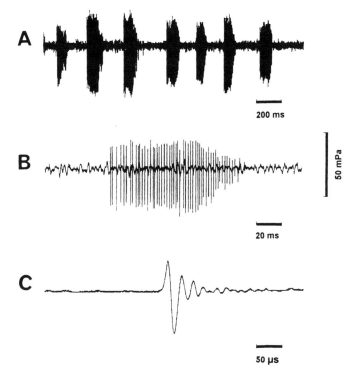

Figure 1. Stridulation signal of the ponerine ant *Megaponera foetens*. Scraping a file on the 4th abdominal tergite against a scraper on the third tergite produces bursts of airborne sound of 100 ms duration, which are repeated 3–5 times per second (A). Each single burst (B) consists of a sequence of high-frequency pulses (C) repeated at a rate of 1 kHz. Bars indicate time and intensity scales. After Hölldobler et al. (1994).

drumming is particularly widespread in arboreal species of the formicine genera *Camponotus* and *Polyrhachis*. Carpenter ants, *Camponotus herculeanum* and *Camponotus ligniperda*, drum with the mandibles and the gaster on the substrate (Markl and Fuchs, 1972; Fuchs, 1976a, b). The Dolichoderine ant *Dolichoderus thoracicus* produces sound by scraping with the mandibles (Rohe, 1991).

Many termites produce vibrations by banging their heads against the ground and ceiling of their galleries. The resulting vibrations have been studied in the North American damp wood termite *Zootermopsis* (Howse, 1964a, 1965; Stuart, 1963, 1988; Kirchner et al., 1994) and in the African fungus-growing termite *Macrotermes* (Kirchner et al., 1995). Drumming signals (Fig. 2) induce primarily substrate vibrations. However, in most cases there is also an audible airborne sound emitted by the vibrating surfaces. Similar to stridulatory sounds, drumming sounds consist of short pulses, but at lower pulse repetition rates, 20 Hz or less, determined by the

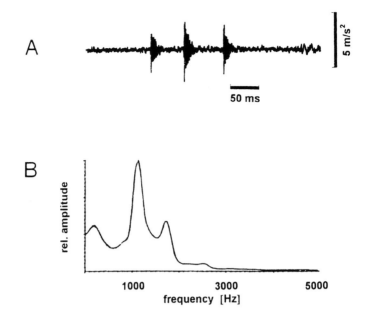

Figure 2. Drumming signal of a soldier of the termite *Zootermopsis nevadensis*. Banging the head against the wooden nest material produces series of short pulses of vibrations of the nest which are repeated at a rate of 20 Hz (A). The carrier frequency is about 1 kHz (B). Bars indicate time and intensity scales. After Kirchner et al. (1994).

frequency of drumming. The carrier frequencies lie in the range of 1 to 5 kHz, and are, in contrast to those of stridulatory signals, highly dependent on the substrate. The amplitude of substrate vibration can reach values of up to 10 m/s² (acceleration, RMS).

Airborne sound through wings movements

The wingstroke of flying insects induces airborne sound. In social insects, the carrier frequencies of these sounds, which are identical with the wing-beat frequencies, vary among species between 100 Hz and 1 kHz (Armbruster, 1922; Hansson, 1945; Abrol and Kapil, 1989; Unwin and Corbet, 1984). In honeybees there are, in addition to these low frequency sounds, spectral components in the ultrasonic frequency range (Rose et al., 1948; Spangler, 1986).

Sound emitted by the wings is used by several social insects for the purpose of communication. In honeybees, Esch (1961) and Wenner (1962a) independently discovered that acoustical signals are emitted as airborne sound during waggle dances. Sound emission is caused by dorsoventral vibrations of the wings (Michelsen et al., 1987). Vibrations of the comb on

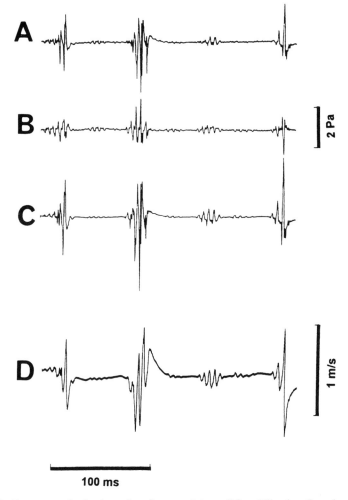

Figure 3. Dance sound of a honeybee forager, *Apis mellifera*. Vibrating the wings dorso-ventrally produces pulses of airborne sound. Simultaneous microphone recordings above (A) and below (B) the plane of the wings revealed high sound pressure gradients close to the wings (C), leading to intense air currents across the edge of the wings (D). The carrier frequency is about 250 Hz, and the pulses are repeated at a rate of 15 Hz. Bars indicate time and intensity scales. After Michelsen et al. (1987).

which the bees are dancing seem not to be induced by the dancers (Michelsen et al., 1986a). The frequency of the dance sounds ranges from 200 Hz to 300 Hz, depending on the dancer's body temperature, and on the distance and profitability of the collected food (Stabentheiner and Hagmüller, 1991; Spangler, 1991; Waddington and Kirchner, 1992). The dance sounds are strictly coupled to the tail wagging movements of the dancing bees (Esch, 1961). A temporal separation of about 0.5 s between the onset of tailwagging and the onset of the dance sounds, recently reported by Griffin and Taft (1992), has been shown to be based on a technical artefact of the recording system (D.R. Griffin, personal communication). The sounds consist of short pulses at a repetition rate of about 20 Hz. The sound pressure level, measured a few mm behind the dancer, is about 94 dB.

Whereas in normal far-field sound there is a certain fixed relationship between sound pressure and the corresponding air particle movement, these relationships are more complex in the acoustic near-field of a sound emitter. In the bee's round dances (Kirchner et al., 1988), as well as in waggle dances (Michelsen et al., 1987), air particle oscillations close to the dancing bee are about 200 times more intense than expected for the sound pressure amplitudes measured. The peak velocity of air particle movement close to a dancing bee was found to be about 1 m/s. The vibrating wings act as dipole sound emitters: sound pressure below and above the wings are 180° out of phase and the corresponding large pressure gradients around the edges of the vibrating wings cause oscillating air currents around the abdomen of the dancer, which decrease rapidly with distance from the dancing bee (Fig. 3). Similar dance sounds have been found in the Asian honeybee species, *Apis cerana* (Towne, 1985) and *Apis dorsata* (Kirchner and Dreller, 1993).

The dance language is not the only context in which honeybees emit airborne sound signals generated by wing movement. In response to disturbance, the entire colonies of Asian and western bees emit airborne sounds (Fuchs and Koeniger, 1974; Spangler, 1986), presumably aimed at vertebrate predators. Similarly, bumble bee colonies, *Bombus terrestris*, emit loud hissing sounds on disturbance of the nest (Haas, 1961; Schneider, 1972, 1975; Röschard and Kirchner, 1995).

Substrate vibration through flight muscle activity

The flight muscles are often used to induce vibration without actual wing activity. Honeybees (Michelsen et al., 1986a, b; Kirchner, 1993b), bumble bees (Röschard and Kirchner, 1995) and stingless bees (Lindauer and Kerr, 1957, 1960; Esch et al., 1965; Esch, 1967b; Kerr, 1994; Nieh and Roubik, 1995) can press the thorax against the substrate, vibrate the flight motor without any correlated movement of the wings, and transmit vibrational signals directly into the substratum. Nevertheless, there is often an audible sound emitted at the same time, produced not directly by the insect, but rather by the vibrating substrate.

As these signals are generated by the flight muscles, the carrier frequency is close to the wing beat frequency, or slightly higher. Depending on the mechanical properties of the substrate there might be some higher harmonics in the substrate vibrations and the audible sound. The temporal structure of these signals is highly variable among species and among different behavioural contexts within single species.

Other possible mechanisms of sound production

In addition to the four mechanisms of sound production described above, several authors have proposed wind instrument-like mechanisms of sound production in social insects. It has been speculated that the tracheal system and the stigmata are involved in sound production as in some solitary insects (Busnel and Dumortier, 1959; Nelson, 1979). Woods (1956) proposed that honeybee queens produce sound by pressing air through the spiracles, but Simpson (1964) has shown that this is not the case. To date, there is no evidence that any social insect uses such a mechanism to produce acoustical signals.

Mechanisms of sound perception in social insects

Because acoustical signals in social insects include either airborne sounds or substrate-borne vibrations, or both, two different mechanisms of perception must be considered, namely perception of airborne sound (hearing), and perception of substrate vibrations.

Perception of airborne sound

Until recently, all social insects have been assumed to be entirely deaf to airborne sound – or nearly so. Several early studies on sound perception in ants and honeybees led to the conclusion that these insects respond to substrate vibrations, but not to airborne sounds (Fielde and Parker, 1904; Haskins and Enzmann, 1938; von Frisch, 1923; Kröning, 1925; Hansson, 1945; Heran, 1959). However, Autrum (1936) reported that some ant species perceive airborne sound if it is of high particle velocity. In anatomical studies, a variety of sensory structures were discussed as possibly serving as ears (McIndoo, 1922; Ishay and Shimony, 1986).

The first direct evidence for a sense of hearing in social insects was provided by Towne and Kirchner (1989) who showed that honeybees can be trained to associate an airborne sound signal at a feeding site with a mild electric shock. Using an operant learning paradigm, Kirchner et al. (1991) determined the sensitivity and the frequency range of hearing in bees. They found that bees can hear sound of low pitch up to 500 Hz (Fig. 4A), and showed that the bee's "ears" are not pressure-sensitive, but rather particle-velocity sensitive, and must therefore consist of structures that can be bent or deflected by air currents, such as hairs or antennae. Dreller and

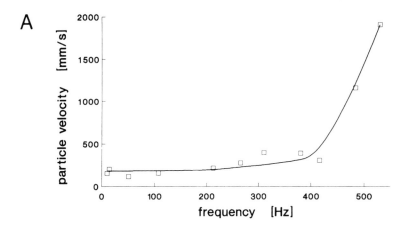

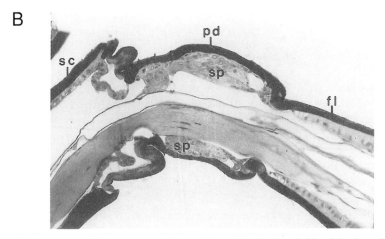

Figure 4. Perception of airborne sound in honeybees. (A) The thresholds of the honeybees' sense of hearing determined in behavioural experiments are plotted versus frequency. Bees can hear low frequency sound up to 500 Hz. (Data from Kirchner et al., 1991). (B) The auditory sense organ is Johnston's organ, a ring of mechanosensory scolopidia (sp) in the pedicel (pd) of the bee's antenna, sensing minute movements of the intersegmental membrane between the pedicel and the flagellum (fl) induced by sound deflecting the flagellum. To the left, the proximal segment of the antenna, the scapus (sc). Modified from Dreller and Kirchner (1995).

Kirchner (1993a) showed that the bee's ear is located in the distal joint of the pedicel of the antenna. Sensory hairs on the head which, according to a physiological study by Eskov (1975), respond to acoustical stimulation, are not involved. The stimulus perceived in the distal joint of the pedicel is the deflection of the antennal flagellum by the air current.

The bee's ear consists of a gorup of scolopidia, forming a ring-shaped sense organ within the pedicel, called Johnston's organ (Fig. 4B). Johnston's organ has already been shown to be used by some flies and mosquitoes to perceive airborne sound (Ewing, 1989). The sensory cells are attached to the intersegmental membrane of the joint and are thus stimulated by any movement of this membrane, induced by deflection of the flagellum. The vibrations at the tip of the antenna are about 100 times smaller than the vibrations of the air molecules, irrespective of frequency (studied up to 1000 Hz), amplitude and direction of the sound (Kirchner, 1994). On the behavioural level, bees can be trained to discriminate between sounds of different frequencies (Towne and Kirchner, 1989; Kirchner et al., 1991; Towne, 1995).

Behavioural responses to airborne sound have not only been found in the western honeybee, *Apis mellifera*, but also in two Asian species of honeybees, *Apis florea* and *Apis dorsata* (Dreller and Kirchner, 1994). To date, no social insect other than the honeybee has been shown to communicate intraspecifically through airborne sound. The significance of reactions of some ant species to intense air particle oscillations (Autrum, 1936) is unclear.

Perception of substrate vibrations

On the behavioural level, a vibration sense has been described in honeybees by Hansson (1945), Frings and Little (1957), and Abramson (1986), and quantified by Michelsen et al. (1986a). In ants, sensitivity to substrate vibrations has been investigated by Markl (1970) and Markl and Fuchs (1972), in termites by Howse (1962, 1964a, b) and Kirchner et al. (1994). Physiologically, the vibration sense was first studied in a variety of social and solitary insects by Autrum and Schneider (1948). They found vibration-correlated neural activity in the leg nerves of insects, and measured and compared thresholds among species.

It is assumed that the most sensitive vibrational sense organ of most insects is the subgenual organ, located in the tibia. Autrum (1942) proposed a theory of how the vibrational stimuli are converted into haemolymph turbulences which are then sensed by the subgenual organ. More recent studies (Markl, 1970; Menzel and Tautz, 1993; Sandeman et al., 1996) confirmed vibration sensitivity of the legs. However, the subgenual organ is not the only vibrational sense organ. In leaf-cutting ants, campaniform sensillae also respond to substrate vibrations (Markl, 1970).

The physiological as well as the behavioural studies revealed that social insects are sensitive to vibrations in a frequency range up to several kHz. However, there has been some confusion about the frequencies to which they are most sensitive. In honeybees, Autrum and Schneider (1948) reported that the lowest thresholds (i.e., the highest sensitivities) of some 10 nm of displacement amplitude had been found at a frequency of 2500 Hz. Frings and Little (1957) and Little (1962) found that the highest sensitivity is at 500 Hz. They induced vibrations of the combs by loud

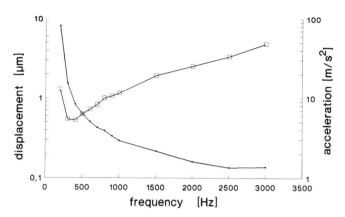

Figure 5. Perception of substrate vibration in honeybees. The thresholds of the honeybees' sense of vibration determined in behavioural experiments are plotted versus frequency. The same set of data is expressed as displacement amplitudes (solid symbols) and as acceleration amplitudes (open symbols). Data from Michelsen et al. (1986a).

airborne noise and made use of the so-called "freezing response" of worker bees (see below) to determine the frequency range to which the bees respond behaviourally. It remained unclear why the physiological threshold was lowest at 2500 Hz, but the behavioural one at 500 Hz, until Michelsen et al. (1986a) showed that both results were correct. If displacement amplitudes are considered, the threshold is lowest at high frequencies, but if thresholds are expressed as acceleration of the substrate – which is proportional to the sound pressure used by Frings and Little (1957) as a measure of intensity – the best frequency is at 300–400 Hz (Fig. 5).

The functions of acoustical signals in social insects

Acoustical signals serve a variety of functions in social insects. Some of them are non-communicatory but nevertheless adaptive for single individuals. Other serve in intraspecific communication to transmit alarm or recruit nestmates. In ants and bees, some acoustical signals are produced by reproductive individuals. Sound signals can also modulate the response to pheromones and might synchronize colony activity. Finally, acoustical signals are involved in interspecific interactions.

Non-communicatory functions of sound production

Bumble bees emit intense buzzes of sound when they forage on certain pollen providing plants, especially on solanaceans, but also on some rosaceans. The pollen freed from the anthers by vibration is then collected, at the same time pollinating the flower (Buchmann 1983, 1985, 1986; King, 1993; King and Lengoc, 1993). Buzz pollination by bumble bees is increasingly used commercially in greenhouses for pollination of tomatoes.

In the leaf-cutter ant *Atta cephalotes*, it has recently been suggested that stridulation might enhance the efficiency of leaf cutting by a mechanism similar to that technically exploited in vibratomes and in electric kitchen knifes (Tautz et al., 1995).

Acoustical alarm communication

In the leaf cutter ant genus *Atta*, stridulatory signals are part of a rescue alarm system (Markl, 1965, 1967, 1968). The signals are emitted when part of the colony is buried by a cave-in of the nest. In response to the signals, which are transmitted as vibrations of the soil, medium-size workers start to dig and search for their nestmates. A similar system has been proposed for *Pogonomyrmex badius* (Wilson, 1971).

Vibrational alarm communication seems to be widespread among termites. Conspicuous mechanical signals are produced by workers and soldiers in many termite species by drumming the head against the substratum. This behaviour, described by König as early as 1779, can be elicited by almost any disturbance of the termite colony such as sudden bright light, air currents, water, or strong vibrations. Emerson and Simpson (1929) and Howse (1964a) argued that drumming serves as an alarm signal, but Stuart (1963) pointed out that the evidence for this function is weak. The only termite genus in which vibrational communication has been studied in some detail is the New World damp-wood termite *Zootermopsis*, which lives in rotten wood in temperate climates (Howse 1964a, 1964b; Stuart 1963, 1988). Workers and soldiers react to disturbances by drumming their heads against the substratum. Kirchner et al. (1994) examined the behavioural responses of these termites to substrate vibrations and found that the animals respond preferentially to temporal patterns similar to those of the natural conspecific signals, indicating that drumming serves for intraspecific communication.

The only other termite genus in which the alarm function of vibrational communication has been demonstrated is the African fungus-growing termite genus *Macrotermes* (Kirchner et al., 1995). In this higher termite genus, vibrational alarm signals are transmitted within the nest and the gallery system over distances of many meters, because the soldiers respond by drumming not only to disturbance, but also to the drumming signals of nearby soldiers, thereby amplifying the intensity of the signals. Whereas the signal of a single soldier can be perceived only up to a distance of about 20 cm, this chain reaction ensures that alarm can reach remote areas of the nest.

Similar systems of alarm communication might occur in other termite species that have not been investigated so far.

Acoustical recruitment communication

Recruitment in social insects has been defined by Wilson (1971) as any communication that brings nestmates to some point in space where work is required. Many social insect species recruit nestmates to new nesting sites or food sources by means of pheromones, tactile interactions or acoustical signals.

Some species have been shown to use acoustical signals to recruit workers for tasks to be performed inside the nest. For example, in the hornets *Vespa crabro* (Gontarski, 1941) and *Vespa orientalis* (Ishay and Landau, 1972), hungry larvae scrape their mandibles against the comb wall, inducing vibrations of the comb that recruit workers to the task of feeding the larvae. There are some other acoustical emissions of wasps described in the literature (Esch, 1971; Ishay and Schwarz, 1965; Ishay and Nachshen, 1975; Ishay, 1977; Gamboa and Dew, 1981; Ishay and Sadeh, 1982). However, the above-mentioned "hunger signals" are the only ones in which a functional significance is obvious. Another example of recruitment within the nest, namely the rescue behaviour of the leaf cutter ant *Atta* (Markl, 1967, 1968), has already been mentioned above.

However, the best investigated recruitment communication involves recruitment to foraging sites. Leaf cutter ants stridulate frequently during leaf cutting. As mentioned previously, it has been proposed that stridulation enhances the efficiency of the cutting process (Tautz et al., 1995). Roces et al. (1993) and Roces and Hölldobler (1996), however, demonstrated that the main function of stridulation at food sources is communicatory. The stridulation attracts conspecifics to the sites where their nestmates are cutting leaves. Similarly, ants of the closely related genera *Novomessor* (Markl and Hölldobler, 1978) and *Messor* (Baroni-Urbani et al., 1988) enhance recruitment of nestmates to prey items through stridulatory vibration signals.

Stingless bees (Meliponini) recruit nestmates to food sources by a variety of mechanisms (Lindauer and Kerr, 1958, 1960). In several species, foragers produce sound signals upon returning to the nest. Esch et al. (1965) and Nieh (1996) report that the duration of the sound pulses is correlated with the distance of the feeding site in bees of the genus *Melipona*. Nieh (1996) reports, in addition, that the duration of the interval between sound pulses is correlated with the elevation of the feeding site above the ground. It therefore seems possible that some stingless bee species inform their nestmates about the location of food sources using acoustical signals, as do honeybees. However, so far there is no direct support of this view.

Acoustical recruitment communication in honeybees

Much of our knowledge on recruitment in social insects is based on extensive work on honeybees, *Apis mellifera*. This work was pioneered by Karl

dance signal (airborne sound)

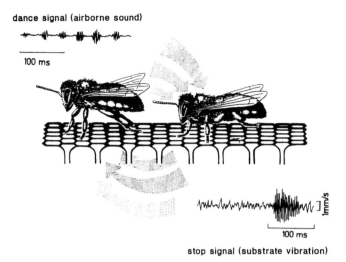

100 ms

1mm/s

100 ms

stop signal (substrate vibration)

Figure 6. The honeybees' dance language is an acoustical communication system. Dancing foragers emit airborne sound signals by dorsoventral oscillations of the wings. These airborne dance signals inform the nestmates about the location of profitable food sources. Recruits produce substrate-borne vibrations, which propagate through the wax comb. These substrate-borne stop signals induce the dancer to stop dancing and offer small samples of food to the recruits. After Kirchner (1993a).

von Frisch who discovered that honeybee foragers perform, upon return from a profitable food source, dances informing their nestmates of the distance and direction of the feeding site (reviewed by von Frisch, 1967). For a long time it remained unclear how the bee ballet is perceived by the dance attenders in the darkness of the hive. Today we know that airborne sound signals emitted during dancing are critical in conveying the information about the location of the food source (Fig. 6). During the straight tail-wagging run, the dancer emits a sound signal (Esch, 1961; Wenner, 1962a). The duration of the sound associated with the wagging run indicates the distance of the food source (Esch, 1964), and the orientation of the dancer's body while emitting the sound indicates the direction to the food (Kirchner et al., 1988). Experiments using an artificial robot dancer (Michelsen et al., 1989, 1992), or wing-mutant bees that produce sounds of atypical frequency and amplitude (Kirchner and Sommer, 1992), as well as experiments using bees manipulated such that they could not perceive the dance sounds (Dreller and Kirchner, 1993b), all confirmed that sounds are crucial for successful recruitment. Thus, the dance language clearly constitutes an acoustical recruitment communication system (see review by Kirchner and Towne, 1994).

Both the dance sound (Waddington and Kirchner, 1992) and the pattern of movement of the dancers (Esch, 1963; Waddington, 1982) correlate not only with the location of a feeding site, but also with the profitability of the food source. It is unclear, however, whether information on profitability is communicated to nestmates by any means other than the duration of dancing (Seeley and Towne, 1992).

The role of vibrational signals in honeybee recruitment communication

In addition to the airborne dance signals, there are also vibrational signals involved in the recruitment communication system of honeybees (Fig. 6). Esch (1964) reported that bees attending the dances of their nestmates from time to time make short squeaking sounds, which were at that time recorded as airborne sounds, but were later shown to be transmitted as vibrations of the comb (Michelsen et al., 1986a). The signals last typically for about 100 ms at a frequency of about 350 Hz and amplitudes of about 1 μm comb displacement. The emitters press the thorax against the comb and by doing so induce substrate vibrations by contraction of the wing muscles. As a response to this signal, the dancer sometimes, but not always, stops dancing and delivers small samples of the collected food to the dance attenders. The signal has therefore been termed begging signal (von Frisch, 1967) or stop signal (Gould, 1976).

Tautz (1996) reported that the recruitment success of bee dances is poorer for bees dancing on capped brood cells than for bees dancing on open empty cells. He suggested that these differences might be caused by differences in the motivation of the recruits or by differences in the transmission of vibrational signals through the comb. It is not clear yet whether differences in the transmission of the dance follower's stop signals contribute to the effect. Tautz et al. (1996) analyzed the pattern of movement of each single leg during the waggle run and found that, although individual legs perform 1–3 steps in the course of the run, for most of the time four or more legs of the dancer hold on to the comb, as opposed to the tripod gait of normally walking bees. The authors also emphasized that dancing bees move quite slowly during the waggle runs, confirming previous data by von Frisch and Jander (1957). The authors concluded that the bees are trying to maximize their stability on the comb during the waggle movements in order to improve transmission of vibrational signals from their legs to the comb, thus enhancing the recruitment effect on the dance attenders. Several earlier findings, however, argue against this hypothesis. First, emission of vibrational signals during the waggle run has, so far, not been shown to occur. Second, waggle dances are effective not only on wax combs, but also on the surface of swarms (Lindauer, 1955), where vibration would have no effect. And, finally, recruitment can be induced even by a robot dancer that does not touch the comb at all, provided that it emits the typical *airborne* wing sound (Michelsen et al., 1989, 1992). Although it cannot be ruled out that such hypothetical vibrational signals emitted by the dancers somehow enhance the recruitment effect of the dances, the simplest explanation of the findings by Tautz et al. (1996) would be that dancers are forced to maximize their stability on the comb due to the intense tail-wagging movements performed during dancing.

Nieh (1993) showed that stop signals are also emitted by so-called "tremble" dancers. Tremble dances can be observed in honeybee colonies when returning nectar foragers must wait for a very long time before they get unloaded by a food-storer bee (Seeley, 1992; Kirchner and Lindauer, 1994). The comb vibrations induced by the tremble dancers are indistinguishable in duration and frequency from those made by dance followers (Kirchner, 1993b). The tremble dance seems to have two different functions. It reduces the duration of the waggle dances of the nestmates and thus the rate of recruitment of further foragers (Kirchner, 1993b), and it recruits more bees for the task of unloading the returning foragers (Seeley et al., 1996).

Evolution of sound-based recruitment communication in honeybees

Most of the Asian honeybee species nest in the open, and thus they are, at least theoretically, able to use visual cues to perceive the dance information (Lindauer, 1956). Still, one of these open-nesting species, the giant bee *Apis dorsata*, produces dance sound similar to those of the western bee, *Apis mellifera* (Kirchner and Dreller, 1993). Two closely related species, however, the Himalayan giant honeybee, *Apis laboriosa* (Kirchner et al., 1996), and the dwarf bee, *Apis florea* (Towne, 1985), do not: they dance "silently". There is some evidence that the silent dance of *Apis florea* represents the ancestral form of dance communication (Kirchner et al., 1996). Acoustical recruitment signals probably evolved when honeybees began to dance under low light intensities. In ancestral tropical honeybees this might have at first been an adaptation to nocturnal foraging as it is found in the modern giant honeybee *Apis dorsata* (Dyer, 1985; Kirchner and Dreller, 1993). A recruitment system that works in the dark would have allowed the ancestors of the cavity-nesting honeybees species to build their nests in more and more sheltered places, as is the case in the modern tropical species *Apis cerana*. Finally, the adaptation to nesting in sheltered places, such as hollow trees, may have allowed the ancestors of the western honeybee to spread from the tropics to temperate climate zones.

The dwarf honeybee, *Apis florea*, can hear airborne sounds (Dreller and Kirchner, 1994), even though this bee does not use sound for recruitment communication. We currently do not know what this bee uses its ears for, but it seems likely that this, and perhaps other honeybee species as well, use acoustical signals not only for their dance communication, but also in other, as yet unknown behavioural contexts. An airborne sound of unknown significance has, for example, been reported by Esch (1967a) in swarming colonies. It seems possible that it serves as a take-off signal.

Acoustical communication in reproductive castes

Young queens of at least two genera of social insects communicate by means of acoustical signals, the myrmecine ant genus *Pogonomyrmex* and the honeybees, genus *Apis*. In *Pogonomyrmex*, young queens stridulate vigorously during copulation as soon as the spermatheca has been filled (Markl et al., 1977). The signal has therefore been termed a "female liberation signal" that seems to communicate the female's non-receptivity to approaching males and to make the males let the females go.

Young honeybee queens emit sounds which have been called queen piping. These signals have been known for centuries. One of the first descriptions was given by Charles Butler (1609) in his book on honeybees, "the feminine monarchie". Using musical notation, he documented two different types of vocalizations produced by bees. Huber (1792) discovered that one of these songs, called tooting, is produced by young queens which have already emerged from the wax cells, whereas the other vocalization, called quacking, is the response of other young queens which are still sitting in their cells. Since then, a number of investigations have addressed the question of the biological significance of these signals, as well as the mechanisms of signal production, transmission, perception and discrimination (Woods, 1956; Wenner, 1962b, 1964; Simpson, 1964; Simpson and Cherry, 1969; Simpson and Greenwood, 1974; Spangler, 1971; Bruisma et al., 1981; Grooters, 1987).

Honeybee queen piping is transmitted within the bees' nest as vibrations of the combs, at amplitudes of 0.1 to 1 μm displacement (Michelsen et al., 1986b). The attenuation of the signals with distance is relatively low, i.e., about 6 dB per 10 cm. The signals (Fig. 7) are more or less pure tones at low frequencies of about 400 Hz. They are produced by rapid contractions of the thoracic muscles and transmitted directly to the substratum. The wings do not vibrate. The temporal structure of the two types of vocalization is different and easily distinguishable. Tooting starts with a first syllable that lasts for more than 1 s and rises in amplitude as well as frequency at the beginning. This sound is then followed by a variable number of syllables lasting for about 250 ms each, that, again, show an initial rise in amplitude. Quacking consists of a number of syllables which are somewhat shorter, typically <200 ms duration, and which lack the initial rise in frequency and amplitude. Although frequency is generally slightly lower in quacking as compared with tooting, there is some overlap and also an age dependence of these frequencies. Young queens reply to tooting more frequently than to quacking. They distinguish the two signals mainly by making use of the differences in the temporal structure. Also worker bees react to queen piping. They immediately stop moving and freeze for the duration of the queen's song (Michelsen et al., 1986b).

The biological significance of queen piping is not yet fully understood. The tooter acquires from the quacking responses information about the

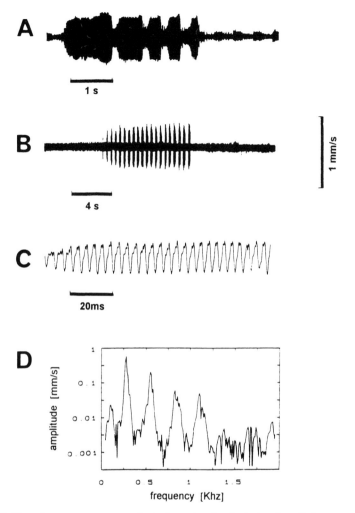

Figure 7. Honeybee queens communicate through vibratioal signals called queen piping. Tooting (A) and quacking (B) differ in their temporal structure. Tooting is characterized by an initial rise in amplitude in each syllable (C). The carrier frequency of a quacking signal (D) shows a fundamental frequency of 300 Hz and some harmonics. Modified from Michelsen et al. (1986b).

presence and location of her competitors. However, why should a young unemerged queen make any sound at all in response to the tooting signals of an emerged young queen? This queen is extremely aggressive. Once she has localized the sound emitter, she will try to open the cell and kill the occupant. Piping thus seems to constitute communication involving not only queens, but also workers. They cluster around the quacking queens' cells, chase the tooter away and seem to confine the queens in their cells.

They also feed them through small slits in the cells. After a couple of days, the tooter will either leave the colony with a second swarm or will eventually succeed in killing all of her competitors.

Although it seems plausible that the quacking signals indeed serve the unemerged queens to draw the workers' attention, this hypothesis has not been proven experimentally. Visscher (1993) has recently put forward an alternative explanation. He argues that the tooting queen might use the quacking response as a measure of the number and viability of her competitors. She might then calculate the risk of fighting and taking over the nest, and compare that with the risk of leaving with a second swarm to found a new colony. She may then stay if the response is weak, but swarm if the response is strong. This hypothesis would explain why the tooters toot and the quackers quack – however, it assumes that queens, rather than workers, decide about swarming, which is not supported by any experimental evidence. Nevertheless, this hypothesis can easily be tested by playback-experiments simulating strong quacking responses, which should, according to the hypothesis, increase the probability of secondary swarms.

There are several reports about a behaviour that was mostly referred to as "worker piping" (Armbruster, 1922; Örösi-Pal, 1932; Ohtani and Kamada, 1980; Schneider and Gary, 1984). It seems to occur rarely and under strange, often quite unnatural conditions. At present, the function of this behaviour is unknown.

Modulatory function of sound

In some species of ants, stridulation seems not to signal anything to the nestmates, unless it is accompanied by the release of chemical signals. In *Novomessor, Messor* and *Leptogenys*, stridulation enhances the effectiveness of pheromones during the recruitment of nestmates to food sources or new nest sites (Markl and Hölldobler, 1978; Maschwitz and Schönegge, 1983, Hahn and Maschwitz, 1985; Baroni-Urbani et al., 1988). In *Leptothorax muscorum*, stridulation has been observed in a variety of further contexts (Stuart and Bell, 1980), indicating again that its function is modulatory.

In honeybees, there is yet another "dance" behaviour given a number of different names: shaking dance (Gahl, 1975), dorso-ventral abdominal vibrations (DVAV) (Fletcher, 1975, 1978a, 1978b), or vibration dance (Schneider, 1987; Schneider et al., 1986a, b; Schneider and McNally, 1991). It is performed in a variety of behavioural contexts. At least when performed directly on the wax comb, it probably induces vibrations of the comb. Although the message of the DVAV to the nestmates and the channel of signal transmission are still unclear, this behaviour may serve a modulatory function similar to that of the stridulation signals mentioned above.

Acoustical signals in social insects and circadian rhythms

Gaius Plinius reported, almost 2000 years ago, that, in the bee colony, one of the bees would awaken the others in the morning by buzzing like a trumpet, like in a military camp. There is no evidence that something like that indeed occurs in honeybees, but there are more recent claims that acoustical "awakening signals" do occur in colonies of bumble bees (Haas, 1961) and wasps (Ishay, 1977). Still, honeybees display circadian changes in the overall level of substrate vibration in the combs, being higher during the day than at night, and affected by the intensity of foraging and recruitment (Minnemann, 1975; Kirchner, 1987). There are some indications that substrate vibrations act as a social synchronizer in the circadian system of the honeybee (Lindauer 1985; Kirchner, 1987). Minnemann (1975) discusses the possibility that they might also be part of a recruitment system that activates foraging in experienced forager bees. Such an effect might also explain the observation reported by Tautz (1996) that the rate of recruitment depends on the mechanical properties of the combs.

Interspecific acoustical communication in social insects

Acoustical signals are also used in interspecific interactions (Masters, 1979, 1980) in which social insects can act as senders as well as receivers. Some social insects emit acoustical warning signals aimed at potential predators. On the other hand, at least one pedatory animal, the antlion larva, exploits vibrations generated by ants. In some cases social insects perceive and respond behaviourally to heterospecific sound signals.

The African ponerine ant *Megaponera foetens* conducts well organized group raids on termites (Longhurst et al., 1978; Longhurst and Howse, 1979). During these raids the ants can often be heard to stridulate. It had been speculated that these signals are involved in maintaining group cohesiveness by transmitting either substrate borne or airborne sound from ant to ant. However, a recent study (Hölldobler et al., 1994) showed that the ants react neither to airborne sound nor to substrate vibrations, suggesting that the signals are directed towards another species. The signals are produced when the ants are disturbed. Experimentally, weak stridulation responses could be elicited by strong vibrations and strong air currents, and strong stridulatory responses were induced by just breathing against the columns of ants. Furthermore, the response to air currents was much stronger with air that was artificially enriched with CO_2. This finding strongly suggests that stridulation is normally a response to vertebrate predators and is used as an aposematic warning signal.

Similarly, bumble bees, genus *Bombus*, respond to disturbance at their nests by a conspicuous hissing sound (Schneider, 1972, 1975; Röschard and Kirchner, 1995). This hissing can be elicited best by vibrations and by breathing against the nest, indicating that it serves as a warning signal aimed at vertebrate predators, especially rodents, which often live in the very same system of burrows and tunnels in the soil as do bumble bees. Similar to *Megaponera, Bombus terrestris* responds most strongly to air that is enriched with CO_2. The hissing sound contains a high proportion of ultrasonic frequencies and it clearly differs from all other kinds of vocalization in bumble bees. Mice (*Mus domesticus*) react strongly to playback of the bumble bees' signals and escape immediately (Röschard and Kirchner, 1995). This finding, as well as the lack of evidence for any intraspecific significance of the hissing sounds, strongly suggest that they are aposematic warning signals as in *Megaponera*.

Honeybee colonies emit hissing sounds on disturbance at the nest. These sounds are audible in western bees, but are much more intense in the Asian honeybee species *Apis cerana* and *Apis dorsata* (Fuchs and Koeniger, 1974; Koeniger and Fuchs, 1972, 1975; Schneider and Kloft, 1971). Like the aposematic sounds of *Megaponera* and *Bombus*, they have a high intensity in the ultrasonic frequency range (Spangler, 1986).

The alarm sounds made by termites on disturbance of the colony might also have some effect on potential predators.

Sound emissions of social insects can also be exploited by heterospecific animals. The antlion larva (*Myrmeleon formicarius*), a neuropteran insect, digs pits in sandy soil and waits for prey at the bottom of the pit (Devetak, 1985). As soon as a prey animal is detected, the predator throws sand at the victim. More than half of the prey are ants. The antlion larva is extremely sensitive to vibrations of the ground generated by the prey; an ant of the species *Formica rufa*, for example, can be detected from a distance of about 6 cm.

Finally, some sound signals of heterospecific animals are aimed at social insects and modify the latter's behaviour. For example, the larvae of myrmecophilous riodinid and lycaenid butterflies emit stridulatory signals that enhance the recruitment of ants towards these larvae which produce secretions on which the ants feed (DeVries 1990, 1992; DeVries et al., 1993; Schurian and Fiedler, 1991; Wieland, 1995). The death head hawk moth *Acherontia atropos* emits sound signals by a unique mechanism (Busnel and Dumortier, 1959). These sounds have been reported to be quite similar to the piping sounds of honeybee queens. As this species of hawk moth does not collect nectar at flowers but rather honey in honeybee colonies, it had been speculated that the acoustical emissions might mimic a honeybee queen and somehow appease the workers bees (Huber, 1792). However, recently it has been shown that *Acherontia* never uses these sounds within the beehive, but rather exclusively on disturbance by potential predators during daytime. Within the bees' nest this moth is camouflaged not by beelike sounds, but by volatile chemicals similar to those emitted by bees (Moritz et al., 1991).

Sensory ecology of acoustical communication systems in social insects

We have seen that acoustical signals are used by a variety of social insects in various behavioural contexts. Most of these acoustical communication systems use substrate-borne vibrations. Airborne sound is significant in only two cases, the dance language of honeybees, and aposematic signals aimed at vertebrate predators. The latter are not restricted to social insects. A variety of solitary insects and other animals, including snakes and several bird and mammal species, produce similarly broad-banded high frequency sounds upon disturbance (Masters, 1979, 1980). The acoustical warning signals of social hymenopterans and poisonous snakes seem to have evolved as a Muellerian mimicry system. Some other hissing animals, such as birds, mammals and solitary insects may be considered as Batesian imitators, exploiting the former system. As the signals are aimed at vertebrates, which are generally much less sensitive to substrate vibrations than are insects, but usually have a well developed sense of hearing, airborne sound is far better suited for the purpose than are vibrations.

In the case of the honeybees' dance language, there are two possible reasons for the use of airborne sound, rather than substrate vibrations. One is the extremely short reach of the signals. Only a few bees in the immediate vicinity of the dancer can perceive the message of the dance. As there are sometimes 50 or more bees dancing at the same time and advertising different feeding sites, it might have been necessary to restrict the reach of the signals. Airborne near-field sound might therefore be the most suitable channel for information transfer. The second reason is that the dances are used not only within the bees' nest to advertise food sources, but also on the surface of clustering swarms to advertise new nesting sites (von Frisch, 1967). In the absence of a substrate suitable for transmitting vibrations, acoustical communication is restricted to airborne sound. The substrate-borne stop signals, on the other hand, are not required in swarming colonies, as the nest-site dancers have no food to offer. Nor is there a need for tremble dancing in swarming bees.

All of the other acoustical signals found in social insects are broadcast through substrate vibrations. In some cases, it seems obvious that substrate vibrations are the best or even the only possible channel of signal transmission, as, for example, in the case of a group of leaf cutter ants which is buried in the ground (Markl, 1983). However, there are other cases in which pheromones might do as well as vibrational communication. Social insects use pheromones as a source of information in many situations (Hölldobler and Wilson, 1990; Free, 1987). Alarm communication, for example, is pheromonal rather than vibrational in many species. In ants, there seems to be a tendency to use vibrations in species that live in the soil, but pheromones in species that nest in plants (Markl, 1973). The fragile comb structures of wasps and bees seem to be an ideal ground on which vibrational communication could evolve. Vibrations of the nest restrict signals to the privacy of the colony, reducing the risk of being exploited by predators.

The privacy in which intraspecific communication occurs might also be a reason why the significance of acoustical signals in social insects has been underestimated. Wilson (1975), for example, compared the relative importance of sensory channels in a variety of organisms. He classified social insects (like moths and microorganisms) as species that rely nearly exclusively on their chemical senses. In recent years it became increasingly clear that vibrations and sound are widely used in social insects' communication systems. However, the acoustical signals of social insects are, except for those that are defensive sounds, not meant for human listeners. They are subtle signals, not easy to detect – neither by predators, nor by scientists.

References

Abramson, C.I. (1986) Aversive conditioning in honey bees (*Apis mellifera*). *J. Comp. Psychol.* 100:108–116.

Abrol, D.P. and Kapil. R.P. (1989) Morphometrics and wing stroke frequency of some bees. *Proc. Indian Nat. Sci. Acad.* 55:369–376.

Armbruster, L. (1922) Über Bienentöne, Bienensprache und Bienengehör. *Archiv für Bienenkunde* 4:221–259.

Autrum, H. (1936) Über Lautäusserungen und Schallwahrnehmung bei Arthropoden. I. Untersuchungen an Ameisen. Eine allgemeine Theorie der Schallwahrnehmung bei Arthropoden. *Z. vergl. Physiol.* 23:332–373.

Autrum, H. (1942) Schallempfindung bei Mensch und Tier. *Naturwissens.* 30:69.

Autrum, H. and Schneider, W. (1948) Vergleichende Untersuchungen über den Erschütterungssinn der Insekten. *Z. vergl. Physiol.* 31:77–88.

Baroni-Urbani, C., Buser, M.W. and Schillinger, E. (1988) Substrate vibration during recruitment in ant social organization. *Ins. Soc.* 35:241–250.

Bruinsma, O., Kruijt, J.P. and Dusseldorp, W. van (1981) Delay of emergence of honey bee queens in response to tooting sounds. *Proc. Kon. Ned. Akad. Wet. C* 84:381–387.

Buchmann, S.L. (1983) Buzz pollination in angiosperms. *In*: C.E. Jones and R.J. Little (eds): *Handbook of experimental pollination biology*. Van Nostrand Reinhold, New York, pp 73–114.

Buchmann, S.L. (1985) Bees use vibration to aid pollen collection from non-poricidal flowers. *J. Kans. Entomol. Soc.* 58:517–525.

Buchmann, S.L. (1986) Vibratile pollination in *Solanum* and *Lycopersicon*. A look at pollen chemistry. *In*: W.G. D'Arcy (ed.): *Solanaceae: biology and systematics*. Columbia University Press, New York, pp 237–252.

Busnel, R.G. and Dumortier, B. (1959) Vérification par des méthodes d'analyse acoustique des hypothèses sur l'origine du cri du sphinx *Acherontia atropos* (Linné). *Bull. Soc. Entomol. France* 64:44–59.

Butler, C. (1609) *The feminine monarchie*. Barnes, Oxford.

Devetak, D. (1985) Detection of substrate vibrations in the antilion larva, *Myrmeleon formicarius* (Neuroptera: Myrmeleonidae). *Biol. Vestn.* 33:11–22.

DeVries, P.J. (1990) Enhancement of symbiosis between butterfly caterpillars and ants by vibrational communication. *Science* 248:1104–1106.

DeVries, P.J. (1992) Singing caterpillars, ants and symbiosis. *Sci. Am.* 268(10):56–62.

DeVries, P.J., Thomas, J.A. and Cocroft, R. (1993) A comparison of acoustical signals between Maculinea butterfly caterpillars and their obligate host ant species. *Biol. J. Linn. Soc.* 49:229–238.

Dreller, C. and Kirchner, W.H. (1993a) Hearing in honeybees: localization of the auditory sense organ. *J. Comp. Physiol. A* 173:275–279.

Dreller, C. and Kirchner, W.H. (1993b) How bees perceive the information of the dance language. *Naturwissens.* 80:319–321.

Dreller, C. and Kirchner, W.H. (1994) Hearing in the Asian honeybees, *Apis dorsata* and *Apis
florea. Ins. Soc.* 41:291–299.

Dreller, C. and Kirchner, W.H. (1995) The sense of hearing in honeybees. *Bee World* 76:6–17.

Dyer, F.C. (1985) Nocturnal orientation by the Asian honey bee, *Apis dorsata. Anim. Behav.*
33:769–774.

Emerson, A.E. and Simpson, R.C. (1929) Apparatus for the detection of substratum communication among termites. *Science* 69:648–649.

Esch, H. (1961) Über die Schallerzeugung beim Werbetanz der Honigbiene. *Z. vergl. Physiol.*
45:1–11.

Esch, H. (1963) Über die Auswirkung der Futterplatzqualität auf die Schallerzeugung im
Werbetanz der Honigbiene (*Apis mellifica*). *Zool. Anz. (Suppl.)* 26:302–309.

Esch, H. (1964) Beiträge zum Problem der Entfernungsmessung in den Schwänzeltänzen der
Honigbiene: *Z. vergl. Physiol.* 48:534–546.

Esch, H. (1967a) The sounds produced by swarming honey bees. *Z. vergl. Physiol.* 56:
408–411.

Esch, H. (1967b) Die Bedeutung der Lauterzeugung für die Verständigung der stachellosen
Bienen. *Z. vergl. Physiol.* 56:199–220.

Esch, H. (1971) Wagging movements in the wasp *Polistes versicolor vulgaris* Bequaert.
Z. vergl. Physiol. 72:221–225.

Esch, H., Esch, I. and Kerr, W.E. (1965) Sound: an element common to communication of
stingless bees and to dances of the honey bee. *Science* 149:320–321.

Eskov, E.K. (1975) Phonoreceptors of honey bees (in Russian). *Biofizika* 20:646–651.

Ewing, A.W. (1989) *Arthropod bioacoustics.* Comstock, Ithaca.

Fielde, A.M. and Parker, G.H. (1904) The reactions of ants to material vibrations. *Proc. Acad.
Nat. Sci. Philadelphia* 56:642–650.

Fletcher, D.J.C. (1975) Significance of dorsoventral abdominal vibration among honey-bees
(*Apis mellifera* L). *Nature* 256:721–723.

Fletcher, D.J.C. (1978a) The influence of vibratory dances by worker honeybees on the activity
of virgin queens. *J. Apic. Res.* 17:3–13.

Fletcher, D.J.C. (1978b) Vibration of queen cells by worker honey bees and its relation to
the issue of swarms with virgin queens. *J. Apic. Res.* 17:14–26.

Free, J.B. (1987) *Pheromones of social bees.* Cornell University Press, Ithaca, NY.

Frings, H. and Little, F. (1956) Reactions of honey bees in the hive to simple sounds. *Science*
125:122.

Frisch, K. von (1923) Über die Sprache der Bienen – eine tierpsychologische Untersuchung.
Zool. Jb. (Physiol.) 40:1–186.

Frisch, K. von (1967) *The dance language and orientation of bees.* Harvard Univ Press, Cambridge, MA.

Frisch, K. von and Jander, R. (1957) Über den Schwänzeltanz der Bienen. *Z. vergl. Physiol.*
40:239–263.

Fuchs, S. (1976a) The response to vibrations of the substrate and reactions to the specific
drumming in colonies of carpenter ants (*Camponotus*, Formicidae, Hymenoptera). *Behav.
Ecol. Sociobiol.* 1:155–184.

Fuchs, S. (1976b) An informational analysis of the alarm communication by drumming
behavior in nests of carpenter ants (*Camponotus*, Formicidae, Hymenoptera). *Behav. Ecol.
Sociobiol.* 1:315–336.

Fuchs, S. and Koeniger, N. (1974) Schallerzeugung im Dienst der Verteidigung des Bienen-
volkes (*Apis cerana* Fabr.). *Apidologie* 5:271–289.

Gahl, R.A. (1975) The shaking dance of honey bee workers: evidence for age discrimination.
Anim. Behav. 23:230–232.

Gamboa, G.J. and Dew, H.E. (1981) Intracolonial communication by body oscillations in the
paper wasp, *Polistes metricus. Ins. Soc.* 28:13–26.

Gontarski, H. (1941) Lautäusserungen bei Larven der Hornisse (*Vespa crabro*). *Natur und Volk*
71:291.

Gould, J.L. (1976) The dance-language controversy. *Quart. Rev. Biol.* 51:211–244.

Griffin, D.R. and Taft, L.D. (1992) Temporal separation of honeybee dance sounds from waggle
movements. *Anim. Behav.* 44:594–595.

Grooters, H.J. (1987) Influences of queen piping and worker behaviour on the timing of emergence of honey bee queens. *Ins. Soc.* 34:181–193.

Haas, A. (1961) Das Rätsel des Hummeltrompeters: Lichtalarm. *Z. Tierpsychol.* 18:129–138.

Hahn, M. and Maschwitz, U. (1985) Foraging strategies and recruitment behaviour in the european harvester ant *Messor rufitarsis* (F.). *Oecologia* 68:45–51.

Hansson, A. (1945) Lauterzeugung und Lautauffassungsvermögen der Bienen. *Opusc. Entomol. Suppl.* 6:1–124.

Haskins, C.P. and Enzmann, E.V. (1938) Studies of certain sociological and physiological features in the Formicidae. *Ann. New York Acad. Sci.* 37:97–162.

Heran, H. (1959) Wahrnehmung und Regelung der Flugeigengeschwindigkeit bei *Apis mellifica* L. *Z. vergl. Physiol.* 42:103–163.

Hölldobler, B. and Wilson, E.O. (1990) *The ants*. Harvard University Press, Cambridge, MA.

Hölldobler, B., Braun, U., Gronenberg, W., Kirchner, W.H. and Peeters, C. (1994) Trail communication in the ant *Megaponera foetens* (Formicidae, Ponerinae). *J. Ins. Physiol.* 40:585–593.

Howse, P.E. (1962) The perception of vibration by the subgenual organ in *Zootermopsis angusticollis* Emerson and *Periplaneta americana* L. *Experientia* 18:457–458.

Howse, P.E. (1964a) The significance of the sound produced by the termite *Zootermopsis angusticollis* (Hagen). *Anim. Behav.* 12:284–300.

Howse, P.E. (1964b) An investigation into the mode of action of the subgenual organ in the termite *Zootermopsis angusticollis* Emerson and the cockroach *Periplaneta americana* L. *J. Ins. Physiol.* 10:409–424.

Howse, P.E. (1965) On the significance of certain oscillatory movements of termites. *Ins. Soc.* 12:335–346.

Huber, F. (1792) *Nouvelles observations sur les Abeilles*. Paschoud, Geneve

Ishay, J. (1977) Acoustical communication in wasp colonies (Vespinae). *Proc. Int. Congr. Entomol.* 15:406–435.

Ishay, J. and Landau, E.M. (1972) *Vespa* larvae send out rhythmic hunger signals. *Nature* 237:286–287.

Ishay, J. and Nachshen, D. (1975) On the nature of the sounds produced within the nest of the wasp *Paravespula germanica* F. *Ins. Soc.* 22:213–218.

Ishay, J. and Sadeh, D. (1982) The sounds of honey bees and social wasps are always composed of a uniform frequency. *J. Acoust. Soc. Am.* 72:671–675.

Ishay, J. and Schwarz, J. (1965) On the nature of the sounds produced within the nest of the oriental hornet, *Vespa orientalis* F. *Ins. Soc.* 12:383–388.

Ishay, J. and Shimony, T. (1986) Tympanic organ and social wasps (Vespinae). *Mon. Zool. Ital.* 20:381–400.

Kerr, W.E. (1994) Communication among *Melipona* workers (Hymenoptera: Apidae). *J. Ins. Behav.* 7:123–128.

King, M.J. (1993) Buzz foraging mechanism of bumble bees. *J. Apic. Res.* 32:41–49.

King, M.J. and Lengoc, L. (1993) Vibratory pollen collection dynamics. *Am. Soc. Agric. Eng.* 36:135–140.

Kirchner, W.H. (1987) *Tradition im Bienenstaat – Kommunikation zwischen den Imagines und der Brut der Honigbiene durch Vibrationssignale*. PhD Thesis, Würzburg.

Kirchner, W.H. (1993a) Acoustical communication in honeybees. *Apidologie* 24:297–307.

Kirchner, W.H. (1993b) Vibrational signals in the tremble dance of the honeybee. *Behav. Ecol. Sociobiol.* 33:169–172.

Kirchner, W.H. (1994) Hearing in honeybees: the mechanical response of the bee's antenna to near field sound. *J. Comp. Physiol.* A 175:261–265.

Kirchner, W.H. and Dreller, C. (1993) Acoustical signals in the dance language of the giant honeybee, *Apis dorsata. Behav. Sociobiol.* 33:67–72.

Kirchner, W.H. and Lindauer, M. (1994) The causes of the tremble dance of the honeybee, *Apis mellifera. Behav. Ecol. Sociobiol.* 35:303–308.

Kirchner, W.H. and Sommer, K. (1992) The dance language of the honeybee mutant *diminutive wings. Behav. Ecol. Sociobiol.* 30:181–184.

Kirchner, W.H. and Towne, W.F. (1994) The sensory basis of the honeybee's dance language. *Sci. Am.* 270(6):74–80.

Kirchner, W.H., Lindauer, M. and Michelsen, A. (1988) Honeybee dance communication: Acoustical indication of direction in round dances. *Naturwissens.* 75:629–630.

Kirchner, W.H., Dreller, C. and Towne, W.F. (1991) Hearing in honeybees: operant conditioning and spontaneous reactions to airborne sound. *J. Comp. Physiol.* A 168:85–89.

Kirchner, W.H., Broecker, I. and Tautz, J. (1994) Vibrational alarm communication in the damp wood termite *Zootermopsis nevadensis*. *Physiol. Entomol.* 19:187–190

Kirchner, W.H., Korb, J. and Leuthold, R.H. (1995) Vibratorische Alarmkommunikation bei Termiten der Gattung *Macrotermes*. *Proc IUSSI Utrecht* 1995: 15.

Kirchner, W.H., Dreller, C., Grasser, A. and Baidya, D. (1996) The silent dances of the Himalayan honeybee, *Apis laboriosa*. *Apidologie* 27:331–339.

König, J.C. (1779) Naturgeschichte der sogenannten weissen Ameisen. *Beschäftigungen der Berliner Gesellschaft für Naturforschung* 4:1–28.

Koeniger, N. and Fuchs, S. (1972) Kommunikative Schallerzeugung von *Apis cerana* Fabr. im Bienenvolk. *Naturwissens.* 59:169.

Koeniger, N. and Fuchs, S. (19775) Zur Kolonieverteidigung der asiatischen Honigbienen. *Z. Tierpsychol.* 37:99–106.

Kröning, F. (1925) Über die Dressur der Biene auf Töne. *Biol. Zbl.* 45:496–507.

Lindauer, M. (1955) Schwarmbienen auf Wohnungssuche. *Z. vergl. Physiol.* 37:263–323.

Lindauer, M. (1956) Über die Verständigung bei den indischen Bienen. *Z. vergl. Physiol.* 38:521–557.

Lindauer, M. (1985) The dance language of honeybees: the history of a discovery. *Fortschr. Zool.* 31:129–140.

Lindauer, M. and Kerr, W.E. (1958) Die gegenseitige Verständigung bei den stachellosen Bienen. *Z. vergl. Physiol.* 41:405–434.

Lindauer, M. Kerr, W.E. (1960) Communication between the workers of stingless bees. *Bee World* 41:29–71.

Little, F. (1962) Reactions of the honey bee, *Apis mellifera* L., to artificial sounds and vibration of known frequencies. *Ann. Entomol. Soc. Am.* 55:82–89.

Longhurst, C. and Howse, P.E. (1979) Foraging, recruitment and emigration in *Megaponera foetens* (Fab.) (Hymenoptera: Formicidae) from the Nigerian Guinea Savanna. *Ins. Soc.* 26:204–215.

Longhurst, C., Johnson, R.A. and Wood, T.G. (1978) Predation by *Megaponera foetens* (Fabr.) (Hymenoptera: Formicidae) on termites in the Nigerian Southern Guinea Savanna. *Oecologia* 32:101–107.

Markl, H. (1965) Stridulation in leaf-cutting ants. *Science* 149:1392–1393.

Markl, H. (1967) Die Verständigung durch Stridulationssignale bei Blattschneiderameisen I. Die biologische Bedeutung der Stridulation. *Z. vergl. Physiol.* 57:299–330.

Markl, H. (1968) Die Verständigung durch Stridulationssignale bei Blattschneiderameisen II. Erzeugung und Eigenschaften der Signale. *Z. vergl. Physiol.* 60:103–150.

Markl, H. (1970) Die Verständigung durch Stridulationssignale bei Blattschneiderameisen III. Die Empfindlichkeit für Substratvibrationen. *Z. vergl. Physiol.* 69:6–37.

Markl, H. (1973) The evolution of stridulatory communication in ants. *Proc. VIIth Int. Congr. IUSSI* (London), pp 258–265.

Markl, H. (1983) Vibrational communication. *In*: F. Huber and H. Markl (eds): *Neuroethology and behavioral physiology*. Springer-Verlag, Heidelberg, pp 332–353.

Markl, H. and Fuchs, S. (1972) Klopfsignale mit Alarmfunktion bei Roßameisen (*Camponotus*, Formicidae). *Z. vergl. Physiol.* 76:204–225.

Markl, H. and Hölldobler, B. (1978) Recruitment and food-retrieving behavior in *Novomessor* (Formicidae, Hymenoptera). *Behav. Ecol. Sociobiol.* 4:183–216.

Markl, H., Hölldobler, B. and Hölldobler, T. (1977) Mating behavior and sound production in harvester ants (*Pogonomyrmex*, Formicidae). *Ins. Soc.* 24:191–212.

Maschwitz, U. and Schönegge, P. (1983) Forage communication, nest moving recruitment, and prey specialization in the oriental poneringe *Leptogenys chinesis*. *Oecologia* 57:175–182.

Masters, W.M. (1979) Insect disturbance stridulation: its defensive role. *Behav. Ecol. Sociobiol.* 5:187–200.

Masters, W.M. (1980) Insect disturbance stridulation: characterization of airborne and vibrational components of the sound. *J. Comp. Physiol.* 135:259–268.

Masters, W.M., Tautz, J., Fletcher, N.H. and Markl, H. (1983) Body vibration and sound production in an insect (*Atta sexdens*) without specialized radiating structures. *J. Comp. Physiol.* A 150:239–249.

McIndoo, N.E. (1922) The auditory sense of the honey-bee. *J. Comp. Neurol.* 34:173–198.

Menzel, J.G. and Tautz, J. (1993) Vibration sensitivity in the ant *Camponotus*. *Proc. 21st Göttingen Neurobiology Conference*. Thieme, Stuttgart, New York, p 229.

Michelsen, A., Kirchner, W.H. and Lindauer, M. (1986a) Sound and vibrational signals in the dance language of the honeybee, *Apis mellifera. Behav. Ecol. Sociobiol.* 18:207–212.

Michelsen, A., Kirchner, W.H., Andersen, B.B. and Lindauer, M. (1986b) The tooting and quacking vibration signals of honeybee queens: a quantitative analysis. *J. Comp. Physiol.* A 158:605–611.

Michelsen, A., Towne, W.F., Kirchner, W.H. and Kryger, P. (1987) The acoustic near field of a dancing honeybee. *J. Comp. Physiol.* A 161:633–643.

Michelsen, A., Andersen, B.B., Kirchner, W.H. and Lindauer, M. (1989) Honeybees can be recruited by a mechanical model of a dancing bee. *Naturwissens.* 76:77–280.

Michelsen, A., Andersen, B.B., Storm, J., Kirchner, W.H. and Lindauer, M. (1992) How honeybees perceive communication dances, studied by means of a mechanical model. *Behav. Ecol. Sociobiol.* 30:143–150.

Minnemann, D. (1975) Experimentelle Untersuchungen über die kommunikative Bedeutung der Substratvibration bei *Apis mellifera* L. *forma et functio* 8:9–18.

Moritz, R.F.A., Kirchner, W.H. and Crewe, R.M. (1991) Chemical camouflage of the death's head hawk moth (*Acherontia atropos* L.) in honeybee colonies. *Naturwissens.* 78:179–182.

Nelson, M.C. (1979) Sound production in the cockroach, *Gromphadorhina portentosa*: the sound-producing apparatus. *J. Comp. Physiol.* 132:27–38.

Nieh, J.C. (1993) The stop signal of honey bees: reconsidering its message. *Behav. Ecol. Sociobiol.* 33:51–56.

Nieh, J.C. (1996) A stingless bee, *Melipona panamica*, may use sounds to communicate the distance and canopy height of a food source. *Proc. 10th. Int. Meeting on Insect Sound and Vibration, Woods Hole*.

Nieh, J.C. and Roubik, D.W. (1995) A stingless bee (*Melipona panamica*) indicates food location without using a scent trail. *Behav. Ecol. Sociobiol.* 37:63–70.

Örösi-Pal, Z. (1932) Wie tütet die Arbeitsbiene? *Zool. Anz.* 98:147–148.

Ohtani, T. and Kamada, T. (1980) Worker piping: The piping sounds produced by laying and guarding worker honeybees. *J. Apic. Res.* 19:154–163.

Plinius, G.S. (1875) *Historia naturalis*. C. Mayhoff (ed.): Teubner, Leipzig.

Roces, F. and Hölldobler, B. (1996) Use of stridulation in foraging leaf-cutting ants: mechanical support during cutting or short-range recruitment signal? *Behav. Ecol. Sociobiol.* 39:293–299.

Roces, F., Tautz, J. and Hölldobler, B. (1993) Stridulation in leaf-cutting ants. *Naturwissens.* 80:521–524.

Rohe, W. (1991) *Systematik and Biologie westmalayischer* Dolichoderus-*Arten (Hymenoptera, Formicidae)*. PhD Thesis, Mainz.

Röschard, J. and Kirchner, W.H. (1995) Akustische Signale von Erdhummeln (*Bombus terrestris*) im Nest und deren biologische Relevanz. *Proc. IUSSI Utrecht 1995*, p 14.

Rose, M., Savorning, J. and Casanova, J. (1948) Sur l'émission d'ondes ultra-sonores par les Abeilles domestiques. *C. R. Acad. Sci.* 227:912–913.

Sandeman, D.C., Tautz, J. and Lindauer, M. (1996) Transmission of vibration across honeycombs and its detection by bee leg receptors. *J. Exp. Biol.* 199:2585–2594.

Schneider, P. (1972) Akustische Signale bei Hummeln. *Naturwissens.* 59:168–169.

Schneider, P. (1975) Versuche zur Erzeugung des Verteidigungstones bei Hummeln. *Zool. Jb. Physiol.* 79:111–127.

Schneider, P. and Kloft, W. (1971) Beobachtungen zum Gruppenverteidigungsverhalten der östlichen Honigbiene *Apis cerana* Fabr. *Z. Tierpsychol.* 29:337–342.

Schneider, S.S. (1987) The modulation of worker activity by the vibration dance of the honeybee, *Apis mellifera. Ethology* 74:211–218.

Schneider, S.S. and Gary, N.E. (1984) 'Quacking': A sound produced by worker honeybees after exposure to carbon dioxide. *J. Apic. Res.* 23:25–30.

Schneider, S.S. and McNally, L.C. (1991) The vibration dance behavior of queenless workers of the honey bee, *Apis mellifera* (Hymenoptera: Apidae). *J. Ins. Beh.* 4:319–332.

Schneider, S.S., Stamps, J.A. and Gary, N.e. (1986a) The vibration dance of the honey bee. I. Communication regulating foraging on two time scales. *Anim. Behav.* 34:377–385.

Schneider, S.S., Stamps, J.A. and Gary, N.E. (1986b) The vibration dance of the honey bee. II. The effects of foraging success on daily patterns of vibration activity. *Anim. Behav.* 34: 386–391.

Schurian, K.G. and Fiedler, K. (1991) Einfache Methoden zur Schallwahrnehmung bei Bläulings-Larven (Lepidoptera: Lycaenidae). *Entomol. Z.* 101:393–412.

Seeley, T.D. (1992) The tremble dance of the honey bee: message and meanings. *Behav. Ecol. Sociobiol.* 31:375–383.

Seeley, T.D. and Towne, W.F. (1992) Tactics of dance choice in honey bees: do foragers compare dances? *Behav. Ecol. Sociobiol.* 30:59–69.

Seeley, T.D., Kühnholz, S. and Weidenmüller, A. (1996) The honey bee's tremble dance stimulates additional bees to function as nectar receivers. *Behav. Ecol. Sociobiol.* 39:419–427.

Simpson, J. (1964) The mechanism of honey-bee queen piping. *Z. vergl. Physiol.* 48:277–282.

Simpson, J. and Cherry, S.M. (1969) Queen confinement, queen piping and swarming in *Apis mellifera* colonies. *Anim. Behav.* 17:271–278.

Simpson, J. and Greenwood, S.P. (1974) Influence of artificial queen-piping sound on the tendency of honeybee, *Apis mellifera*, colonies to swarm. *Ins. Soc.* 21:283–288.

Spangler, H.G. (1967) Ant stridulation and their synchronization with abdominal movement. *Science* 155:1687–1689.

Spangler, H.G. (1971) Effects of recorded queen pipings and of continuous vibration of the emergence of queen honey bees. *Ann. Entomol. Soc. Am.* 64:50–51.

Spangler, H.G. (1986) High-frequency sound production by honeybees. *J. Apic. Res.* 25: 213–219.

Spangler, H.G. (1991) Do honey bees encode distance information into the wing vibrations of the waggle dance? *J. Ins. Behav.* 4:15–20.

Stabentheiner, A. and Hagmüller, K. (1991) Sweet food means "hot dancing" in honeybees. *Naturwissens.* 78:471–473.

Stuart, A.M. (1963) Studies of the communication of alarm in the termite *Zootermopsis nevadensis* (Hagen), Isoptera. *Physiol. Zool.* 36:85–96.

Stuart, A.M. (1988) Preliminary study on the significance of head-banging movements in termites with special reference to *Zootermopsis angusticollis* (Hagen) (Isoptera: Hodotermitidae). *Sociobiol.* 14:49–60.

Stuart, R.J. and Bell, P.D. (1980) Stridulation by workers of the ant *Leptothorax muscorum* (Nylander) (Hymenoptera, Formicidae). *Psyche* 87:199–210.

Tautz, J. (1996) Honeybee waggle dance: recruitment success depends on the dance floor. *J. Exp. Biol.* 199:1375–1381.

Tautz, J., Roces, F. and Hölldobler, B. (1995) Use of a sound-based vibratome by lead-cutting ants. *Science* 267:84–87.

Tautz, J., Rohrseitz, K. and Sandeman, D.C. (1996) One-strided waggle dance in bees. *Nature* 382:32.

Towne, W.F. (1985) Acoustical and visual cues in the dances of four honey bee species. *Behav. Ecol. Sociobiol.* 16:185–187.

Towne, W.F. (1995) Frequency discrimination in the hearing of honey bees (Hymenoptera: Apidae). *J. Ins. Behav.* 8:281–286.

Towne, W.F. and Kirchner, W.H. (1989) Hearing in honeybees: detection of air-particle oscillations. *Science* 244:686–688.

Unwin, D.M. and Corbet, S.A. (1984) Wingbeat frequency, temperatur and body size in bees and flies. *Physiol. Entomol.* 9:115–121.

Visscher, P.K. (1993) A theoretical analysis of individual interests and intracolony conflict during swarming of honey bee colonies. *J. Theor. Biol.* 165:191–212.

Waddington, K.D. (1982) Honey bee foraging profitability and round dance correlates. *J. Comp. Physiol.* 148:297–301.

Waddington, K.D. and Kirchner, W.H. (1992) Acoustical and behavioral correlates of profitability of food sources in honey bee round dances. *Ethol.* 92:1–6.

Wenner, A.M. (1962a) Sound production durin the waggle dance of the honeybee. *Anim. Behav.* 10:79–95.

Wenner, A.M. (1962b) Communication with queen honey bees by substrate sound. *Science* 138:446–448.

Wenner, A.M. (1964) Sound communication in honeybees. *Sci. Am.* 210:116–124.

Wieland, A. (1995) *Untersuchungen zum Einfluss des von Raupen erzeugten Substratschalls auf Bläulings-Ameisen-Assoziationen.* Diploma Thesis, Würzburg.

Wilson, E.O. (1971) *The insect societies.* Belknap Press of Harvard University Press, Cambridge, MA.

Wilson, E.O. (1975) *Sociobiology.* Belknap Press of Harward University Press, Cambridge, MA.

Woods, E.F. (1956) Queen piping. *Bee World* 10:185–194, 216–219.

Orientation and Communication in Arthropods
ed. by M. Lehrer
© 1997 Birkhäuser Verlag Basel/Switzerland

Acoustic communication and orientation in grasshoppers

D. von Helversen

Institut für Zoologie der Universität Erlangen, Staudtstr. 5, D-91058 Erlangen, Germany

Summary. Phonotactic orientation of *Chorthippus biguttulus,* a behaviourally well studied acridid species, serves as an example to outline the function, capacities and limitations of the grasshopper's auditory performance. Results of behavioural, electrophysiological and anatomical studies are combined to reveal the neural pathway of acoustic information processing, and the mechanisms involved in acoustical communication and orientation. In *Chorthippus,* both males and females stridulate. The male approaches the female in small, distinct steps, each being associated with a turn towards the side from which the sound arrives, thus bringing the sound source to a lateral position on the opposite side. This lateralization behaviour is based on the interaural intensity difference. For species-specific signal recognition and for sex discrimination, the acoustic inputs from both sides are summed, whereas directional information is processed in parallel on either side via separate channels ascending to the brain. Although in both the task of signal recognition and the task of sound localization a substantial part of information processing occurs in the first stage of synaptic interactions in the metathoracical ganglion (TG3), the final step of song recognition, as well the final decision on the direction of turning, take place in the brain.

Introduction:
Chorthippus biguttulus as a model system for orientation mechanisms in grasshoppers

Mate finding by means of acoustical communication involves two different, but closely related tasks, namely (i) recognition and discrimination of signals, and (ii) localization of the sound source. The very extensive work on these two processes has been reviewed in general by Ewing (1989) and Bailey (1991), for crickets by Huber et al. (1989) and Huber (1990), and for bushcrickets (katydids) by Bailey and Rentz (1990). So far, however, there has been no comprehensive review of orientation mechanisms in grasshoppers. The present chapter is meant to provide an updated review of the behavioural results, as well as of several relevant neuro-anatomical and -physiological findings on the grasshopper's auditory system.

In the present review, *Chorthippus biguttulus* is used as a model system for two reasons. Firstly, it readily displays phonotactic behaviour in the laboratory, where acoustic stimuli can be varied systematically. Secondly, the results of anatomical and electrophysiological studies on *Locusta* (e.g., Boyan, 1984, 1992; Römer, 1976; Römer and Marquart, 1984; Marquart, 1985a; Römer et al., 1988; Halex et al., 1988), which does not show easily quantifiable orientation behaviour, can also be applied to *Chorthippus* (Ronacher and Stumpner, 1988).

Sound production and phonotactic behaviour in *Ch. biguttulus*

In *Ch. biguttulus*, as in many other gomphocerine grasshopper species, mating is preceded by acoustical duetting of the two potential partners (Fig. 1A). Animals of both sexes stridulate by rubbing a file on the inner side of the femur of the hindlegs against a protruding vein of the forewing (Fig. 1, inset). Recognition of the species-specific signal relies on those song parameters that are common to both sexes, such as syllable structure (Fig. 1C) (von Helversen, 1972; von Helversen and von Helversen, 1997). Sex recognition, on the other hand, is based on those parameters in which male and female songs differ from each other (Fig. 1B, C). For males to recognize a female signal, the syllables must consist of short, distinct pulses of a ramped shape, separated by gaps. Females, on the other hand, reject such gappy syllables and respond only to long continuous syllable blocks (von Helversen and von Helversen, 1997). The song preferences of males and females differ, in addition, with respect to the carrier frequency spectrum of the signals (Fig. 1D). Females respond preferentially to signals containing both high and low-frequency components, whereas males only orient to signals of a low carrier frequency.

A male in search of a female sings at fairly regular time intervals. A female ready to mate responds to the song, but, because competition for females is high among males, it is up to the male to localize the female and to take the risk of approach. Upon hearing a female response, the male turns abruptly towards the side from which the female's signal comes. It moves forward a short distance (about 5–15 cm), sings again and turns anew after the next answer of the female. Thus, the male approaches in a stepwise manner, making successive decisions (Fig. 2). The turning angle is usually larger than that needed to bring the sound source to the frontal auditory field. Instead, the turn brings the sound source in a lateral position on the previously sound-contralateral side. This behaviour, termed lateralization, results in a zig-zag course of the approach (see Fig. 2). Only when a female response is perceived in the frontal auditory field, the male may react by jumping straight forward towards the sound source.

The intensity of the signals produced by *Ch. biguttulus* males, as measured in the field at a distance of 10 cm, is between 60 and 66 dB SPL (von Helversen and von Helversen, 1994). Female signals are less intense by about 10 dB. Because density of vegetation and the height above ground influence sound propagation (Michelsen, 1978), the effective communication distance is typically within a range of 50 cm, but in favourable surroundings it may extend to more than 1 m (Gilbert and Elsner, 1995).

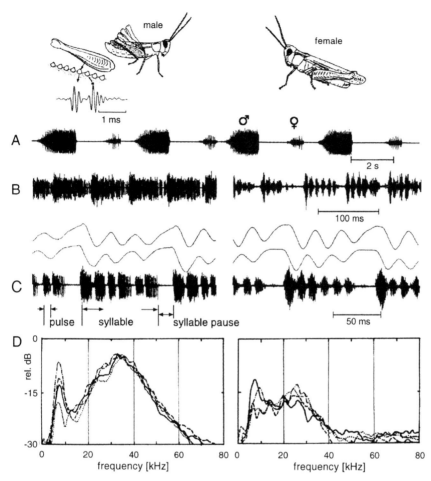

Figure 1. Song pattern of male and female *Chorthippus biguttulus*. Insets (top) show the technique of song production. Both males and females stridulate by rubbing a file on the inner side of the hindfemora against the elytra. (A) Duetting of male and female during phonotaxis. (B) Magnified part of phrases of males and females stridulating with both hindlegs to demonstrate the species specific syllable pattern common to both sexes, and the sex-specific differences between male and female song. Note the continuous syllables in the song of males, and the distinct pulses in the syllables of the female song. (C) Generation of the patterns presented in (B). The upper two traces represent the movement patterns of the stridulating legs. Each down- and upward movement produces one short sound pulse, as shown for a single stridulating leg in the third line (the other leg was made soundless by cutting the elytron). Note the different shape of male and female pulses. In the syllables of males, the short gaps at the reversal points of movements are camouflaged by the phase shift between the two stridulating legs, as shown in (B) (Elsner, 1974), whereas female syllables maintain their gappyness even with both legs active. (D) Sound frequency spectra of four males (left-hand panel) and four females (right-hand panel).

Figure 2. Phonotactic approach paths of male *Ch. biguttulus* to loudspeakers broadcasting a female response song following every calling song of the male. Open arrows: orientation of the male's longitudinal body axis when calling; black arrows: orientation of the male's longitudinal body axis after the turning response. Small empty arrowheads mark the male's head position in successive frames of the filmed approaches (24 frames per s). Occasionally, the male does not move forward after a turn, but it still keeps singing repeatedly (dotted arrows). In some cases, it would jump forward, in which case no arrowheads are shown. Inset at bottom shows definition of target angle and turning angle, also used in Figure 3.

Orientation performance as revealed by behavioural experiments

The male's behavioural response can be easily elicited in the laboratory by playback of model female songs to a freely moving male. For stimulation from left and right, sound stimuli were delivered via two loudspeakers mounted on a hand-held frame (see insets in Figs 7 and 9–14). The two speakers could be positioned equidistant from the animal's longitudinal body axis by aligning the frame accordingly. In experiments involving song recognition, the response of the male was evaluated in terms of the proportion of stimuli that elicited a turning response, whereas in the experiments concerning lateralization, the response was expressed as turns to the left or to the right in percentage of the total number of turns observed.

Magnitude of the turning angle in relation to the angle of sound incidence

The female's reply song does not only initiate the male's turning response. It may provide the male with additional directional information during the turn itself (closed-loop conditions). In order to create open-loop conditions, the simulated female signals (of normally 1 s duration) were cut down to 400 ms-stimuli, which still reliably induced turning behaviour, but terminated before the male initiated its turning response. In Figure 3A, the turning angles recorded under these conditions are plotted as a function of the target position (with respect to the animal's midline). It is evident that there is no correlation between target angle and turning angle at target positions between 30° and 150°. However, males turn consistently towards the louder side: no errors occur at target angles larger than ±10°. Thus, the male does not estimate the exact target angle. Rather, the directional response is restricted to lateralization of the sound source. This strategy results in a considerable scatter – between 40° and 80° – of the turning angles. With a female signal of normal duration (1 s) (Fig. 3B), the scatter is smaller, because, under these closed-loop conditions, the male often corrects immediately for some of the overshooting. But even in this case, it is evident that the male makes little use of the actual target angle. With these longer signals, however, the frequency of jumps (Fig. 3C) is increased, as compared with that obtained with short signals (not shown). Males tend to jump (between 5 and 30 cm) at target angles in the frontal auditory field (position 0° in Fig. 3C). This behaviour seems to be elicited when the signal lasts sufficiently long, and when the intensity of the auditory inputs is equal on both sides. Jumps also occur when the sound comes from behind (position 180° in Fig. 3C), showing that the insect cannot discriminate between sound coming from the front and from the rear.

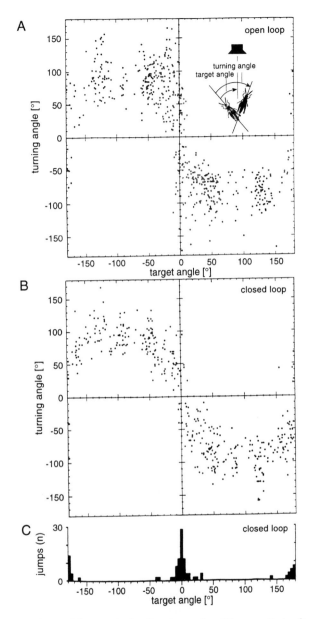

Figure 3. Independence of the magnitude of turning angles of the magnitude of target angle, as revealed by frame-by-frame evaluation of filmed turning responses. (A) Turning angles of male *Ch. biguttulus* as a function of target angle, under open-loop conditions, (B) the same under closed-loop conditions. In (A), the replayed natural female signal was artificially shortened to 400 ms, and therefore it terminated before the male initiated a turn (that occurs, on the average, only 560 ms after the onset of the female signal). In (B), the female signal was of normal length (1.2 s). The normal signal released not only turns, but, in addition, jumps (C) when the sound came from ahead or from the rear. (D. von Helversen and J. Rheinlaender, unpublished data.)

The role of visual feed-back

The magnitude of turning angles is independent of sound intensity (D. von Helversen, unpublished data), but it is influenced by optic feed back. In a homogenous optical surrounding, 67% of the turning angles were larger than 60°, whereas in a striped cylinder (stripe width 18°) they were smaller than 45° in 80% of the recorded responses (D. von Helversen, unpublished data). Thus, in the presence of optomotor inputs (which is usually the case in the animal's natural environment), the magnitude of the turning response is reduced (see also Kirschfeld, this volume). A model calculation, minimizing the number of phonotactic steps in proportion to the magnitude of a given turning angle (Ronacher et al., 1989), suggests that angles around 40° are, indeed, optimal. The model suggests, in addition, that the strategy of stepwise approach by means of lateralization is robust against handicaps, such as defective hearing on one side, or acoustic asymmetry, and tolerates a broad range of ambiguity in the frontal auditory field (Ronacher et al., 1989). Thus, the magnitude of the turning angle seems to have been selected such as to be optimal in natural conditions.

The effective frequency range of phonotaxis

Males of *Chorthippus biguttulus* respond best to frequencies between 6 and 8 kHz (Fig. 4B), suggesting that the inputs for lateralization are mainly provided by those groups of receptors tuned to 6–8 kHz (Fig. 4A), a finding that agrees well with the low-frequency peak of the female song spectrum (see Fig. 1D). Moreover, high-frequency components, when added to a replayed female song, reduce the turning probability significantly (von Helversen and von Helversen, 1997). Thus, high-frequency components, among others, may constitute an important cue to distinguish male and female song and to prevent males from approaching other singing males, thereby reducing the risk of predation.

The finding that turning is released only by low frequency signals does not imply that grasshoppers are unable to localize high-frequency sounds. Indeed, male locusts are perfectly capable of localizing high-frequency sounds as well, as revealed by the negative phonotactic response of tethered flying locusts to high-frequency signals (Robert, 1989).

Directional cues provided by the tympanal organs

The ear of orthopteran insects, the so-called tympanal organ, is similar to the vertebrate ear in that it mediates sound perception via oscillations of a specialized membrane. Still, at least at first sight, it might seem paradoxical that small grasshoppers such as *Chorthippus* should perform excellent

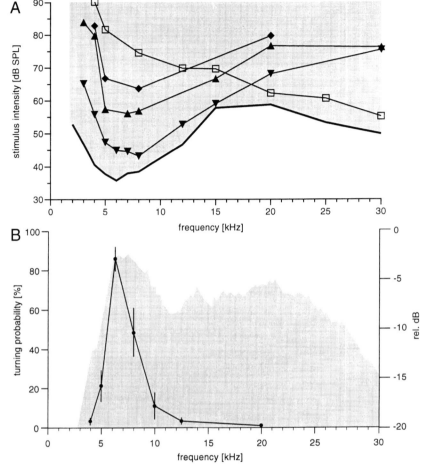

Figure 4. (A) Threshold curves of single receptor elements of four types, and the summed action potential of the tympanal nerve (thick line) recorded in males of *Ch. biguttulus* (A. Stumpner, unpublished data). (B) Spectral range of phonotaxis in males. Mean turning probability and standard deviation (left ordinate) of six males to a replayed female signal, consisting of syllables with six ramped pulses presented at 55 ± 3 dB SPL. The spectrum was 1/3 octave-noise presented at mid-frequencies as indicated on the abscissa (from von Helversen and von Helversen, 1997). Grey area: mean spectrum of the songs of five females (right ordinate, see also Fig. 1 D). Note that the phonotactic response is mediated only by low-frequency receptors.

phonotaxis even though the directional cues used by vertebrates (being much larger animals) are not available to them. Because the wavelengths contained in the signal are usually longer than their body size, insects cannot make use of interaural intensity differences caused by direction-dependent diffraction of sound. Moreover, insects are not expected to make use of the time differences between sound arrival at the two tympanal membranes, because, due to the short distance between the two ears

(1–2 mm in *Chorthippus*), these differences would be too small to be detected by the nervous system. As will be shown below, insects have found some ingenious solutions to overcome these difficulties.

Pressure-difference receivers

The tympanal organ of insects, as opposed to the vertebrate ear, allows sound to act simultaneously on both sides of the tympanal membrane. The vibration amplitude is thus determined by the difference between the external and the internal sound pressures, the tympanal organ thus acting as a pressure-difference receiver. Because the phase difference of sound waves arriving at the two sides of the tympanal membrane varies with the angle of sound incidence, the resulting vibration amplitude may provide information on the location of the sound source.

This general principle of insect hearing was proposed by Autrum as early as 1940, but only in recent years have the phase-shifting properties of particular structures of the orthopteran tympanal organs been investigated in detail (Michelsen and Nocke, 1977; Lewis, 1974; Stephen and Bailey, 1982; Stephen and Bennet-Clark, 1982; Breckow and Sippel, 1985; Larsen et al., 1989; Stumpner and Heller, 1992; Michelsen et al., 1994a, b; Michelsen and Rohrseitz, 1995).

A single pressure-difference receiver would not suffice to localize the sound source, except in successive steps (see Schildberger and Kleindienst, 1989). For instantaneous measurement of sound direction, the vibration amplitudes of the two membranes must be compared with each other. These differences between the oscillation amplitudes can be increased more effectively by reducing the membrane vibrations on the side opposite to the sound than by increasing the vibration amplitude on the side exposed to the sound (Larsen et al., 1989).

The grasshopper's ear: Morphology and sensory elements

In grasshoppers, the two tympanal organs are located on the two sides of the first abdominal segment (Fig. 5A), with air-filled tracheal sacs between them, thus allowing sound waves to travel through one tympanal membrane and act on the inside of the other (Fig. 5B). The tympanal membrane, framed by a sclerotized cuticular rim, consists of a larger thin membrane and a smaller membrane of thicker cuticle (Fig. 5C). The two regions are separated by a structure known as Müller's organ, consisting of sclerites that serve as the attachment sites of four groups of receptor elements, the a-, b-, c-, and d-cells (Fig. 5C) (Schwabe,1906; Gray, 1960). The four morphologically defined receptor groups differ with respect to their sensitivity and frequency response characteristics, due to the frequency-dependent vibration modes of the tympanal membrane (Michelsen, 1971a), which cause the attached Müller's organ to oscillate in a complex manner (Stephen and Bennet-Clark, 1982; Breckow and Sippel, 1985). In *Ch. biguttulus*, the a-, b-, and c-cells are tuned to low frequencies, with best responses between 6 and 8 kHz. The d-receptors are sensitive to frequencies above 12 kHz, best frequency being 30 kHz, with thresholds that are 10 to 15 dB higher than the most sensitive low frequency receptors (Fig. 4A, A. Stumpner, personal communication).

The receptors respond to acoustical stimuli tonically over a dynamic range of about 20–25 dB with spike rates up to 200–300 Hz (Rehbein, 1976; Römer, 1976; Ronacher and Römer, 1985; Stumpner and Ronacher, 1991).

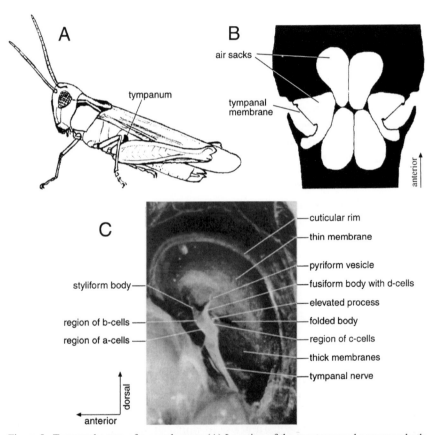

Figure 5. Tympanal organ of a grasshopper. (A) Location of the two tympanal organs on both sides of the first abdominal segment. (B) Schematic drawing of a horizontal section through the tympanal organ of the grasshopper species *Oedipoda* (after Michelsen and Rohrseitz, 1995, modified). Three pairs of airsacs allow sound to travel through one tympanal membrane and to act on the inner side of the other. (C) Lateral view of the inner side of the tympanal membrane and Müller's organ in the grasshopper *Locusta* (courtesy of A. Stumpner). See text for further details.

Information on rapid or slow amplitude modulations, or on the presence of small gaps (of 1 to 2 ms), that cannot be detected by a single receptor element, can be encoded by synchronous or asynchronous spiking of many elements (Ronacher and Römer, 1985; Krahe and Ronacher, 1993; Lang, 1996).

Directional characteristics

Because the adequate cue for the lateralization response is the difference between the vibration amplitudes of the tympanal membranes on the two sides, it is the tympanal directional characteristics that finally determine lateralization acuity. Tympanal directional sensitivity was determined several times in a number of *Locusta, Schistocerca* and *Chorthippus* species, using behavioural (von Helversen, 1984) and electrophysiological methods (Autrum et al., 1961; Römer, 1976; Adam, 1977a; Miller, 1977; Wolf, 1986; von Helversen and Rheinlaender, 1988; Werner and Elsner, 1995) and, more recently, by direct measurements of membrane vibrations (Michelsen and Rohrseitz, 1995). The results obtained differ quantitatively, partly because species of different size and stimuli with different frequency spectra were used in the different studies. Qualitatively, however, they all lead to the same conclusions, namely that sound on the side opposite to the source is attenuated (by 6 to 15 dB) as compared to sound on the source-ipsilateral side, and that there is a steep change in sensitivity as the source crosses the animal's midline.

Figure 6 A shows directional characteristics recorded for *Ch. biguttulus* using the female song spectrum (von Helversen and Rheinlaender, 1988). It can be seen that there is little change in directional sensitivity on the sound-ipsilateral side when the speaker is moved on that side. However, crossing to the other side results in a drastic change in sensitivity, as is indeed expected from an animal orienting by lateralization of the sound source.

The effective intensity difference for eliciting the lateralization response

To determine the magnitude of intensity differences between the right and left tympanal organs, the directional sensitivity curve shown in Figure 6 A (solid line) was subtracted from its mirror image (dashed line). The curve thus obtained (Fig. 6 B) shows that the grasshopper experiences rapidly increasing intensity differences in the frontal auditory field (between plus and minus 15°), whereas between 30° and 150° the interaural dB-difference remains fairly constant at about 8–9 dB. Thus, intensity differences cannot provide a cue to discriminate between different angles in the lateral acoustical field, a result that is in agreement with the results shown in Figure 3.

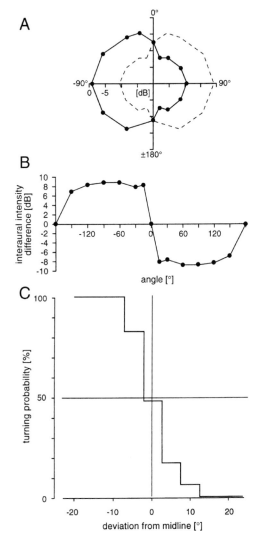

Figure 6. (A) Mean directional characteristics of the left and right tympanal organ (solid and dotted line, respectively) as measured for four *Ch. biguttulus* males using the spectrum of the female signal. At various angles of sound incidence, the threshold of the tympanal nerve was monitored by headphones (data from von Helversen and Rheinlaender, 1988). (B) Interaural intensity difference as a function of the angle of sound incidence, calculated by subtracting the directional characteristic from its mirror image (see A). Note the rapid change in intensity differences in the frontal auditory field, and the constant intensity differences in the lateral field. (C) Probability of correct lateralization (turns to the left in % of all turns) of three *Ch. biguttulus* males as a function of the target angle. The percentage of left turns was calculated for classes of angles corrected for each male's individual deviation from the midline (see text). (D. von Helversen, unpublished data.)

In Figure 6C, the probability of turning to the louder side is plotted as a function of sound incidence, on the basis of behavioural data such as those presented in Figure 2A, using sound intensities of 50–55 dB. The percentage of correct lateralizations increases steeply as soon as the sound source has crossed the midline, and we can read from Figure 6B that, at target angles larger than 10°, the intensity difference provided by the system is sufficient for error-free turns towards the louder side.

The intensity-difference curve shown in Figure 6B was derived from a tympanal characteristic tuned to the most sensitive receptor elements (with peak responses at 7 kHz, see Fig. 4A). In experiments by Wolf (1986), using the same sound intensity as that used in Figure 6C, the course of the tympanal characteristic was less steep, suggesting that even smaller intensity differences than indicated above would suffice to elicit correct turns. Therefore, the interaural intensity difference necessary to elicit lateralization was investigated in yet another behavioural experiment, using two loudspeakers simultaneously presenting (non coherent) female signals from the two sides (von Helversen and Rheinlaender, 1988). The intensity of one speaker was kept constant, while that of the other was varied in steps of 1 dB from being less loud than the reference speaker to being louder than the latter. The mean turning probability of 12 males to the reference speaker is shown in Figure 7A. Clearly, males turn very reliably towards the louder speaker even at intensity differences as small as 1–2 dB between the two speakers.

Directional cues at the neural level

Our current understanding of the neural mechanisms involved in the encoding and processing of acoustical information in the grasshopper's auditory system is based on results of behavioural as well as neurophysiological studies. For better understanding of the conclusions to be drawn below, a brief survey of the neural pathway along which acoustical signals travel to finally elicit a behavioural response is presented.

The central auditory pathway

The acoustic afferents originating in the tympanal organs enter the metathoracic ganglion (TG3) via the tympanal nerve, arborising exclusively in the soma-ipsilateral hemiganglion (Rehbein, 1976; Halex et al., 1988; Riede et al., 1990). In the TG3, they synapse with first-order interneurons that connect the two hemiganglia of the TG3. These interneurons receive polysynaptic inputs (Römer and Marquart, 1984) and exert either excitation or inhibition on the contralateral side, or distribute it to both sides (Marquart, 1985a, b; Römer et al., 1988; Sokoliuk et al., 1989; Boyan, 1992). The

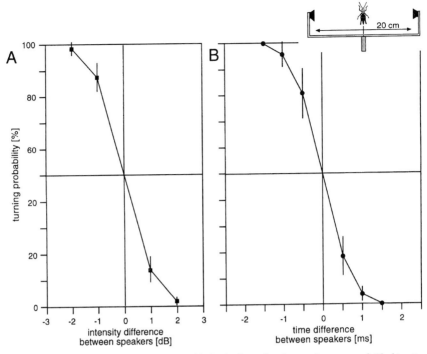

Figure 7. The preference for the louder side in the lateralization performance of *Ch. biguttulus*. (A) Mean probability of turns (in % of all turns) of 12 males towards a reference speaker of a constant sound intensity presented on one side, as a function of intensity differences between this speaker and a second one, presented on the other side, the intensity of which was either higher or lower than that of the reference speaker (negative and positive values on the abscissa, respectively). Data from von Helversen and Rheinlaender (1988). (B) Probability of turns of six males towards a reference speaker as a function of the interaural time difference between the female signal from the reference speaker and an equally loud female signal presented by another speaker on the other side. The latter signal was either delayed with respect to the reference signal, or it preceded the latter (negative and positive values on the abscissa, respectively). Data from von Helversen and Rheinlaender (1988).

neural activities depend on the temporal relations of excitation and inhibition, and vary with stimulus frequency and intensity (Römer and Dronse, 1982; Römer and Marquart, 1984; Marquart, 1985a, b; Römer et al., 1988). Although these complex interactions are not fully understood so far, it is clear that, already at this first stage, complex processing of acoustic information occurs. The TG3 can be regarded as the site at which the acoustical signals are encoded with respect to temporal pattern, frequency content, intensity, amplitude modulation, and direction (Römer and Marquart, 1984; Marquart, 1985a, b; Römer et al., 1988; Stumpner, 1988; Stumpner and Ronacher, 1991; Boyan, 1992).

From the TG3, at least 16 neurons ascend to the brain on each side. The ascending neurons (ANs) are individually characterized by their morphological and physiological properties. Each type of AN is represented by one individual in each of the two hemiganglia (Stumpner, 1989), the input regions being, with only a few exceptions, on the soma side. In all cases, the axons ascend on the soma-contralateral side and terminate, as far as is known, in the brain (Boyan, 1983; Hedwig, 1986a; A. Stumpner, unpublished data).

Neural codes used for orientation: Spike rate vs. latency-induced time difference

The different vibration amplitudes of the tympanal membranes may affect the excitation in the receptor cells in two ways: (i) the receptors will respond with a lower spike rate on the less intense side than on the louder side (Mörchen, 1980), and (ii) the onset of excitation will be delayed on the less intense side, due to the intensity-dependence of latency (Mörchen et al., 1978; Mörchen, 1980). There is convincing neurophysiological evidence, to be presented below, that both the rate of binaural spike discharges and latency difference encode directional information.

Neuronal evidence for spike-rate coding
To test whether or not the louder side is determined by comparing the spike rates elicited in the receptor cells on the two sides, Römer and Rheinlaender (1983) recorded the response of an ascending neuron (AN1) to a 16 kHz sound pulse, while on the contralateral side the spike rate of tympanal fibers was varied by stimulating the tympanal nerve electrically with short impulses, each of which gave rise to a single spike in the receptor fibers. This procedure allowed the spike frequency to be varied without changing latency. In these experiments, with increasing spike rate of the electrically stimulated tympanal fibers, the response induced by the acoustic stimulus was cancelled out. From this result the authors concluded that spike rate alone is sufficient for determining direction.

Neuronal evidence for time coding
Encoding sound intensity in the spiking rate is very suitable for comparing long continuous signals that can be integrated over time. However, for encoding short signals that induce only one or a few spikes upon each stimulation, the temporal relations between the inputs from the two sides would be more reliable. Rheinlaender and Mörchen (1979) and Rheinlaender (1984), recording from the AN1, varied the temporal interval (from 2 to 40 ms) between two short mechanical stimuli ("clicks") applied to either tympanal membrane (Fig. 8). When a click presented on the opposite side of the recording preceded the ipsilateral click, the response of the AN1 was suppressed. When the two stimuli were presented simultaneously, both neurons responded. A delay of 2 ms or more of the click on the con-

tralateral side had no effect, provided that the two stimuli were of equal intensity (Rheinlaender and Mörchen, 1979; Rheinlaender, 1984). When the contralateral click was louder (by 10 dB), it inhibited the response of the recorded AN1, even when presented 4 ms later than the ipsilateral click. Conversely, when the contralateral click was less intense, the neuron responded, although the contralateral click was still leading by 2–4 ms (Rheinlaender, 1984). Because each of these short clicks generated only one spike, the shift in the onset of the response (Fig. 8) cannot be due to compensation by a higher spike rate, but rather exclusively to the intensity-induced latency difference between the two sides.

Behavioural evidence for the use of time differences for sound localization
The finding that interaural time differences are perceived at the neuronal level does not automatically imply that such are actually used in the lateralization behaviour. To test whether or not grasshoppers make use of interaural time differences for sound localization, a grasshopper male was stimulated from both sides with two female signals of equal intensities, one of which was delayed with respect to the other (von Helversen and Rheinlaender, 1988). The point of symmetry (i.e., of equal stimulation from both sides) was determined by adjusting the intensities such that, at $\Delta t = 0$ (no time delay), the animal turned equally often to either side. At this inten-

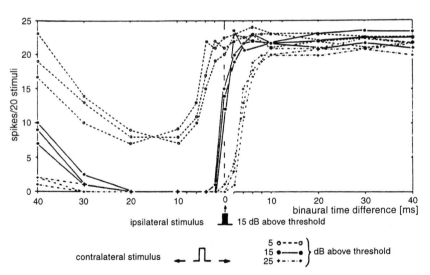

Figure 8. Effects of interaural time differences on the spiking activity of an auditory inter-neuron in the locust. To decouple the two tympanal organs, the tympanal membranes were stimulated with short mechanical contact-stimuli, the time delay between which was varied. As long as the stimulus contralateral to the recording site was leading, the recorded neuron was inhibited. At simultaneous presentation of the two clicks, the neuron responded. Note that the onset of the response could be shifted by varying the intensity of the contralateral click relative to that of the ipsilateral click. Modified from Rheinlaender (1984).

sity-equilibrium, the signals on one side were shifted in steps of 0.5 ms with respect to the other side. The results (Fig. 7B) reveal that a time difference of only 0.5 ms is sufficient to elicit a significant proportion of turns towards the leading side. When the interaural time difference was 1 ms, 100% of the turns were towards the leading side, showing that latency differences are sufficient for identifying the louder stimulus.

The use of interaural time differences for sound localization is as expected when one considers that (i) the natural female pulses, 80% of which are shorter than 8 ms, elicit just one spike per pulse, and (ii) sufficiently large interaural time differences (1–2 ms for stimulation with rectangularly modulated pulses, and 3–4 ms for ramp-shaped pulses) are available to the animal over a fairly wide range of intensities (Krahe and Ronacher, 1993, for *Locusta*), which alone may ensure errorless lateralization at all lateral angles larger than ±10° (see Fig. 6B).

"Trade-off" between intensity and latency

The electrophysiological results shown in Figure 8 suggest that an increase in intensity compensates for an increase in time difference. The magnitude of increase in intensity that is necessary to compensate for an induced time delay was now determined in behavioural experiments by increasing the sound intensity on the delayed side until the male reversed the direction of its turn, now preferring the delayed, but louder side. An example of such a set of "compensation curves" of one individual male is shown in Figure 9A. The point of intersection of each curve with the abscissa indicates the magnitude of intensity increase necessary for compensating for the time delay. For example, to compensate for a delay of 1 ms, it was necessary to increase the intensity on the delayed side by 2 dB.

In Figure 9B, the increase of intensity of the delayed speaker necessary for compensation is plotted for various time differences (D. von Helversen, unpublished data). Up to a delay of 2 ms, the course of the trade-off curve is fairly linear, which can also be derived from the data shown in Figure 7 when intensity differences (Fig. 7A) and time differences (Fig. 7B) are plotted against each other at corresponding turning probabilities (Fig. 9C). Longer delays, however, are compensated by a relatively smaller increase of intensity (Fig. 9B). The fact that compensation curves of both the behavioural and electrophysiological measurements (from Fig. 8, linearly interpolated by the solid line in Fig. 9C) are less steep than the intensity-dependent latency curve of the summed action potential (SAP) of the tympanal nerve (dashed line in Fig. 9C) means that smaller increases in intensity than those predicted by the SAP-curve were sufficient to compensate for delay differences. Therefore, it can be concluded that, in addition to latency differences, the effect of spike-rate-coding and the number of activated units be-

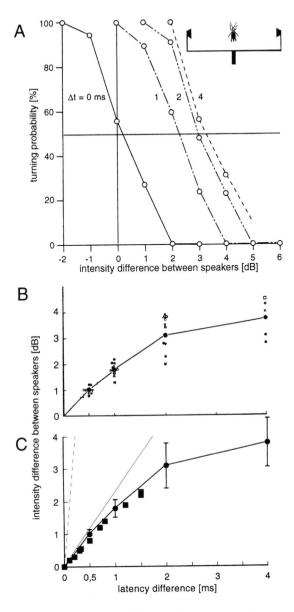

Figure 9. Trade-off between intensity and latency: Compensation of interaural time differences by increasing the intensity of the speaker on the delayed side. (A) Four curves of one male presented with various delays between the two speakers. Turning probability to the leading speaker is plotted as a function of the intensity difference between the two speakers (abscissa). Each data point is based on 20–30 turning responses. The tendency to turn towards the leading stimulus is taken to be compensated by the increased intensity of the delayed stimulus whenever the animal turns equally often to either side (50%-line, ordinate) (D. von Helversen, unpublished data). (B) Intensity increase of the delayed speaker (ordinate) necessary to compensate the attractiveness of the leading speaker (abscissa). Individual males are denoted by different symbols, means are represented by the large symbols (D. von Helversen, unpublished data). In

comes increasingly important when the intensity difference between the two sides is increased.

The role of inhibition

When trains of spikes travel along the two tympanal nerves continuously, as is the case during a natural song, the single acoustic events cannot easily be assigned unequivocally to corresponding spikes on the two sides, i.e., the auditory system may not be able to lateralize correctly on the basis of latency differences. The problem of correct assignment is eluded in systems that respond phasically. Indeed, in the experiment by Rheinlaender and Mörchen (1979) (Fig. 8), the inhibiting influence of the leading side was immense; the response of the recorded AN1 was suppressed even when the leading click (on the opposite side) had occurred 40 ms prior to the one on the recorded side.

However, this result might have been due to the unnatural electro-mechanical stimulation used. Therefore, the duration of inhibition induced by short clicks was investigated again, this time in a behavioural experiment. In this experiment, turning was induced by replaying a female song from the rear, which caused 50% turns to either side (Fig. 10, shaded disc). When 2 ms clicks were added from one side, the turning probability to that side was increased to about 93% (filled disc). The duration of inhibition induced by the click was measured by presenting a second click on the opposite side at various delays with respect to the leading click (Fig. 10, abscissa). Up to intervals of about 10 ms between the two clicks, males continued to turn preferentially to the side of the leading clicks. At larger time intervals, however, turns to either side occurred equally often. When the interval between the clicks was further increased, the delayed click became leading with respect to the one on the opposite side, and the turning tendency was reversed. Thus, inhibition lasts for about 10 ms (D. von Helversen, unpublished data).

Integration over several units possessing different thresholds

For a single receptor fibre, latency and spike rate change with increasing intensity and reach a saturation level at about 20 to 25 dB above threshold

◄───

(C), the behaviourally measured trade-off curve (see B) is compared with intensity and time differences between speakers at corresponding turning probabilities, as measured in Figure 7 A,B, here represented by squares, as well as with the trade-off curves recorded electrophysio-logically in the AN1 by Rheinlaender (1984) (see Fig. 8), here represented by the thin solid line. The intensity-dependent change of latency of the summed action potential (SAP) of the tympanal nerve (see Fig. 4, thick line) (J. Schul, unpublished data) is represented by the dotted line. For interpretation, see text.

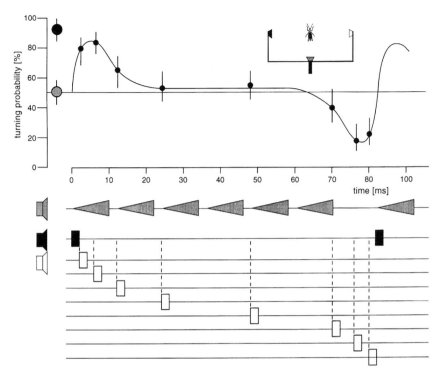

Figure 10. Duration of the inhibition of a laterally presented click on a second click presented on the other side at various time delays (mean turning frequencies of five males). In order to induce turning behaviour, a female signal was presented from behind (inset, grey speaker), causing an equal probability of turns to either side (large grey symbol in the top panel). When a click was added from laterally every 80 ms (inset, black speaker), the turning probability was shifted to that side (large black symbol in top panel). A second click from the other side (inset, white speaker) became effective at delays longer than 10 ms with respect to the first click, equalizing the attractiveness of the leading click. When the second click was shifted by more than 70 ms with respect to the first click, thus approaching the next click on the previously leading side, it became the leading click, causing the animal to reverse the direction of turns. Intensity of the clicks was 10 dB above the intensity of the ramped pulses of the female signal. (D. von Helversen, unpublished data.)

(Mörchen et al., 1978). The expectation, therefore, was that lateralization performance would be better at low intensities (at which the receptors are in their dynamic range) than at higher intensities.

In behavioural experiments, however, the interaural intensity difference necessary to obtain a given turning probability was found to be constant over the whole range of intensities in which the turning response could be induced (D. von Helversen, unpublished data). In Figure 11 A, the turning probability measured for two individuals at intensities close to threshold, and at 15 and 21 dB above threshold, is plotted as a function of intensity difference between the two lateral speakers. The curves are very similar to one another, indicating that, for lateralization, the same interaural intensity difference is necessary at all intensities. The same conclusion can be drawn from the results obtained from eight animals, each tested at a low as well as at a high intensity (Fig. 11B). The data points scatter about the bisection line, indicating that, contrary to the expectation, lateralization probability is fairly independent of absolute intensity.

This result makes sense when the summed action potentials (SAPs) of the tympanal nerve are considered. The amplitude of the SAP (which depends on the number of units involved and on the degree of synchronization) changes fairly linearly over a range of about 40 dB (Adam, 1977b), and so does the SAP's latency (J. Schul, unpublished data). Thus, at least some first-order interneurons attain a dynamic range that is broader than that of single receptor elements by integrating over several units possessing different thresholds.

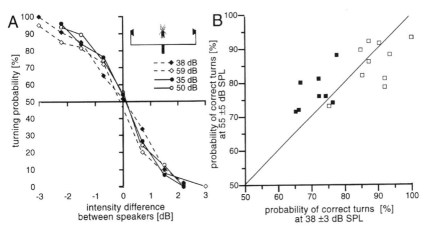

Figure 11. Independence of the probability of correct lateralization of absolute sound intensity in *Ch. biguttulus*. (A) Turning probabilities as a function of intensity differences between two speakers presenting a natural female song (as specified in the inset) near threshold (filled symbols) and at 15 or 21 dB above threshold (empty symbols). Shown are response curves of two males (broken and solid lines, respectively). (B) For eight males, the percentage of correct turns at high intensities (ordinate) is plotted against the percentage of correct turns at lower intensities (abscissa). The turning probabilities towards the reference speaker were measured at intensity differences of 0.75 dB and 1.5 dB (filled and empty symbols, respectively) with respect to the reference speaker. (D. von Helversen, unpublished data.)

The function of ramps

If, as concluded above, the grasshopper is able to exploit time differences between onsets of excitation on the two sides, then lateralization probability is expected to increase with increasing interaural time differences. Adam (1977b) was the first to suggest that ramp-shaped pulses might improve the percentage of correct lateralizations, because, for pulses with slowly rising ramps, the interaural time difference would be increased by the amount of time needed to reach the threshold on the sound-contralateral side. Indeed, Krahe and Ronacher (1993) demonstrated that Adam's hypothesis is valid for single receptor elements: latency differences between the two sides were considerably larger for pulses with slow rise times than for rectangularly modulated pulses. In behavioural experiments, however, it was impossible to compare directly between responses to ramped pulses and to such with steep onsets, because males do not respond to the latter (von Helversen, 1993). To overcome this difficulty, Ronacher and Krahe (1997) conducted experiments presenting, via two lateral speakers, a natural female song pattern in which they modified only the first pulse of each syllable. First pulses consisted of either a 10 ms ramp with steep cut-off, or of the mirror-imaged pulse with steep onset and ramp-shaped decline. The intensity of this first pulse was varied on one side relative to that on the other side, while pulses 2 to 6 were natural pulses presented with equal intensity on both sides (to obtain a 50% turning probability to either side). Under these conditions, the increase in intensity difference (between speakers) needed for a given probability of correct

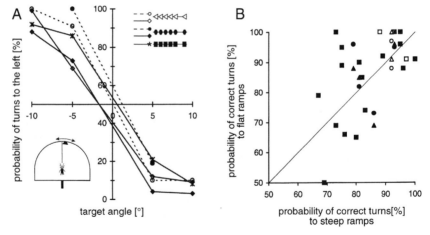

Figure 12. Influence of risetime of pulses on the probability of correct lateralization of *Ch. biguttulus*. Signals differing only with respect to the shape of the pulses (3 ms and 10 ms ramps, and rectangularly modulated pulses, top inset) are compared with respect to the probability of correct turns at a given angle of sound incidence (bottom inset). Stimulus pattern: Six pulses of 10 ms duration separated by 2 ms gaps, syllable interval 12 ms. (A) Response curves of two males (broken and solid lines, respectively) tested with steeply rising (empty symbols, see inset) and slowly rising pulses (filled symbols). One of the two males was also tested with rectangularly modulated pulses (asterisks). (B) At target angles of 5° and 10° (filled and empty symbols, respectively), the turning probabilities using flat ramps were plotted against those using steep ramps (squares: 3 ms vs. 10 ms, circles and triangles: rectangularly modulated pulses vs. 3 and 10 ms ramps, respectively). (D. von Helversen, unpublished data.)

choices was larger by 3 dB for rectangularly modulated pulses than for such with slowly rising ramps.

When the intensity difference involved all of the pulses equally (as is the case under natural conditions or when a sound source is moved around the animal), no significant difference was found between stimuli consisting of pulses which differed only with respect to the steepness of ramps (Fig.12 A) (D. von Helversen, unpublished data). In Figure 12 B, the turning probabilities for steep ramps are plotted against those obtained with less steep ramps. The proportion of correct turns is slightly but not significantly higher with the flat ramp pulses, as compared with the steeper ones. Thus, rectangularly modulated pulses are not substantially less useful for sound localization than are ramped pulses. It is possible that the improved synchronization of spikes expected at the receptor level for steep onsets of stimuli compensate for the effect of the ramp-induced time delay.

Separate processing for pattern recognition and lateralization

In the task of recognizing the species-specific (and sex-specific) song pattern, directional sensitivity is not required, whereas in the sound localization task it is. Therefore, these two pieces of information are expected to be processed in parallel in two separate neural channels.

Internal summation of information for pattern recognition: Split song experiments

The results of the following behavioural experiments show that song recognition is based on the summation of inputs from the two tympanal organs (von Helversen, 1984). Because the summation destroys directional information, these results provide evidence that the song recognition mechanism is distinct from that underlying the lateralization behaviour.

In a first set of experiments, *Ch. biguttulus* females were presented with two song patterns, offered simultaneously via two speakers placed at variable angles between them. Each of the two patterns was behaviourally ineffective when presented alone. However, when combined, the two patterns constitute the adequate male song. Indeed, when the two speakers were placed laterally close to each other, the female responded with a reply song.

The expectation now was that the female would fail to respond when the angle between the two speakers was made large enough for the two sounds to be detected from two different directions, each thus presenting an ineffective pattern. This expectation was based on results obtained from crickets (Pollack, 1986): female crickets fly towards the conspecific song when given a choice between it and a heterospecific song presented simultaneously from a different direction.

Contrary to this expectation, our *Ch. biguttulus* females responded consistently to all appropriate composite patterns, irrespective of the angle between the two speakers, even at a separation of 180°. It follows that, in grasshoppers, the mechanism underlying song recognition pools sound signals from all directions (von Helversen, 1984).

This finding does not imply that the female is unable to localize the sound source. As a matter of fact, if a female is sufficiently motivated, it can make its own way to a singing male, but this behaviour is difficult to quantify. Therefore, in a second set of experiments, the turning response of *males* was exploited.

As already mentioned above, *Ch. biguttulus* males readily turn towards stimuli, the syllables of which are composed of six ramped pulses separated by short gaps (Fig. 13a). Such a pattern can be split up into two components, each consisting of three ramped pulses with every second pulse omitted, neither of which is effective when presented alone (Fig. 13b). When the two patterns are presented on one side, the grasshopper will take them as an effective signal and will reliably turn towards that side (not shown). However, the male responds even when the two patterns are presented one on each side (Fig. 13d) (von Helversen and von Helversen, 1990, 1995). Thus, similar to the female, the male is unable to discriminate between sound sources placed at different directions. One might argue that the male responded to a signal generated by superposition of the strong ipsilateral pulses with the 8 dB attenuated pulses (as predicted by the tympanal directional characteristics, see Fig. 6) arriving at the same ear from the opposite side. However, when two signals differing by 8 dB were offered simultaneously on one side, they induced no turning response (Fig. 13c). The conclusion is that the signals from the two sides are summed, irrespective of their positions in the auditory field.

Separation of information on pattern and position

Because directional information cannot be derived from inputs that are summed over the entire auditory field, the analysis of sound direction must be accomplished, in parallel, by a separate channel. If so, then it should be possible to make the animal turn towards a pattern that is ineffective for song recognition, but that provides, instead, strong directional cues.

Males turn readily to a pattern with slowly rising ramps (Fig. 14a), but hardly ever to one with rectangularly modulated pulses (Fig. 14b). However, when the same two patterns are presented simultaneously from the two sides, turns are always directed away from the effective pattern and towards the one with the stronger directional cues, (i.e., louder and leading with respect to the other) (Fig. 14d, von Helversen and von Helversen, 1995). The possibility that the turns are in fact a response to an effective pattern being generated by superposition of the ipsi- and contralateral stimulus at

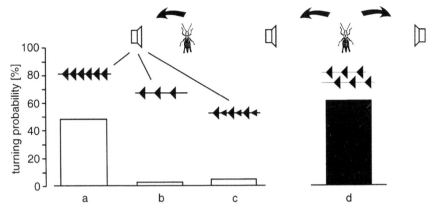

Figure 13. Summation of inputs from the two sides in the pattern recognition task. Shown is the turning probability of *Ch. biguttulus* males towards an effective pattern consisting of six ramped pulses (a), and towards a pattern with every second pulse omitted (b). When two ineffective patterns (as the one shown in b) that complement each other to form an effective pattern are presented from the two sides, they induce turns in either direction with equal probabilities (d). A control experiment (c) demonstrating that the pattern generated by superposition of the two speakers is not effective when one is louder than the other and both are presented on one side. Modified from von Helversen and von Helversen (1990).

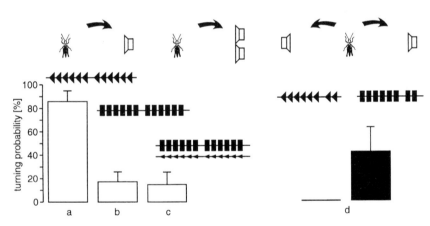

Figure 14. Separation of information on song pattern and direction. Turning probabilities of *Ch. biguttulus* males to signals of different efficiency presented singly or simultaneously. Inset above each bar (a, b, c) shows two syllables of the sound stimulus used. The pattern in (a) is effective, the one in (b) is not, because of the steep onsets of pulses. When the two patterns are presented simultaneously, one from either side (d), the males do not turn very frequently, but in case they do, the turns are always directed towards the pattern that provides the better directional cues, although it is not the effective (species-specific) pattern. A control experiment (c) demonstrates that a pattern that might have become attractive by the superposition of two speakers, both presented on one side, does not elicit turning. Modified from von Helversen and von Helversen (1990).

the tympanal organ, was ruled out as follows: when such a superposed pattern is presented from one side, turning probability is as low as with the rectangular stimulus alone (Fig. 14c). The conclusion is that the cues serving for sound localization are independent of those that are used for pattern recognition. This conclusion is corroborated by the results of the experiments using lateral clicks (see Fig. 10), in which it was possible to reverse the male's turning response.

Localization of the centres for pattern recognition and directional hearing in the central nervous system

In the early stages of research on acoustical communication, filter mechanisms for recognition of the species-specific song were thought to be located rather peripherally. It was surprising, therefore, to find interneurons mediating, in addition to species-specific signals, signals that are not species-specific. In recent years, much evidence has accumulated suggesting that the recognition mechanism is located in the brain. It is only there that neurons have been found that not only copy the song pattern, but are, in addition, tuned to specific properties of the species-specific song. Schildberger (1984) was the first to show, in the cricket, that particular bandpass neurons in the brain respond preferentially to the same syllable rates that elicit the animal's phonotactic behaviour.

Neurons with complex response properties and very specific filter characteristics were also found in the thoracic ganglia of grasshoppers (e.g., Römer and Marquart, 1984; Stumpner and Ronacher, 1991; Boyan, 1992), suggesting that the *Innate Releasing Mechanism* might consist of a chain of filtering processes at various stations in the auditory pathway. However, because the filtering processes occurring in the TG3 are common to all grasshoppers investigated so far (Ronacher and Stumpner, 1988), including species that do not communicate acoustically, pattern recognition is expected to occur at a later stage of processing, probably within the brain. The behavioural experiments described below were designed to determine the location of the recognition mechanism.

Selective heating of the brain

In grasshoppers, as in all other orthopterans, the rate of stridulation movements strongly depends on the ambient temperature. Song rates vary by up to a factor of four within the range of temperatures at which grasshoppers normally sing. In *Ch. parallelus*, the female's preference for the species-specific song shifts with temperature, i.e., it is tuned to the song rate that the conspecific male would produce at the female's body temperature (Bauer and von Helversen, 1987). Thus, a female responds better to a high song rate at a high temperature, whereas at low temperatures low song rates are preferred (Fig. 15 B, open symbols). It should be possible then, by selective heating (Fig. 15 A), to determine the location within the central nervous system at which pattern recognition occurs.

Females of *Chorthippus parallelus* whose head and thorax temperature differed from each other by 5 to 9°C were tested with various song models, the rate of which corresponded to that of males recorded at different temperatures. The results (Fig. 15 B, filled symbols) show that the preferred pattern corresponds to that of a male singing at the temperature of the female's head. This result suggests that the recognition system is located in the brain, a conclusion that is corroborated by earlier results (Elsner and

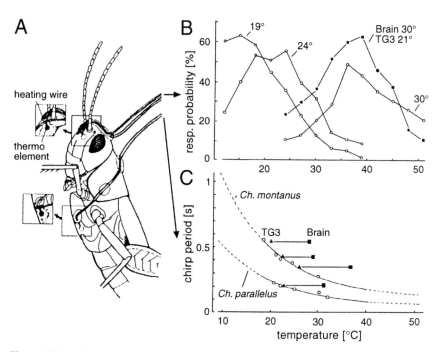

Figure 15. Localization of the neural center for pattern recognition, and that for sound production, by selectively heating the brain. (A) Experimental set-up for heating the head of *Chorthippus parallelus* females with a thin wire twisted around the insect's head, with two thermocouples recording the temperatures near the brain and the thoracic ganglia. (B) Response curves of an untreated female (open circles) indicate that the preferred song patterns shift with ambient temperature. In the patterns that became the most attractive, the wavelengths corresponded to the songs produced by the male at that temperature. When the head was selectively heated, the female preferred the song pattern that corresponded to the head temperature (closed symbols), showing that the neural circuitry for song recognition resides in the brain. (C) Wavelengths of song patterns of untreated females of *Ch. parallelus* and *Ch. montanus* as a function of temperature (open circles). The songs produced by the treated females match the pattern wavelengths corresponding to the thoracic temperature (TG3), showing that the center for song production resides in the thorax. Modified from Bauer and von Helversen (1987).

Huber, 1969) showing that brainless male grasshoppers do not respond to the female signal. The results show, in addition, that the song pattern is generated in the thorax. The syllable rate of the female's response stridulation corresponds to that produced at the thorax temperature (Fig. 15C) (Bauer and von Helversen, 1987).

Surgical manipulations

By means of surgical manipulations it is possible to delimit the paths of information flow and thus to determine the sites where particular steps of information processing occur. This method has often been applied to trace the information channels which are necessary and sufficient to elicit particular motor responses and to control their coordination (Huber, 1963; Elsner and Huber, 1969; Hedwig, 1986a, 1995; Ronacher, 1989, 1991; Elsner, 1994). When such manipulations are used to investigate the paths of sensory inputs, one must be careful not to impair the function of the motor pathway, because it is the motor response that will finally provide the answer to the question under consideration. Therefore, in the surgical experiments to be described below, one chain of connectives had to be left intact to allow for a normal singing response. Nevertheless, several conclusions on information routes and stations, three of which will be reviewed here, could be drawn from these experiments.

The neuronal summation of pattern information takes place in the TG3
This conclusion was drawn from the stridulation response of animals with one connective chain dissected between TG2 and TG3, with the tympanal nerve cut on the same or on the opposite side. With respect to their stridulatory response, males and females so treated behaved like intact animals, regardless of whether the tympanal nerve was cut on the side of the dissected connective, or opposite to it, and irrespective of the side on which the sound stimulus was presented (Ronacher et al., 1986). This result was as expected on the basis of the finding that pattern information from the two sides is summed (see above). Because the cut was between TG2 and TG3, the result indicates that summation occurs at the level of TG3.

*An essential step of directional information processing takes place
in the TG3*
This conclusion was drawn from experiments comparing the turning responses of intact males (Fig. 16A) with that of males that were deafened on one side (Fig. 16B), and such with one connective chain cut between TG2 and TG3 (Fig. 16C), or between the subesophageal ganglion (SOG) and the brain (Fig. 16D) (Ronacher et al., 1986). As expected, males with one tympanal nerve dissected always behaved as if the loudspeaker was on the side of the intact ear, regardless of whether the loudspeaker was on the

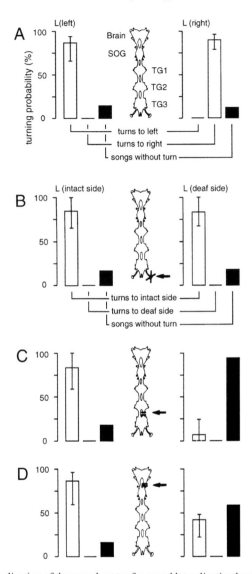

Figure 16. Localization of the neural center for sound lateralization by surgical manipulations. Turning probabilities of male *Ch. biguttulus* to female signals presented by a loudspeaker on one or the other side of the animal. The responses of intact males (A) are shown for comparison with animals unilaterally deafened (B), or with one connective chain transected between TG2 and TG3 (C), or between SOG and the brain (D). Ordinates: percentage of total responses (= singing without turning plus turns, which always include response singing as well) for each of the categories indicated on the abscissa. Modified from Ronacher et al. (1986).

right or on the left (Fig. 16 B). However, the responses of males with one connective cut (Fig. 16 C, D) depended on the position of the speaker. When stimulated from the intact side, they behaved like untreated males, but when stimulated from the operated side, they usually failed to turn, although they still sang in response. The few turns that occurred in this situation were all directed to the intact side; no turns to the operated side were observed.

The possibility that the absence of turns towards the operated side was due to some damage caused to command fibers for turning can be excluded, because, when, in addition to the female sound, an optical stimulus was offered on the operated side (by moving a female glued to a small stick), males did perform (rare but typical) turns towards the operated side. Consequently, the failure to turn towards the operated side was due to the absence of appropriate directional information on that side. This finding corroborates earlier results (see above) demonstrating the reciprocal inhibition that the two sides exert on each other. When the loudspeaker is on the dissected side, directional information on the intact side is suppressed, and, at the same time, the signal on the operated side cannot ascend, due to the cut. Therefore, the animal fails to respond. Obviously, essential steps of directional information processing are accomplished in the TG3.

Channels for both pattern recognition and directional information must ascend to the brain
Because the interface between sensory input and motor output is located in the brain, the information needed for pattern recognition as well as for lateralization must make its way through the brain. The finding that animals with a cut below the turning apparatus (retaining the connection between descending commands from the brain and the motor circuitry in the TG1/TG2) never turn to the operated side (Fig. 16 C) shows that the directional information indeed ascends side-specifically to the brain. Were the brain only required to give a directionally non-specific "permission" for a turn, the information for the direction of the turn being delivered by the TG3 directly to the motor centres in the TG1-TG2 complex, then the animals would be expected to turn to both sides. The males, however, fail to turn to the operated side regardless of whether or not the motor apparatus has been left intact together with the site of directional processing in TG3 (cut between SOG and brain, Fig. 17 D), or together with the brain, the location from where the turning commands descend (cut between TG2 and TG3, Fig. 17 C). Clearly, the final decision on the direction of the turn is made in the brain.

These results suggest that the patterns of excitation arriving in the TG3 are processed in (at least) two ways. For species-specific (and sex specific) song recognition they undergo a neuronal summation and then ascend to the brain without further crossing. The recognition processes of species-specific (and sex-specific) signals are not well understood so far, but it is clear that they drive descending command fibers which trigger different types of songs, e.g., courtship song, calling song or copulatory jerks (Hedwig, 1995; Hedwig and Heinrich, 1997), and may therefore be regarded as the outputs of the *Innate Releasing Mechanisms*. In a parallel pathway, for processing directional information, the inputs of the two sides undergo reciprocal inhibition that enhances the interaural intensity difference. The processed information then ascends side-specifically to the brain.

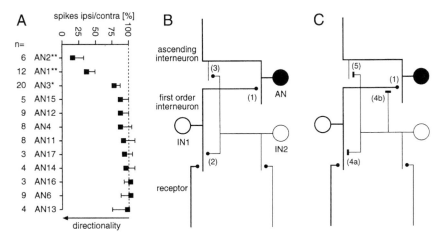

Figure 17. Model of neural processing in the auditory system of grasshoppers. (A) Rank order of directionality for 12 ascending auditory neurons of *Chorthippus biguttulus,* as determined from the spike rates elicited by ipsilateral and contralateral sound stimulation. Number of spikes elicited by soma-ipsilateral stimulation are expressed in % of spike number obtained with contralateral stimulation. Because most ascending neurons produce more spikes when stimulated contralaterally (i.e., from the side on which the axon ascends), a low value of the percentage indicates a higher degree of directionality. Bars denote 99% confidence intervals, numbers refer to number of investigated specimens; asterisks denote significant deviation from the 100% value. From Ronacher and Stumpner (1993), modified. (B, C) Schematic drawing of possible connections resulting in summation (B) and directional enhancement by inhibition (C) of left and right inputs. Thick lines represent the basic path of information: the receptors are restricted to their soma side and are excitatory presynaptic to first order interneurons. These interlinking neurons (IN) transmit the information to the opposite side, where they are presynaptic (interaction (1)) to ascending neurons (AN) that cross back and ascend on the side of sound incidence (Römer et al., 1988). (B) Summation of inputs can be achieved by a local neuron of the other side (IN2, thin lines) mediating excitation to a local neuron (2), which is activated also by ipsilateral receptor inputs, or directly to an ascending neuron (AN) with receiving structures in both hemiganglia (3). (C) Enhancement of directionality is achieved either by inhibition of the local neuron transmitting excitation to the ascending neuron (integrating path (4a) and presynaptic (4b)), or by directly inhibiting the soma-ipsilateral ascending neuron (5). Note that the activity pattern of the ascending neuron will depend on the time relations of excitation and inhibition.

Evidence from anatomical and physiological data

The data available so far on the anatomy, the connections and the physio-logical properties of the neuronal elements are in accordance with the behavioural results. The postulated "loop through the brain" of the recog-nition pathway, as well as of the directionally-sensitive pathway, has its anatomical correlates in the ascending neurons that project to the brain (Boyan, 1983; Hedwig, 1986b; A. Stumpner, personal communication). The most likely candidates to transfer directional information are the AN1 and AN2 (Fig. 17A), because of their pronounced inhibition when the sound source is on the opposite side of the ascending axon (Römer et al., 1988; Stumpner and Ronacher, 1991). In surgical experiments, Ronacher et al. (1993) showed that both of these neurons are involved in sound localization, the AN2 being sufficient for correct turning when the AN1 is dissected.

Further, there is strong evidence for pooling of information from the two sides as early as at the level of the TG3. The majority of interneurons, amongst which at least the AN4 and AN12 are involved in recognition of species-specific song (Stumpner et al., 1991), respond equally strongly to stimuli from both sides and are thus likely to sum the activity of side-specific local interneurons (Stumpner and Ronacher, 1991, 1994; Ronacher and Stumpner, 1993).

A scheme of the possible information flow within the TG3 is presented in Figure 17 B,C. The basic path of acoustic information implies a twofold crossing of the midline before the information finally ascends on the side facing the sound source. The receptor cells are presynaptic to interlinking local neurons (IN1), which transmit the information to the opposite side, where it is picked up by ascending neurons (interaction (1) in Fig. 17B), which in turn cross the midline and ascend on the side facing the sound source (thick lines in Fig. 17B) (Römer et al., 1988). Summation of audi-tory inputs from both sides can be achieved by a further interneuron of the other side (IN2) mediating excitation either via the interneuron activat-ing the ascending neuron (2), or by a direct interaction (3) with the as-cending neuron (thin lines in Fig. 17B).

The inhibition exerted by the interlinking neuron IN2 (thin lines in Fig. 17C) enhances directionality by acting either on the excitation-media-ting interneuron (IN1) at its input region (4a), or by presynaptic inhibition of its endings at the ascending neuron (4b). Direct inhibition of the ascend-ing neuron (5) would also result in enhancement of directionality if the ascending neuron is ipsilateral to the IN2 (i.e., has its soma on the same side).

A candidate neuron corresponding to the ascending neuron in the scheme is, as already mentioned above, AN1. Candidates corresponding to the interlinking neurons IN1 and IN2, respectively, are the BSN1, a first-order-neuron that directly excites AN1 (Marquart 1985a, b; Boyan, 1992) and is itself directionally sensitive due to inhibitory input from the other side, and the SN1, a local interneuron that receives monosynaptic input from receptors and mediates inhibition on the other side (Marquart, 1985b).

The connectivity proposed in Figure 17A, B predicts that not only differences in spike rates between the two sides, but also latency differences at the level of the tympanal nerves would determine the activity of ascending neurons, thus encoding directional information. On the side facing the sound source, excitation will precede inhibition mediated from the other side, which would result in phasic excitation of the ascending neuron on that side (Fig. 17C), while on the other side excitation will be delayed relative to inhibition. Thus, the ascending neuron on the side facing the sound source will be excited, whereas its contralateral counterpart will be inhibited. The effects of temporal interactions between excitatory and inhibitory postsynaptic potentials on the activity of ascending neurons was demonstrated by Römer et al. (1981) and Römer and Dronse (1982).

Comparison with bushcrickets and crickets

The present review has focused on the acoustical communication and orientation performance of an acridid insect, here representing the caeliferan group of orthopteran insects. It would have been beyond the scope of this review to include all that is known on communication and orientation in the ensiferan species (crickets and bushcrickets) investigated so far. Still, we do not wish to close this chapter without providing a short comparison between the findings described in this chapter and those obtained from crickets and bushcrickets.

Biophysics and directional sensitivity of the tympanal organs

In all of the three major orthopteran groups, crickets, bushcrickets and grasshoppers, in principle, the tympanal organs operate as pressure-difference receivers (at least at low sound-wave frequencies). However, the manner in which the internal and external forces acting on the tympanal membranes summate to generate the effective driving force differs among the three groups. In grasshoppers, as we have seen, the two tympanal membranes are coupled by air sacs, and the phase shifts between sound acting on the inner and outer surfaces of the tympanal membrane are generated by particular structural specializations of the membrane and Müller's organ (Michelsen, 1971b; Stephen and Bennet-Clark, 1982; Breckow and Sippel, 1985; Michelsen and Rohrseitz, 1995). In the crickets, a great part of the phase shift is accomplished with the help of a specialized membrane (termed septum) intercalated in the transverse acoustic trachea (Löhe and Kleindienst, 1994). In many bushcricket species, the pressure-receiver properties of the tympanal organ are achieved by the relatively high gain of the acoustic trachea (Michelsen et al., 1994b).

The diversity of pressure-difference receivers among the various orthopteran groups is mirrored in the diverse and unique tympanal directional characteristics, which in turn determine and limit the orientational capacities of each group. In grasshoppers, as has been described above, the tympanal directional characteristic is reflected in the animal's reliable lateralization capacity in the frontal auditory field, and in the absence of discrimination capacity among directions of stimuli presented laterally. As a consequence, the grasshopper's phonotactic strategy is either to turn towards the louder side, the magnitude of the turn being independent of target angle, or to jump forward when the sound source is ahead and both inputs are equally loud. Crickets, on the other hand, possess a rather flat tympanal directional sensitivity in the frontal auditory field (Boyd and Lewis, 1983; Larsen et al., 1989; Löhe and Kleindienst, 1994). Therefore, they are less accurate in the lateralization performance (Rheinlaender and Blätgen, 1982), which is reflected in their meandering course of locomotion during phonotactic approach (Weber et al., 1981). The magnitude of the cricket's turning angle is not correlated with the angle of sound incidence (Murphy and Zaretzky, 1972) even under closed-loop conditions (Oldfield, 1980). However, the strength of turning increases with deviation angle (Schmitz et al., 1982; Stabel et al., 1989). In contrast, bushcrickets are capable of angle-dependent corrections during phonotaxis even under open-loop conditions (Hardt, 1988; Rheinlaender and Römer, 1990).

The cues used for sound localization

Neural encoding of the interaural intensity differences is based on the differences in the activity of the receptors on the two sides, manifested as (i) the spike rates of individual receptor fibres, (ii) the overall activity of the left and right receptor populations, or (iii) latency differences between the onset of the response on the two sides. Grasshoppers are, as we have seen, capable of using each of these three cues for lateralization. In crickets, as opposed to grasshoppers, lateralization based on time differences alone has not been demonstrated so far, but it is likely that the powerful mutual inhibition of the omega-1-cells (Kleindienst et al., 1981; Wiese and Eilts, 1985) is based on the spike rate in combination with the latency differences of the presynaptic receptor elements. Some bushcricket species that are able to localize very short female clicks seem to make use of the number of receptor units activated on each side (Hardt, 1988).

The neural pathway and neural mechanisms for song recognition and sound localization

The three groups of orthopteran insects differ, in addition, with respect to the organisation of the neuronal elements underlying acoustic information processing. In crickets and bushcrickets, the neural auditory pathway is by far less complex than it is in grasshoppers, at least in as far as the number of neurons and connectivities at the first station of information processing is concerned. In grasshoppers, excitatory and inhibitory interactions of several segmental first-order interneurons result in at least 16 second-order interneurons (each with specific filter properties) ascending on each side. In crickets, the two important ascending neurons, AN1 and AN2, synapse directly with receptor cells (Hennig, 1988) and are inhibited by the con-tralateral omega-neuron-1 (Kleindienst et al., 1981; Wohlers and Huber, 1982; Selverston et al., 1985; Horseman and Huber, 1994). Whether or not the T-fiber units described by Atkins and Pollack (1987) are involved in orientation is not known in crickets, whereas in bushcrickets at least the T1 seems to be involved in directional analysis (Rheinlaender and Römer, 1980; see also Schul, 1997).

A further major difference between acrid and ensiferan insects is that, in the former, most of the ascending neurons pool acoustical information from the two sides and are consequently directionally insensitive. Only two neurons, AN1 and AN2, appear to transmit directional information to the brain. In crickets, on the other hand, the two ascending neurons, AN1 and AN2, encode, at the same time, information about direction *and* song pattern. Thus, there is a strict separation of left and right acoustic inputs, so that two different song patterns can be represented simultaneously on each side (Pollack, 1986), which enables the cricket to choose between two simultaneously presented sound patterns and to track a conspecific in the presence of other simultaneously singing males. The consequence of the difference in organization is revealed in behavioural experiments. Pattern recognition and orientation are processed in parallel in grasshoppers, whereas in crickets they are processed in the same channel (at least up to the brain). The serial processing of information on song pattern and sound location implies that the outputs of the two recognition filters modulate the phonotactic response (Stabel et al., 1989; Wendler, 1990; von Helversen and von Helversen, 1995).

Evolutionary aspects

The principal differences in acoustic information processing outlined above may reflect the differences in the origin of acoustic communication between the ensiferan and caeliferan insects. In bushcrickets and crickets, hearing and stridulation have most probably co-evolved in the context of intraspecific communication (Sharov, 1971; Bailey, 1991). As wing vibra-tion (originally serving for warming up, or to fan odour, or to generate vibratory communication signals) became noisy, and scolopidial cells became sensitive to airborne sound by attaching to the walls of air filled tracheae, the active space of communication was enlarged considerably.

Thus, from the very beginning, the neuronal organisation of the auditory system was selected for recognition and localization of one particular signal, the species-specific song. Only much later in the course of evolution, probably with the appearance of the first insectivorous bats, did the auditory system adapt to the task of predator avoidance (Hoy, 1992). This idea is supported by the finding that, in crickets, the tympanal organs are fully functional only in adult animals (Ball and Young, 1974), when hearing becomes important for mate finding (Ball and Hill, 1978).

Among the caelifera, stridulation appears to have evolved several times independently (as indicated by the many different stridulation modes found in this group), but the same tympanal organs persisted in all families, including those that do not stridulate (Dirsh, 1975; Riede, 1987). This finding and the fact that hearing capacity is fairly well developed in larvae (Petersen et al., 1982) support the idea that hearing in grasshoppers has originally evolved in the context of predator avoidance, the neuronal substrate having been designed to detect possible predators and to distinguish them from insignificant background noise. Summation of the inputs arriving from different directions is advantageous for enhancing the signal-to-noise ratio, though at the cost of the capacity to discriminate between two simultaneously presented sound sources. Thus, a highly developed auditory system might have constituted a pre-disposition to evolve the high diversity of communication signals observed in the repertoire of one species, as well as among different species.

Acknowledgements
For intensive, informative and inspiring discussions I am grateful to Norbert Elsner, Otto von Helversen, Andreas Stumpner and Wolfgang Wickler. I am thankful to Andreas Stumpner and Johannes Schul for providing me with unpublished data, and to Theo Weber for his skilful preparation of the figures. My thanks are also due to Rohini Balakrishnan, Henry Bennet-Clark, Franz Huber, and Miriam Lehrer, who all gave valuable advice and comments on the manuscript and improved the English.

References

Adam, L.J. (1977a) Vorzugsrichtungen nervöser Intensitätsfilter; Hörneuronen im Gehirn der Heuschrecke *Locusta migratoria* (Acridiae). *Zool. Jb. Physiol.* 81:250–272.

Adam, L.J. (1977b) The oscillatory summed action potential of an insect's auditory nerve (*Locusta migratoria*, Acrididae). I. Its original form and time constancy. *Biol. Cybern.* 26: 241–247.

Atkins, G. and Pollack, G.S. (1987) Correlations between structure, topographic arrangement, and spectral sensitivity of sound-sensitive interneurons in crickets. *J. Comp. Neurol.* 266: 398–412.

Autrum, H. (1940) Über Lautäusserungen und Schallwahrnehmungen bei Arthropden. II. Das Richtungshören von *Locusta* und Versuch einer Hörtheorie für Tympanalorgane vom Locustidentyp. *Z. vergl. Physiol.* 28:326–352.

Autrum, H., Schwartzkopff, J. and Swoboda, H. (1961) Der Einfluß der Schallrichtung auf die Tympanal-Potentiale von *Locusta migratoria* L. *Biol. Zentralblatt* 80:385–402.

Bailey, W.J. (1991) *Acoustic Behaviour of Insects*. Chapman & Hall, London.

Bailey, W.J. and Rentz, D.C.F. (1990) *The Tettigoniidae. Biology, Systematics and Evolution.* Springer-Verlag, Berlin, Heidelberg, New-York.

Ball, E.E. and Hill, K.G. (1978) Functional development of the auditory system of the cricket *Teleogryllus commodus. J. Comp. Physiol.* 127:131–138.

Ball, E.E. and Young, D. (1974) Structure and development of the auditory system in the prothoracic leg of the cricket *Teleogryllus commodus* (Walker). II. Postembryonic development. *J. Z. Zellforsch.* 147:313–324.

Bauer, M. and von Helversen, O. (1987) Separate localisation of sound recognizing and sound producing neural mechanisms in a grasshopper. *J. Comp. Physiol.* 161:95–101.

Breckow, J. and Sippel, M. (1985) Mechanics of the transduction of sound in the tympanal organ of adults and larvae of locusts. *J. Comp. Physiol.* 157:619–629.

Boyan, G.S. (1983) Postembryonic development in the Auditory System of the Locust. Anatomical and physiological characterisation of interneurons ascending to the brain. *J. Comp. Physiol.* A 151:499–512.

Boyan, G.S. (1984) Neural mechanisms of auditory information processing by identified interneurons in Orthoptera. *J. Insect. Physiol.* 30:27–41.

Boyan, G.S. (1992) Common synaptic drive to segmentally homologous interneurons in the locust. *J. Comp. Neurology* 321:544–554.

Boyd, P. and Lewis, B. (1983) Peripheral auditory directionality in the cricket *(Gryllus campestris L., Teleogryllus oceanicus* Le Guillou). *J. Comp. Physiol.* 153:523–532.

Dirsh, V.M. (1975) A preliminary revision of the families and subfamilies of Acridoidea (Orthoptera, Insecta). *Bull. Brit. Mus. (Nat. Hist.)* 10:349–419.

Elsner, N. (1974) Neuroethology of sound production in gomphocerine grasshoppers. I. Song patterns and stridulatory movements. *J. Comp. Physiol.* 88:67–102.

Elsner, N. (1994) The search for the neural centres of cricket and grashopper song. *In*: K. Schildberger and N. Elsner (eds): *Neural Basis of Behavioural Adaptations. Progress in Zoology* 39:167–193.

Elsner, N. and Huber, F. (1969) Die Organisation des Werbegesanges der Heuschrecke *Gomphocerippus rufus* L. in Abhängigkeit von zentralen und peripheren Bedingungen. *Z. vergl. Physiol.* 65:389–423.

Ewing, A.W. (1989) *Arthropod Bioacoustics: Neurobiology and Behaviour*. Edinburgh University Press, Edinburgh.

Gilbert, F. and Elsner, N. (1995) Directional hearing of the grasshopper *Chorthippus biguttulus* in natural habitats. *Proceedings of the 23rd Göttingen Neurobiology Conference* Vol. II: 276.

Gray, E.G. (1960) The fine structure of the insect ear. *Phil. Trans. Roy. Soc. Lond.* B 243:75–94.

Hardt, M. (1988) *Zur Phonotaxis von Laubheuschrecken: Eine vergleichende verhaltensphysiologische und neurophysiologisch/neuroanatomische Untersuchung*. Ph.Thesis, Ruhr-Universität, Bochum.

Halex, H., Kaiser, W. and Kalmring, K. (1988) Projection areas and branching patterns of the tympanal receptor cells in migratory locusts, *Locusta migratoria* and *Schistocera gregaria. Cell Tiss. Res.* 253:517–528.

Hedwig, B. (1986a) On the role in stridulation of plurisegmental interneurons of the acridid grasshopper *Omocestus viridulus* L.: I. Anatomy and physiology of descending cephalothoracic interneurons. *J. Comp. Physiol.* A 158:413–427.

Hedwig, B. (1986b) On the role in stridulation of plurisegmental interneurons of the acridid grasshopper *Omocestus viridulus* L.: II. Anatomy and physiology of ascending and T-shaped interneurons *J. Comp. Physiol.* A 158:429–444.

Hedwig, B. (1995) Kontrolle des Stridulationsverhaltens von Feldheuschrecken durch deszendierende Hirnneurone. *Verh. Dtsch. Zool. Ges.* 88.2:181–190.

Hedwig, B. and Heinrich, R. (1997) Identified descending brain neurons control different stridulatory motor patterns in an acridid grasshopper. *J. Comp. Physiol.* A 180:285–294.

Hennig, R.M. (1988) Ascending auditory interneurons in the cricket *Teleogryllus commodus* (Walker): comparative physiology and direct connections with afferents. *J. Comp. Physiol.* A 163:135–143.

Horseman, B.G. and Huber, F. (1994) Sound localization in crickets. I. Contralateral inhibition of an ascending auditory interneuron (AN1) in the cricket *Gryllus bimaculatus. J. Comp. Physiol.* A 175:389–398.

Hoy, R.R. (1992) The evolution of hearing in insects as an adaptation to predation from bats. *In*: D.B. Webster, R.R. Fay and A.N. Popper (eds): *The Evolutionary Biology of Hearing*. Springer-Verlag, New York, Berlin, pp 115–129.

Huber, F. (1963) The role of the central nervous system in Orthoptera during co-ordination and control of stridulation. *In*: R.G. Busnel (ed.): *Acoustic Behaviour of Animals*. Elsevier Publ. Company Amsterdam, London, New York, pp 440–488.

Huber, F. (1990) Cricket neuroethology: Neural basis of intraspecific acoustic communication. *Advances in the Study of Behavior* 19:299–356.

Huber, F., Moore, T.E. and Loher, W. (1989) *Cricket behavior and neurobiology*. Cornell University Press, Ithaca, New York.

Kleindienst, H.-U., Koch, U.T. and Wohlers, D.W. (1981) Analysis of the cricket auditory system by acoustic stimulation using a closed sound field. *J. Comp. Physiol.* 141:283–296.

Krahe, R. and Ronacher, B. (1993) Long rise times of sound pulses in grasshopper songs improve the directionality cues received by the CNS from auditory receptors. *J. Comp. Physiol.* A 173:425–434.

Lang, F. (1996) Noise filtering in the auditory system of *Locusta migratoria*. *J. Comp. Physiol.* A 179:575–585.

Larsen, O.N., Kleindienst, H-U. and Michelsen, A. (1989) Biophysical aspects of sound reception. *In*: F. Huber, T. Moore and W. Loher (eds): *Cricket Behavior and Neurobiology*. Cornell Univ. Press, New York, pp 364–390.

Lewis, D.B. (1974) The physiology of the tettigoniid ear. I. The implications of the anatomy of the ear to its function in sound reception. *J. Exp. Biol.* 60:821–837.

Löhe, G. and Kleindienst, H-U. (1994) The role of the medial septum in the acoustic trachea of the cricket *Gryllus bimaculatus*. *J. Comp. Physiol.* A 174:601–606.

Marquart, V. (1985a) *Auditorische Interneuronen im thorakalen Nervensystem von Heuschrecken. Morphologie, Physiologie und synaptische Verbindungen*. Ph.Thesis, Universität Bochum.

Marquart, V. (1985b) Local interneurons mediating excitation and inhibition onto ascending neurons in the auditory pathway of grasshoppers. *Naturwissenschaften* 72:42–44.

Michelsen, A. (1971a) The physiology of the locust ear. II. Frequency discrimination based upon resonances in the tympanum. *Z. Vergl. Physiol.* 71:63–101.

Michelsen, A. (1971b): The Physiology of the Locust Ear. III. Acoustical properties of the intact ear. *Z. vergl. Physiol.* 71:102–128.

Michelsen, A. (1978) Sound reception in different environments. *In*: M.A. Ali (ed.): *Sensory Ecology*. Plenum, New York, London, pp 345–373.

Michelsen, A. and Löhe, G.(1995) Tuned directionality in cricket ears. *Nature* 375:639.

Michelsen, A. and Nocke, H. (1977) Biophysical aspects of sound communication in insects. *Adv. Insect Physiol.* 10:247–296.

Michelsen, A. and Rohrseitz, K. (1995) Directional sound processing and interaural sound transmission in a small and a large grasshopper. *J. Exp. Biol.* 198:1817–1827.

Michelsen, A., Popov, A.V. and Lewis, B. (1994a) Physics of directional hearing in the cricket *Gryllus bimaculatus*. *J. Comp. Physiol.* A 175:153–164.

Michelsen, A., Heller, K-G., Stumpner, A. and Rohrseitz, K. (1994b) The gain of the acoustic trachea in bushcrickets, determined with a new method. *J. Comp. Physiol.* A 175:145–151.

Miller, L.A. (1977) Directional hearing in the locust *Schistocerca gregria* Forskal (Arididae, Orthoptera): *J. Comp. Physiol.* 71:102–128.

Mörchen, A. (1980) Spike count and response latency: Two basic parameters encoding direction in the CNS of Insects. *Naturwissenschaften* 67:469.

Mörchen, A., Rheinlaender, J. and Schwartzkopff, J. (1978) Latency shift in insect auditory nerve fibers. *Naturwissenschaften* 65:656–657.

Murphy, R.K. and Zaretzky, M. (1972) Orientation to calling song by female crickets, *Scapsipedus marginatus* (Gryllidae). *J. Exp. Biol.* 56:335–352.

Oldfield, B.P. (1980) Accuracy of orientation in female crickets, *Teleogryllus oceanicus* (Gryllidae): dependence on song spectrum. *J. Comp. Physiol.* 147:461–469.

Petersen, M., Kalmring, K. and Cokl, A. (1982) The auditory system in larvae of the migratory locust. *Physiol. Entomol.* 7:43–54.

Pollack, G.S. (1986) Discrimination of calling song models by the cricket, *Teleogryllus oceanicus*: the influence of sound direction on neural encoding of the stimulus temporal pattern and on phonotactic behavior. *J. Comp. Physiol.* A 158:549–561.

Rehbein, H. (1976) Auditory neurons in the ventral cord of the locust: morphological and functional properties. *J. Comp. Physiol.* 110:233–250.

Rheinlaender, J. (1984) *Das akustische Orientierungsverhalten von Heuschrecken, Grillen und Fröschen: Eine vergleichende neuro- und verhaltensphysiologische Untersuchung.* Habilitationsschrift, Universität Bochum.

Rheinlaender, J. and Blätgen, G. (1982) The precision of auditory lateralization in the cricket, *Gryllus bimaculatus. Physiol. Entomol.* 7:209–218.

Rheinlaender, J. and Mörchen, A. (1979) Time-intensity trading in locust auditory interneurones. *Nature* 281:672–674.

Rheinlaender, J. and Römer H. (1980) Bilateral coding of sound direction in the CNS of the bushcricket *Tettigonia viridissima* (Orthoptera, Tettigoniidae). *J. Comp. Physiol.* 140: 101–111.

Rheinlaender, J. and Römer, H. (1990) Acoustic cues for sound localisation and spacing in orthopteran insects. *In*: W.J. Bailey and D.C.F. Rentz (eds): *The Tettigoniidae. Biology, Systematics and Evolution.* Springer-Verlag, Berlin, Heidelberg, New York, pp 248–264.

Riede, K. (1987) A comparative study of mating behaviour in some neotropical grasshoppers (Acridoidea). *Ethology* 76:265–296.

Riede, K., Kämper, G. and Höfler, I. (1990) Tympana, auditory thresholds, and projection areas of tympanal nerves in singing and silent grasshoppers (Insecta, Acridoidea). *Zoomorphology* 109:223–230.

Robert, (1989) The auditory behaviour of flying locusts. *J. Exp. Biol.* 147:279–301.

Römer, H. (1976) Die Informationsverarbeitung tympanaler Receptorelemente von *Locusta migratoria* (Acrididae, Orthoptera). *J. Comp. Physiol.* 109:102–122.

Römer, H. and Dronse, R. (1982) Synaptic mechanisms of monaural and binaural processing in the locust. *J. Insect. Physiol.* 28:365–370.

Römer, H. and Marquart, V. (1984) Morphology and physiology of auditory interneurons in the metathoracic ganglion of the locust. *J. Comp. Physiol.* A 155: 249–262.

Römer, H. and Rheinlaender, J. (1983) Electrical stimulation of the tympanal nerve as a tool for analysing the response of auditory interneurons in the locust. *J. Comp. Physiol.* 152: 289–296.

Römer, H., Rheinlaender, J. and Dronse, R. (1981) Intracellular studies on auditory processing in the metathoracic ganglion of the locust. *J. Comp. Physiol.* 144:305–312.

Römer, H., Marquart, V. and Hardt, M. (1988) Organization of a sensory neuropile in the auditory pathway of two groups of Orthoptera. *J. Comp. Neurol.* 275:201–215.

Ronacher, B. (1989) Stridulaton of acrid grasshoppers after hemisection of thoracic ganglia: evidence for hemiganglionic oscillators. *J. Comp. Physiol.* 164:723–736.

Ronacher, B. (1991) Contribution of abdominal commissures in the bilateral coordination of the hindlegs during stridulation in the grasshopper *Chorthippus dorsatus. J. Comp. Physiol.* A 169:191–200.

Ronacher, B. and Krahe, R. (1997) Long rise times of acoustic stimuli improve directional hearing in grasshoppers. *Naturwissenschaften* 84:168–170.

Ronacher, B. and Römer, H. (1985) Spike synchronization of tympanic receptor fibres in a grasshopper (*Chorthippus biguttulus* L., Acrididae): A possible mechanism for the detection of short gaps in model songs. *J. Comp. Physiol.* A 157:631–642.

Ronacher, B. and Stumpner, A. (1988) Filtering of behaviourally relevant temporal parameters of a grasshopper's song by an auditory interneuron. *J. Comp. Physiol.* A 163:517–523.

Ronacher, B. and Stumpner, A. (1993) Parallel processing of song pattern and song direction by ascending interneurons in the grasshopper *Chorthippus biguttulus. In*: K. Wiese, F.G. Gribakin, A.V. Popov and G. Renninger (eds): *Sensory Systems of Arthropods.* Birkhäuser Verlag, Basel, Boston, Berlin, pp 376–385.

Ronacher, B., von Helversen, D. and von Helversen, O. (1986) Routes and stations in the processing of auditory directional information in the CNS of a grasshopper, as revealed by surgical experiments. *J. Comp. Physiol.* A 158:363–374.

Ronacher, B., von Helversen, D. and von Helversen, O. (1989) Search for the sexual partner in grasshoppers – how to make the best of it. *Proc. 17th Göttingen Neurobiol. Conf:* p 155.

Ronacher, B., Stumpner, A., Sokoliuk, T. and Herrmann, B. (1993) Acoustic communication of grasshopper males after lesions in the thoracic connectives: Correlation with the ascending projections of identified auditory neurons. *Zool. Jb. Physiol.* 97:199–214.

Schildberger, K. (1984) Temporal selectivity of identified auditory neurons in the cricket brain. *J. Comp. Physiol.* A 154:171–185.

Schildberger, K. and Kleindienst, H.-U. (1989) Sound localization in intact and one-eared crickets: Comparison of neural properties and closed-loop behavior. *J. Comp. Physiol.* A 165: 615–626.

Schmitz, B., Scharstein, H. and Wendler, G. (1982) Phonotaxis in *Gryllus campetris* L. (Orthoptera, *Gryllidae*). I. Mechanism of acoustic orientation in intact females. *J. Comp. Physiol.* 148:431–444.

Schul, J. (1997) Neural basis of phonotactic behaviour in *Tettigonia viridissima*: processing of behaviourally relevant signals by auditory afferents and thoracic interneurons. *J. Comp. Physiol.* A 180:573–583.

Schwabe, J. (1906) Beiträge zur Morphologie und Histologie der tympanalen Sinnesapparate der Othopteren. *Zoologica* 20:1–148.

Selverston, A.I., Kleindienst, H.-U. and Huber, F. (1985) Synaptic connectivity between cricket auditory interneurons as studied by selective photoinactivation. *J Neurosci. Res.* 5: 1283–1292.

Sharov, A.G. (1971) Phylogeny of the Orthopteroidea. *Acad. Sci. USSR,* Isr. Prog. Sci. Translat., Jerusalem.

Sokoliuk, T., Stumpner, A. and Ronacher, B. (1989) GABA-like immunoreactivity suggests an inhibitory function of the thoracic low-frequency neuron (TN1) in acridid grasshoppers. *Naturwissenschaften* 76:223–225.

Stabel, J., Wendler, G. and Scharstein, H. (1989) Cricket phonotaxis: localization depends on recognition of the calling song pattern. *J. Comp. Physiol.* A 165:65–177.

Stephen, R.O. and Bailey, W.J. (1982) Bioacoustics of the ear of the bushcricket *Hemisaga* (Saginae). *J. Acoust. Soc. Am.* 72:13–25.

Stephen, R.O. and Bennet-Clark, H.C. (1982) The anatomical and mechanical basis of stimulation and frequency analysis in the locust ear. *J. Exp. Biol.* 99:279–314.

Stumpner, A. (1988) *Auditorische thorakale Interneurone von Chorthippus biguttulus L.: Morphologische und physiologische Charakterisierung und Darstellung ihrer Filtereigenschaften für verhaltensrelevante Lautattrappen.* Ph.Thesis, Universität Erlangen-Nürnberg.

Stumpner, A. (1989) Physiological variability of auditory neurons in a grasshopper. Comparison of twin cells and mirror-image cells. *Naturwissenschaften* 76:427–429.

Stumpner, A. and Heller, K-G. (1992) Morphological and physiological differences of the auditory system in three related bushcrickets (Orthoptera: Phaneropteridae, *Poecilimon*). *Physiol. Entomol.* 17:73–80.

Stumpner, A. and Ronacher, B. (1991) Auditory interneurons in the metathoracic ganglion of the grasshopper *Chorthippus biguttulus*. I. Morphological and physiological characterization. *J. Exp. Biol.* 158:391–410.

Stumpner, A. and Ronacher, B. (1994) Neurophysiological aspects of song and pattern recognition and sound localization in grasshoppers. *Amer. Zool.* 34:696–705.

Stumpner, A., Ronacher, B. and von Helversen, O. (1991) Auditory interneurons in the metathoracic ganglion of the grasshopper *Chorthippus biguttulus*. II. Processing of temporal patterns of the song of the male. *J. Exp. Biol.* 158:411–430.

von Helversen, D. (1972) Gesang des Männchens und Lautschema des Weibchens bei der Feldheuschrecke *Chorthippus biguttulus* (Orthoptera, Acrididae). *J. Comp. Physiol.* 81: 381–422.

von Helversen, D. (1984) Parallel processing in auditory pattern recognition and directional analysis by the grasshopper *Chorthippus biguttulus* L. (Acrididae). *J. Comp. Physiol.* A 154: 837–846.

von Helversen, D. (1993) "Absolute steepness" of ramps as an essential cue for auditory pattern recognition by a grasshopper (Orthoptera; Acrididae; *Chorthippus biguttulus* L.). *J. Comp. Physiol.* A 172:633–639.

von Helversen, D. and Rheinlaender, J. (1988) Interaural intensity and time disrimination in an unrestraint grasshopper: a tentative approach. *J. Comp. Physiol.* A 162:333–340.

von Helversen, D. and von Helversen, O. (1990) Pattern recognition and directional analysis: routes and stations of information flow in the CNS of a grasshopper. *In:* F.G. Gribakin, K. Wiese and A.V. Popov (eds): *Sensory Systems and Communication in Arthropods.* Birkhäuser Verlag, Basel, Boston, Berlin, pp 209–216.

von Helversen, O. and von Helversen, D. (1994) Forces driving coevolution of song and song recognition in grasshoppers. *In*: K. Schildberger and N. Elsner (eds): *Neural Basis of Behavioural Adaptations. Progress in Zoology* 39:253–284.

von Helversen, D. and von Helversen, O. (1995) Acoustic pattern recognition and orientation in orthopteran insects: parallel or serial processing? *J. Comp. Physiol.* A 177:767–774.

von Helversen, D. and von Helversen, O. (1997) Recognition of sex in the acoustic communication of the grasshopper *Chorthippus biguttulus* (Orthoptera, Acrididae). *J. Comp. Physiol.* A 180:373–386.

Weber, T., Thorson, J. and Huber, F. (1981) Auditory behavior of the cricket. I. Dynamics of compensated walking and discrimination paradigms on the Kramer treadmill. *J. Comp. Physiol.* 141:215–232.

Wendler, G. (1990) Pattern recognition and localization in cricket phonotaxis. *In*: F.G. Gribakin, K. Wiese and A.V. Popov (eds): *Sensory Systems and Communication in Arthropods.* Birkhäuser Verlag, Basel, Boston, Berlin, pp 387–394.

Werner, A. and Elsner, N. (1995) Directional hearing of the grasshopper *Chorthippus biguttulus* (L.) in natural habitats. *Proceedings of the 23rd Göttingen Neurobiology Conference* Vol. II, p 276.

Wiese, K. and Eilts, K. (1985) Evidence for matched frequency dependence of bilateral inhibition in the auditory pathway of *Gryllus bimaculatus*. *Zool. Jahrb. Physiol.* 89:181–201.

Wohlers, D.W. and Huber, F. (1982) Processing of sound signals by six types of neurons in the prothoracic ganglion of the cricket *Gryllus campestris* L. *Cell Tiss. Res.* 239:555–565.

Wolf, H. (1986) Response patterns of two auditory interneurons in a freely moving grasshopper (*Chorthippus biguttulus* L.). I. Response properties in the intact animal. *J. Comp. Physiol.* A 158:697–703.

Orientation and Communication in Arthropods
ed. by M. Lehrer
© 1997 Birkhäuser Verlag Basel/Switzerland

Pheromone-controlled anemotaxis in moths

K.-E. Kaissling

Max-Planck-Institut für Verhaltensphysiologie, D-82319 Seewiesen/Starnberg, Germany

Summary. Males of several moth species possess very sensitive chemoreceptors for detecting conspecific female sex pheromone. Behavioural experiments conducted in wind tunnels, as well as electrophysiological experiments with pulsed odor stimuli, revealed that moths are remarkably well adapted to the rapid changes in stimulus concentration they encounter in a natural odor plume. As a walking or a flying moth orients to an odor source, it displays turning responses to individual odor pulses of only a few tens of ms duration. Both receptor cells and central neurons can resolve up to ten odor pules per second. When odor pulses stop coming, a walking moth circles on the spot, and a flying moth selects a path transverse to the odor plume, both types of behaviour being thought to serve for regaining olfactory information. Concentration gradients within the odor plume, however, do not provide reliable information on the location of the odor source. The odor pulses rather act to trigger locomotion upwind (positive anemotaxis), a strategy that will eventually guide the male to the odor source. Moths progress upwind most effectively when the odor pulses are repeated at a rate of about 3/s, regardless of the stimulus intensity. The orientational cues by which the animal adjusts its upwind course of locomotion are provided by mechanical and visual stimuli. Possible mechanisms for odor-elicited anemotaxis are discussed, in particular the cybernetic model system of Kramer (1996) with two feedback loops, that explains anemotactic orientation in flight based solely on the flow of ground patterns over the animal's eyes. A model system for walking requires only one feedback loop. Computer simulations of anemotactic behaviour produce flight and walking tracks similar to those observed in real animals. The odor stimulus modulates several parameters of this model.

Introduction

The orientation of moths to an odor source attracted the interest of numerous researchers over many decades, not least because early reports have described observations on the amazing capacity of males of some species to detect a female over distances of more than a kilometer (e.g., Fabre, 1919; Rau and Rau, 1929; Collins and Potts, 1932). Males of these species, particularly saturniids and lymantriids, possess large antennae with high sensitivity to conspecific sex pheromones (for an extensive literature survey see Priesner, 1973).

The experimental work that followed the early observations was concerned with orientation to *nearby* odor sources, due to the obvious technical difficulties associated with experiments involving long-distance orientation. The largest amount of data was obtained in experiments in which the insects were tested in so-called "wind tunnels", with the odor source at one end, and the male's starting point at the other. Although we do not know whether or not long-distance and short-distance orientation triggered by odor stimuli are governed by the same mechanisms, the studies of short-distance orientation have contributed a lot to our understanding of the possible mechanisms underlying odor-based spatial orientation.

The present chapter is not meant to deal with all of the aspects of odor-controlled orientation behavior in moths and other insects. These have already been addressed in many previous publications (e.g., Priesner, 1973, 1984; Roelofs and Cardé, 1977; Cardé, 1979; Cardé and Charlton, 1984; Mayer and Mankin, 1985; Kaissling, 1986a, 1987; Boeckh and Ernst, 1987; Baker 1989a, b; Kennedy, 1983; Payne et al., 1986; Masson and Mustaparta, 1990; Kaissling and Kramer, 1990; Murlis et al., 1992; Hansson, 1995; Cardé and Mafra-Neto, 1996; Christensen et al., 1996; Hildebrand, 1996). Furthermore, many comprehensive reviews on anemotaxis in moths and other insects are presented in a recent book edited by Cardé and Minks (1996). Therefore, it does not seem necessary to describe again the history of the exciting development achieved in this field of research. Rather, this chapter concentrates on possible mechanisms of anemotaxis and its control by odor. In particular, it outlines a model of causal connections, the anemotactic "nexus" of Kramer (1996), that describes the possible neural mechanisms underlying the anemotactic behavior of moths. All of the theoretical considerations incorporated in the model are based on experimental results to be described below.

Odor pulses elicit anemotaxis

How does a male moth find its female by orienting in an airborne pheromone plume? To answer this question, the properties of the plume must be considered.

The physical properties of the odor plume

A small odor source such as the female pheromone gland releases a slim filament of pheromone-laden air that is distorted due to air turbulence. Within the odor plume so created, diffusion that would produce a uniform gradient of odor concentration can be neglected. The odor plume does not provide a reliable concentration gradient that could guide the male to the odor source. This important aspect of olfactory orientation was first addressed by Wright (1958). It follows that the male moth, once it has detected the presence of the pheromone, must use cues others than the odor gradient to localize the odor source.

Murlis and Jones (1981) and Murlis et al. (1990, 1992) made quantitative estimates of the extremely inhomogeneous odor distribution by replacing the odor source with an ion generator and positioning ion detectors downwind. This arrangement allowed the ion flux to be recorded with high temporal and spatial resolution. At a given downwind site, the ion concentrations varied widely and showed rapid changes. Typically, bursts of peak ion concentrations lasting for 0.1 to 0.4 s alternated with pauses, usually longer, during which no ions were measurable. As the distance from the ion source increased, the time pattern of bursts and pauses did not change significantly if measured within the meandering plume no more than a few tens of meters from the source. The mean concentrations within the ion peaks decreased but were still well above the mean concentration of the ion cloud as a whole (Murlis, 1996), Thus, odor packets can travel a considerable distance and still retain suprathreshold concentrations beyond the mean threshold volume calculated by Bossert and Wilson (1963, see below).

The best strategy is to move upwind: Anemotaxis

A male flying in such a plume will receive brief pulses of pheromone at relatively high concentration whenever it hits a packet of pheromone-laden air. Between the pulses, the pheromone concentration can be very low. What remains for a moth as the best strategy to orient towards the odor source is to move upwind (positive anemotaxis). However, by just averaging over all wind vectors, including those of air without pheromone, the male, although flying due upwind, might easily miss the goal. Therefore, it should orient upwind only if it encounters a pheromone-laden packet of air that has obviously passed the odor source and thus, at least on the average, carries relevant vectorial information.

The adequate stimulus is intermittent

Indeed, male moths are able to respond with turns into the upwind direction to each of several pheromone pulses per second (Fig. 1), as was first reported for males of *Bombyx mori* walking in a locomotion compensator (Kramer, 1986). An optimum upwind orientation was obtained with three pheromone pulses per second, the highest repetition rate used. In a constant odor stream, the orientation was clearly impaired (see Fig. 1). Similar results were obtained using stimulus intensities ranging over four decades of odor concentrations (Kramer, 1986). The conclusion that intermittent odor stimulation improves upwind flight has already been drawn by Kennedy et al. (1980, 1981), and was confirmed, in parallel with Kramer, also for the tortricid *Grapholita molesta* by Willis and Baker (1984) and Baker et al. (1985). The conclusion that only intermittent odor stimuli are effective for olfactory orientation has also been drawn from results obtained in experiments on aquatic crustaceans (see M.J. Weissburg, this volume).

A perfectly constant odor concentration does not elicit upwind orientation even if the concentration is sufficiently high for the receptor cells to

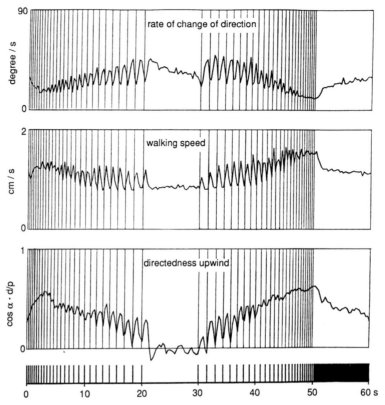

Figure 1. Anemotactic responses of walking male moths of *Bombyx mori* elicited by pulsed stimulation with bombykol (vertical bars, pulse duration 150 ms). Stimulus pulses were first given at increasing intervals followed by a 10 s pause; then pulses were given with decreasing intervals followed by a 10 s period of continuous stimulation. This sequence was repeated ten times for each run. The behavioral responses were measured by means of a locomotion compensator ("Kramer sphere") and averaged over 55 runs of a total of 12 animals. Three different parameters were evaluated with a time resolution of 0.1 s: rate of change of direction, walking speed, and directedness upwind cos $\alpha \cdot$ d/p, where α is the course angle, d the distance between major turns, and p the path length of the walk. The directedness parameter is the length of the upwind vector divided by the path length. The anemotactic orientation performance is best at the highest frequency of bombykol pulses, and poorest during continuous stimulation. Modified from Kramer (1986), with author's permission.

respond. Experimentally, it is difficult to offer a constant odor concentration, because the insect can, by moving its antennae or changing their orientation with respect to the wind direction, modulate the uptake of pheromone molecules even at a constant stimulus concentration in air and a constant wind velocity. A simulation of completely constant pheromone concentration was obtained in experiments on males of *Bombyx mori* by means of the bombykol mimic (Z,E)-4,6-hexadecadiene. Upon brief (10 s) stimulation with this compound, the bombykol receptor cells showed constant firing of nerve impulses for many minutes, but the moths never-theless did not respond anemotactically (Kaissling et al., 1989). However, when the cell's firing was modulated by brief puffs of the inhibitor linalool at a repetition rate of three puffs per s, the moths readily walked upwind (Kramer, 1992).

Receptor properties

As expected on the basis of the behavioral responses, the receptor cells can resolve pheromone pulses at repetition rates up to 10/s (Kaissling, 1986b; Rumbo and Kaissling, 1989; Marion-Poll and Tobin, 1992; Kodadová, 1996). Projection neurons of the antennal lobe display a similarly high temporal resolution (Christensen and Hildebrand, 1988; Christensen et al., 1996). This finding suggests that higher-order neurons in the CNS, prob-ably cells in the mushroom bodies, require a modulated (intermittent) input in order to trigger behavioral responses.

All of these findings show that moths are extraordinarily well adapted to the temporal pattern of intermittent odor stimulation in the natural odor plume. Females of some moth species were found to release odor signals in a pulsed manner (Connor et al., 1980), but even in species in which this is not the case, air turbulences will cause the odor stimulus to become inter-mittent as it travels downwind away from the source.

Responses to termination of odor stimulation

What does a moth do when it loses the odor plume, i.e., when the odor pulses stop coming? The answer to this question is important for under-standing the mechanism underlying odor-dependent orientation to the upwind direction.

Walking animals

When odor pulses stop arriving, walking moths start turning in narrow circles, as was first reported by Kramer (1975) for male *Bombyx mori*. This behavior may be considered as a strategy of waiting at approximately the place where the odor plume had been before and where it may be most likely to return. It is similar to the circling behaviour observed when the male is near the female, also known as a "mating dance" (Schwinck, 1955): circling increases the male's change of meeting the female. Interestingly, *Bombyx* males do not seem to recognize the female visually, as can easily be shown by placing the male upwind of the female, even at distance as short as a few centimeters. Still, upon mechanical contact with almost any object, the male will display copulatory behavior when exposed to the sex pheromone (bombykol).

Flying animals

When flying moths lose the odor plume, they usually exhibit a behavior termed "casting", i.e., lateral cross-wind excursions, alternating between left and right, with increasing amplitudes and, sometimes, with a slight regression downwind (Fig. 2). The casting behavior, as e.g. described by Kennedy and Marsh (1974), can, again, be considered as a strategy to regain the odor plume (David et al., 1983). The offset downwind ("regressive casting", Kuenen and Cardé, 1994) would help to find an odor source that had been missed during the previous upwind flight. Counterturns as observed in casting also occur during the approach flight within the plume, if the moth flies not straight upwind but on an oblique (menotactic) course with alternating changes of direction (zig-zagging). Counterturning in casting and zig-zagging was considered as an internally generated, oscillatory behavior, merely initiated by a chemical stimulus (Kennedy, 1983).

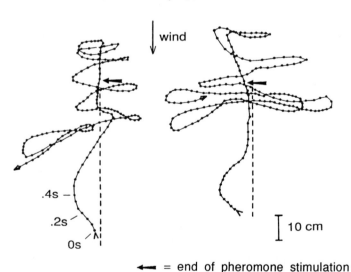

Antheraea polyphemus

wind

.4s —
.2s
0s

10 cm

◄■ = end of pheromone stimulation

Figure 2. Flight tracks of male saturniid moths in a wind tunnel video-taped from above. The pheromone plume was 5 cm in diameter, as calibrated by $TiCl_4$ smoke. Wind velocity 0.5 m/s. Upon termination of the odor plume (arrow), the moths continued upwind flight for 300–500 ms, followed by casting (transverse flight across the plume) with a bias downwind. Modified from Baker and Vogt (1988), with authors' permission.

An internal counterturn generator

Often the counterturns observed during casting or zig-zagging described above occur in a regular manner which Kramer (1975; see also Kennedy and Marsh, 1974) interpreted as indicating the existence of an internal turning tendency (T_0) with changing sign. The value of T_0 is alternatingly added to or substracted from an α-dependent turning tendency (T_α), caused by deviations of the course angle α (between body axis and wind direction) from the exact upwind direction (Kramer, 1975). T_0 may be a constant, whereas T_α depends on the course angle α, but it may also be influenced by other parameters, such as the wind velocity w or the odor concentration c. An oblique (with respect to the wind direction) but straight path occurs if $T_0 >$ zero and the difference T_d between T_0 and T_α,

$$T_d = s \cdot T_0 - T_\alpha \qquad (1)$$

equals zero. When the sign s (either $+1$ or -1) of T_0 changes, a counterturn to the other side, right or left of the wind direction, will occur until the same course angle, but with the altered sign, is obtained. In the absence of wind or odor, T_α is zero and the remaining internal turning tendency ($s \cdot T_0$) causes circling in the walking moth. If the sign s changes, the direction of circling would change. In casting flight, changes of sign would produce counterturns.

A brief odor stimulation triggers a sequence of anemotactic responses

The sequence of events expected to be triggered by a single encounter with an odor-laden air filament during flight is summarized in Figure 3, based on behavioral and electrophysiological data (Kaissling and Kramer, 1990). In this hypothetical example we take the airspeed (a), i.e., the velocity of the animal relative to the surrounding air, to be approximately 60 cm/s, and the wind speed (w) to be 50 cm/s. As the moth crosses an odor filament, in this example 6 mm in diameter, it perceives a stimulus lasting for 10 ms. The receptor cells generate nerve impulses after a latency of at least 10 ms (in our example 25 ms, t_1 in Fig. 3), and reach their maximal discharge rate after about 100 ms. By that time, the flying moth is 6 cm away from its point of encounter with the odor filament. It is only after a behavioral latency (in our example $t_2 = 200$ ms) that the moth begins to turn into the wind and then maintains a course angle α. After further 100 ms (t_3), the receptor cells might stop firing. After some delay ($t_4 = 500$ ms), the moth ceases steering into the wind and begins to fly across the wind line (casting) (t_5). Then a counterturn (t_6) may be elicited by a change of sign of the internal turning tendency, and casting continues to the other side, with increased probability of entering the odor plume. As long as casting goes on, the

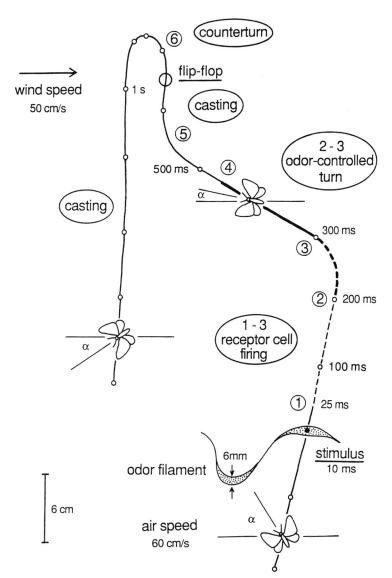

Figure 3. Imaginary flight track of a moth showing the chain of events elicited by crossing of a single odor filament (at time t=0). (1) Receptor cells begin firing; (2) beginning of anemo-tactic manoeuvre, resulting in a small net upwind progress; (3) receptor cell firing ceases; (4) end of upwind surge; (5) downwind turn and onset of casting; (6) counterturn, induced by a change of sign of an internal turning tendency (flip-flop), casting continues. α = course angle. Modified from Kaissling and Kramer (1990).

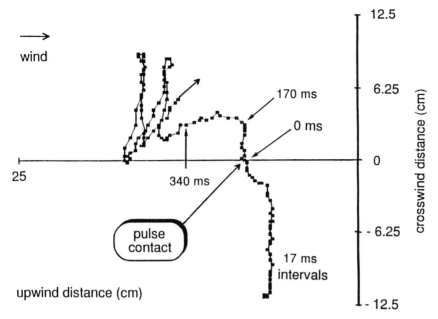

Figure 4. Characteristic change of flight behavior of *Cadra cautella* males in response to contact with a single pheromone pulse, in a wind tunnel. Wind speed was 50 cm/s. Data points represent the position of the moth at 17 ms intervals. Males flying upwind in a fast-pulsed pheromone plume were induced to cast as a result of removal of the plume. Once casting in clean air, the male was exposed to a single pheromone pulse. Contact with pheromone (arrow) was followed by a sharp turn upwind (170 ms after pulse contact), and a faster upwind flight. In the absence of new pheromone pulses, males resumed castig flight (more than 340 ms after pulse contact). From Mafra-Neto and Cardé (1994), with authors' permission.

animal obviously orients to the wind direction, still as a consequence of the previous odor stimulus. Similarly, in a walking silk moth, a single brief bombykol stimulus causes an anemotactic response, a small upwind surge, after which the male starts walking in circles (see above) and exhibits wing vibration for several seconds, sometimes minutes.

Flight tracks like those shown in Figure 3 were, indeed, demonstrated recently in wind tunnel experiments by Cardé and Mafra-Neto (1996) (Fig. 4). The behavioral latencies between bombykol stimulus and wing vibration in *Bombyx mori* males were a few hundred ms at threshold odor concentration, but less than 200 ms with 1000-fold stronger stimuli (Kaissling and Priesner, 1970). Latencies of upwind surge were between 200 and 300 ms for *Cadra cautella* (Cardé and Mafra-Neto, 1996; Mafra-Neto and Cardé, 1996), and 200–500 ms for *Heliothis virescens* (Vickers and Baker, 1994b, 1996; Baker and Vickers, 1996). The duration of the upwind surge in *C. cautella* was 570 ms ± 130 ms for stimuli of 20 ms duration, and 720 ms ± 100 ms for stimuli of 250 ms duration. In *H. virescens*, the surge lasted, on average, for 380 ms. The saturniid moths continued upwind flight for 300–500 ms after termination of the odor stimulation before they started casting (Fig. 2) (Baker and Vogt, 1988).

These examples show that a brief odor stimulus can induce a full sequence of behavioral activities. The oriented locomotion towards an odor source can be regarded as a lining up of several such sequences as odor stimuli are encountered again and again along the animal's course of locomotion. With odor pulses coming in rapid succession, the response sequences to consecutive pulses will overlap. If a new stimulus pulse occurs early enough before casting is initiated, the flight track will straighten, as has been reported for *Grapholita molesta* (Baker et al., 1985; Vickers and Baker, 1994a; Mafra-Neto and Cardé, 1995).

Detection of the adequate stimulus is not only very fast, but also very accurate

The astonishingly high temporal resolution of the olfactory system involved in anemotactic orientation (see above) is even more remarkable when we consider that it is also necessary to discriminate the quality of the pheromone. Only the correct blend of pheromone components elicits an optimal upwind orientation (Linn et al., 1986, 1987; Willis and Baker, 1988; Linn and Roelofs, 1989). Moth pheromones usually consist of a mixture of two or a few pheromone components in species-specific proportions (Arn et al., 1992). Each component is perceived by a specific type of receptor cell (Priesner 1979, 1980; Meng et al., 1989). Cells of a different type can differ also with respect to the temporal pattern of their nerve impulse response, for example with respect to the latency of the first nerve impulse and to the duration of firing following a pheromone stimulus (Kaissling and Kramer, 1990). Thus, recognition of the pheromone blend requires evaluation of a complex pattern of nerve impulse firing.

Wind tunnel experiments showed that the male's flight behavior changes with the composition of the pheromone blend (Palaniswamy et al., 1983; Willis and Baker, 1987, 1988; Witzgall, 1990). Figure 5 shows clear differences in flight tracks for the conspecific female glands, for optimized blends, and for incomplete synthetic pheromone blends, respectively, of two species of moths (Witzgall, 1996). Inhibitory compounds, possibly pheromone components of sympatric species, can also modify the flight pattern, as was reported for *Coleophora laricella* males (Priesner and Witzgall, 1984). These observations suggest that recognition of the pheromone blend occurs upon each of several encounters per second with an air packet containing pheromone.

Finally, the temporal response pattern of a receptor cell depends on stimulus intensity (Kaissling, 1986a), on previous stimulation causing adaptation (Zack, 1979; Kaissling et al., 1987; Baker et al., 1988), and on tem-

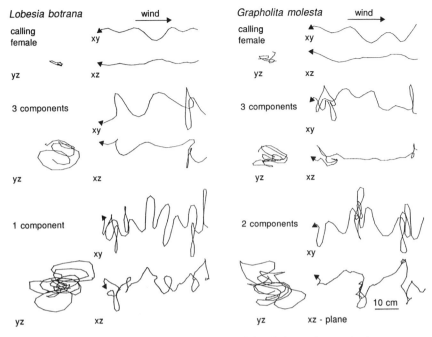

Figure 5. Typical flight tracks of *Lobesia botrana* and *Grapholita molesta* males in response to a natural calling conspecific female (top panels), to optimized (middle panel) and to incomplete synthetic pheromone blends (bottom panels), in a wind tunnel. Each track is shown in three projections (top: xy = horizontal plane; xz = vertical plane, view from side; yz = vertical plane, view from source). The flight tracks towards the natural stimulus were straightest, indicating the importance of the correct pheromone blend. From Witzgall (1996), with author's permission.

perature (Kodadová, 1996; Kodadová and Kaissling, 1996). At high stimulus intensities, i.e., in an adapted state, the responses usually display a shorter latency and last for a shorter time period, i.e., they become more "phasic". This mechanism enables a better resolution of repetitive stimulus pulses, which can be advantageous when the animal is near the odor source (see also M.J. Weissburg, this volume).

The navigation triangle for anemotaxis

Although, as we have seen above, olfactory stimuli elicit and modulate anemotactic orientation, the orientation behavior itself requires the use of other sensory modalities. During walking, the wind direction can be determined by using appropriate mechanoreceptors. In flying, however, the wind drift and involuntary deviations of the animal from its intended direction of locomotion due to the wind need to be compensated (see also K. Kirschfeld, this volume). In this task, a reference to the ground is required, and this reference can be obtained visually ("optomotor anemo-taxis", Kennedy, 1940, 1986; Baker et al., 1984; Willis and Cardé, 1990).

The navigation triangle (Fig. 6), representing movements in a horizontal plane, shows how the wind-based drift causes a flow of visual ground patterns over the moth's eyes. This visual flow can be divided into a trans-verse component T (perpendicular to the animal's longitudinal body axis, to the left positive), and a longitudinal component L (parallel to the ani-mal's longitudinal axis, backwards positive). In order to determine its course angle α or its track angle β, which is the direction of its track over ground with respect to the wind direction ($\beta = \alpha + \delta$), the flyer could eva-luate the ground speed g and the drift angle δ between its body axis and the direction of movement of the ground patterns. Apart from these two para-meters of the visual flow, however, a moth also needs to know the airspeed (a), i.e., its velocity relative to the air. The relations between these angles and the magnitudes of the various velocities are given by the following equations.

$$T = w \cdot \sin \alpha = g \cdot \sin \delta \tag{2}$$

$$L = a - w \cdot \cos \alpha = g \cdot \cos \delta \tag{3}$$

$$w = (T^2 + (a - L)^2)^{1/2} \tag{4}$$

$$g = (T^2 + L^2)^{1/2} \tag{5}$$

$$\sin \beta = \sin \alpha \cdot a/g = aT/[(T^2+L^2)(T^2+(a-L)^2]^{1/2} \tag{6}$$

Flying correctly across the wind (casting) requires adjusting the airspeed (a) to T and L according to

$$a = (T^2+L^2)/L \quad \text{(Kaissling and Kramer, 1990).} \tag{7}$$

According to these considerations, if the animal is to determine its track angle, it needs to measure not only parameters of the visual flow, but, in addition, the airspeed. The airspeed could be measured, for example, by mechanoreceptors capable of sensing air movements, such as have been identified by Gewecke (1975) in locusts, or it could be estimated from the energy consumption of the flight motor, as has been discussed for honeybees (see review by Esch and Burns, 1996), or via an efference copy (see von Holst and Mittelstaedt, 1950). All of these possible mechanisms are, however, rather elaborate and require, in addition, calibration factors, either innate or those acquired by learning.

Is there no simpler neural mechanism for accomplishing successful orientation towards an odor source? In the next section it will be shown that, at least theoretically, there is.

upwind   casting

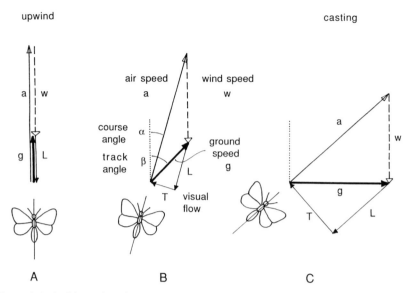

Figure 6. Velocities and angles to be considered in orientation of flying insects. a = airspeed, w = wind speed, g = ground speed. L and T = longitudinal and transverse components, respectively, of the visual flow of ground patterns, in relation to the animal's body axes. (B) The track angle β is always larger than the course angle α, except when both are zero in true upwind flight. Their difference is the drift angle δ (not indicated). For upwind flight (A), the insect needs to minimize T by turning, and to keep L (backwards positive) at a value above zero, by adjusting its airspeed. Precise casting (C) requires measurement of a, T and L (see text).

Kramer's feedback system of odor-controlled anemotaxis

A mechanism of anemotactic orientation that does not require calculations of angles and velocities was first proposed for walking insects by Kramer (1975), and was extended to flying insects by Preiss and Kramer (1984) and Preiss and Spork (1994). In the following, the model for odor-controlled anemotaxis proposed by Kramer (1996) will be described. This model takes into consideration the internal turning tendency ($s \cdot T_0$), as well as the compensation of the α-dependent turning tendency, both of which have been introduced above. It also takes into account the dependence of the animal's speed of motion on the odor concentration, and the experimental results showing that the orientation to wind direction, after it has been initiated by an odor stimulus, persists even in the absence of the odor stimulus (see above). The model produces the observed oriented flight behaviour by solely relying on visual stimuli. It does not require that the airspeed be measured.

Two feedback loops for odor-controlled anemotaxis during flight

According to Kramer's hypothesis, only two feedback loops are sufficient to produce upwind flight under odor stimulation, and casting in the absence of odor. For progressing upwind, it is necessary to keep T (the transverse vector of the visual flow) at a minimum, and L (the longitudinal vector) at a value above zero (see Fig. 6).

One feedback loop controls the component T of the visual flow by eliciting turning about the vertical axis (yaw). When T differs from zero, the animal will turn. In other words, this feedback loop produces a turning command T_d for the flight motor if T_α, the internal representation of T, differs from a set point $s \cdot T_0$ (s standing for the sign $+1$ or -1) (Fig. 7):

$$T_d = s \cdot T_0 - T_\alpha. \tag{8}$$

The flight motor then produces an angular velocity $d\alpha/dt$,

$$d\alpha/dt = k_T (s \cdot T_0 - T_\alpha), \tag{9}$$

(k_T being a proportionality constant).

The second feedback loop controls the component L of the visual flow (see Fig. 6) by adjusting the airspeed (a), i.e., the animal's velocity relative to the air. This loop involves the acceleration (da/dt) of the animal's speed of flight,

$$da/dt = k_L (L_0 - L_a), \tag{10}$$

where $L_0 - L_a = L_d$ is an acceleration command for the flight motor (k_L being a proportionality constant). L_a is an internal variable representing L (see Fig. 6). When L_a differs from a set point L_0, the animal will change its speed of flight: when L_a is small, the animal will accelerate; when L_a is large, it will slow down.

It is important to emphasize that both feedback loops include temporal integration. This is obtained by the flight motor (Fig. 7) producing increasing velocities at a given positive acceleration. If, for example, the wind speed is increased, L and L_a become smaller; L_a then differs from L_0, which leads to an acceleration until L_a is sufficiently increased and the acceleration command L_d returns to zero. Correspondingly, if the course angle α changes, a torque is induced which changes T_α until it becomes equal to $s \cdot T_0$ and the turning command becomes zero.

Using a flight compensator in open-loop conditions, Preiss (1987) measured the torque exerted by a gypsy moth in tethered flight, and found that the torque, indeed, depends on the direction of visual flow in a sinusoidal fashion. Furthermore, the flight speed was shown to be controlled by the movement of the ground patterns (Preiss and Kramer, 1984).

According to the model proposed by Kramer (1996), in wind without odor, if the set points $s \cdot T_0$ and L_0 are exactly zero, the animal will hover on the spot. Its longitudinal axis will point exactly upwind, and its airspeed

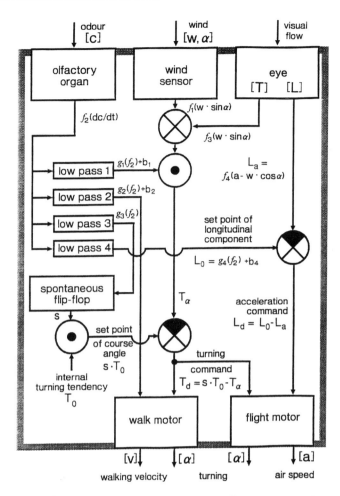

Figure 7. System of interactions underlying the odor-controlled anemotaxis of moths. Arrows symbolize variables and their direction of action, and boxes the relations between the variables. Crossed or dotted circles indicate addition or multiplication, respectively. Black sectors mark a change of sign. The model incorporates two independent feedback loops. Both loops are closed via environmental stimuli (outside the large, grey-bordered box). One loop controls $T = w \cdot \sin \alpha$, internally represented by T_α. In flight, T is the transverse component of the visual flow of ground patterns. For due upwind anemotaxis, the set point of T (and T_α) is zero. As a decisive detail the gain is modified in this loop by the actual olfactory situation (via low-pass filter #1). The term b_1 prevents zero gain in the absence of pheromone pulses, which is necessary to sustain casting flight. The set point of T_α can be >zero to produce menotactic orientation and "counterturning", by the action of the internal turning tendency T_0 and its sign s which is changed by the flip-flop mechanism. The second loop controls the upwind advance in flight with respect to the ground. In this loop, the set point L_0 is determined by the olfactory situation via low-pass filter 4 and a small constant b_4, necessary to sustain casting. The walking speed is controlled by an open loop via low-pass filter 2 and a small constant b_2, necessary to sustain circling. Not shown are the addition of noise to α (α-noise) and to T_0 (internal noise), the proportionality constants k_T and k_L, and a low-pass filter representing the duration of the signal f_2 from the olfactory receptor cells. Modified from Kramer (1996), with author's permission.

will be exactly equal to the wind speed w (see Fig. 6A). When the value of L_0 is slightly above zero, the nexus produces casting across the wind line. This appears to contradict the requirement to measure the airspeed (a) for precise casting (Eq. 8). However, because the value of L_0 is only slightly positive, airspeed and wind speed will be almost equal, and the ground speed vector will remain virtually crosswind. This prediction was validated in a computer simulation in which L_0 was set equal to a small additive constant b_4 (see below). So far, the value of b_4 has been empirically determined for a large range of airstream velocities (30 cm/s – 600 cm/s) in which anemotaxis of insects may, indeed, occur.

Because casting involves anemotactic orientation, the α-dependent turning tendency T_α should not be zero. This requirement is met in the nexus by introducing a small constant b_1; it is added to the odor-dependent signal $g_1(f_2)$ which is zero in the absence of odor.

$$T_\alpha = (g_1(f_2) + b_1) \cdot f_3(T). \tag{11}$$

Two types of casting can be distinguished, depending on the presence or absence of an internal turning tendency T_0. In the presence of T_0, counterturns will be generated if the sign of T_0 and, consequently, the sign of the turning command T_d changes. Because this type of casting is internally controlled, it may be called "endogenous casting" (Fig. 8B). Depending on the frequency of sign changes, endogenous casting may produce regular counterturning. In contrast, when T_0 is zero, sign changes are ineffective, but still the sign of T_d changes whenever the course angle α changes its sign due to fluctuations of the wind direction. This externally caused casting may be called "exogenous casting". It produces irregular counterturns (Fig. 8D). The possibility of casting without internally produced sign changes of T_d was first suggested by Preiss and Kramer (1986a).

How does the odor stimulation control the anemotactic response?

The influence of the odor pulses is to transiently increase L_0 above b_4 resulting in a higher positive value of $(g_4(f_2) + b_4)$. This value depends on the odor concentration (dc/dt). An increase of odor concentration will lead to a higher airspeed a, which is necessary for proceeding upwind. Furthermore, following an odor pulse, the turning tendency T_α is enhanced, which leads to an upwind surge. This results from multiplying (weighting) the internal signal $f_3(T)$ by an odor-dependent function $g_1(f_2)$ (Fig. 7). The small constant b_1 (Eq. 11) prevents T_α from becoming zero when the value of the function $g_1(f_2)$ vanishes after the odor has ceased. Thus, the odor merely modifies parameters of both feedback loops. Consequently, the same causal network may underly the different modes of anemotactic behavior inside and outside the odor plume.

Anemotactic walk

One of the feedback loops of Kramer's anemotaxis system would be sufficient for anemotaxis of the walking animal, which can separately measure the wind speed w and the course angle α using mechanoreceptors. It should be noted that for the walking animal there is practically no wind drift which occurs and needs to be compensated in flying. For upwind walk it is sufficient to control T which depends on $w \cdot \sin \alpha$, i.e., to turn until T_α equals T_0. In walking gypsy moths it has been shown that an increase in either pheromone concentration or wind speed produces smaller course angles α, which are, during walking, identical with track angles β, as is expected from the equation

$$T_\alpha = (g_1(f_2) + b_1) \cdot f_1 (T) \tag{11a}$$

(Preiss and Kramer, 1986b). If the odor stimuli fail to come, T_α becomes very small compared with the internal turning tendency T_0, because the signal from low-pass filter 1 becomes minimal. This produces walking in circles. The same behavior is produced by the model, and is also observed in experiments if odor is present but the wind velocity is zero.

In the model the walking speed v depends on the odor concentration ($v = g_2(f_2) + b_2$). The dependence on odor concentration was implemented in the model based on the finding that male moths of *Bombyx mori* walk faster with increasing pheromone concentration (Kramer, 1975). The small value b_2 guarantees that walking in circles is sustained, after the odor ceased and the low-pass output signal $g_2(f_2)$ becomes zero.

Internal turning tendency and flip-flop

An alternating internal turning tendency has been suggested by Wendler (1975) and Wendler and Scharstein (1986) for the corn weevil, *Calandra granaria (Sitophilus granarius)*. On a vertical plane this beetle orients to gravity at an angle caused by an additive set value, i.e., an internal turning tendency, and hence it walks in circles on a horizontal plane. As Kramer (1996) states, such a mechanism, originally proposed by von Holst and Mittelstaedt (1950), might be common to many insects. If the internal turning tendency $s \cdot T_0$ is large compared with the variations of T_α caused by fluctuations of the wind direction then the course and track angles α and β, respectively, differ from zero (menotactic orientation). The oblique path, together with the changing sign s of T_0, results in zig-zagging of both upwind walking (Kramer, 1975) and upwind flight (Fig. 8A, B). Thus, the sign changes may underly counterturns in both, menotactic zig-zagging and casting; in walking, they may underly zig-zagging and changes in the

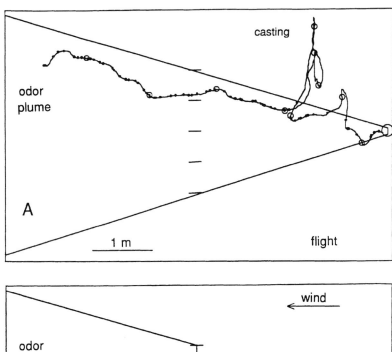

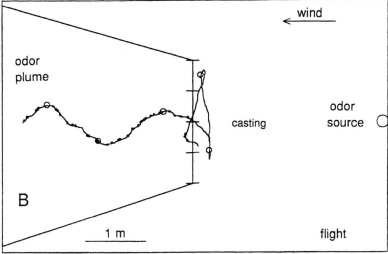

Figure 8. Flight (A,B,D) and walking paths (C) generated with a computer model of the nexus of Figure 7. Within the simulated odor plume, the model animal randomly encounters filaments of odor (dots along the tracks). The change in sign ("flip-flop" of the internal turning tendency (circles along the tracks) occurs every four s in flight, and every eight s in walking. When the odor pulses stop, the animal starts flying at right angles to the average wind direction ("casting", A,B,D), or walking in circles (C). In both cases, the animal maintains its position with respect to the wind for some time. Horizontal dashes in the middle of each panel denote the assumed mean local wind vectors of only those portions of air that had passed the odor source. These vectors do not point exactly to the odor source (empty circle on the right hand in each panel); however, the direction to the odor source was preserved to 15%, and this percentage increased with shorter distance to the odor source (see text). The odor plume is triangular

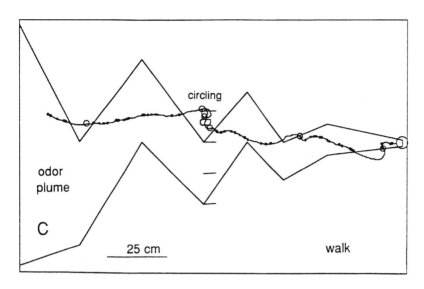

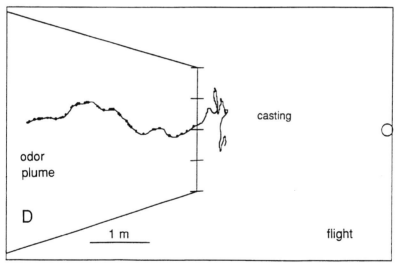

in A, meander-shaped in C, and trapezoidal in B and D. The following parameter settings (see Fig. 7) are used: wind speed 300 cm/s, walking speed 4 cm/s, mean interval between odor pulses 0.3 s. Time constant of the signal from olfactory receptor cells was 0.8 s; time constants of the low-pass filters 1–4 were 0.9, 1.1, 1.1 and 1.3 s, respectively. The turning tendency was 0.05 radians/s in flight (A,B,D), and 5 radians/s in walking (C). Casting occurs with (in A and B), or without an internal turning tendency with regular flip-flop (D). However, without flip-flop, the counterturns appear at randomly varying intervals. Further parameters were α-noise and internal noise, the small constants b_1, b_2 and b_4, and the proportionality constants k_T and k_L. Note that in (B), the model animal happens to move downwind during casting, due to the added noise. An extension of the model would be necessary to produce a systematic offset downwind as observed in flying moths (see Fig. 2).

direction of circling, respectively (Kramer, 1975; Kanzaki, 1996). However, it might be difficult to distinguish the more regular menotactic zig-
zagging (Fig. 8A and B) from irregular zig-zagging, which occurs in the
absence of an internal turning tendency (Fig. 8D).

Counterturning may occur in a very regular fashion, as frequently observed in wind tunnels (Kennedy and Marsh, 1974; Kennedy, 1983; Baker et
al., 1984; David and Kennedy, 1987) (Fig. 9). It continues in wind with as
well as without odor, i.e., during circling or casting, and even in odor without wind (Mankin and Hagstrum, 1995). These observations indicate the
existence of an internal oscillator, or flip-flop element, which changes the
sign of an internal turning tendency, as first suggested by Kramer (1975).
This mechanism represents the "counterturn generator" of Kennedy
(1983). It could certainly also induce irregular counterturning by randomly
switching the flip-flop. In this case, endogenous casting may be difficult to
distinguish from exogenous casting where counterturning is induced by
changes of the sign of α due to fluctuations of the wind direction (Preiss
and Kramer, 1986c).

In the nexus, the frequency of sign changes may be controlled by the
odor via the low-pass filter (3) with the output signal $g_3(f_2)$ (Fig. 7). Odor
stimuli may even elicit single sign changes, as is concluded from recordings from descending interneurons of the neck connectives. These neurons
show a flip-flopping behavior, i.e., they alternatingly initiate or terminate
nerve impulse firing when stimulated by single pheromone pulses, or by
light pulses (Olberg, 1983; Kanzaki et al., 1989, 1994; Olberg and Willis,
1990; Kanzaki, 1996).

Low-pass filters of the nexus

Low-pass filters that control the dynamics of the orientation are important
elements of the nexus. First, the olfactory receptor cells may be considered
as a combination of low-pass and high-pass filters. Upon a brief stimulus
pulse (20 ms), they produce a slow receptor potential followed by a burst
of nerve impulses (see e.g., Kodadová, 1996). The effects of this burst on
the components T_α and L_a, the walking speed, and the flip-flop generator,
must decay with appropriate time constants (low-pass filters 1 to 4, respectively; Fig. 7). In flight, the turning commands and the acceleration commands must be adjusted to each other in an appropriate manner by using
further low-pass filters (not shown in Fig. 7). Partially, these account for
latencies and inertia involved in walking and flight manoeuvers. Furthermore, sources of noise were implemented in the model (not shown in
Fig. 7), to account first for irregularities of the wind direction (α-noise,
added to α) and, second, for imprecisions in the behavior of the animal
(internal noise, added to T_α, see above). How all of these functions representing a network of causal connections are realized in the nervous system

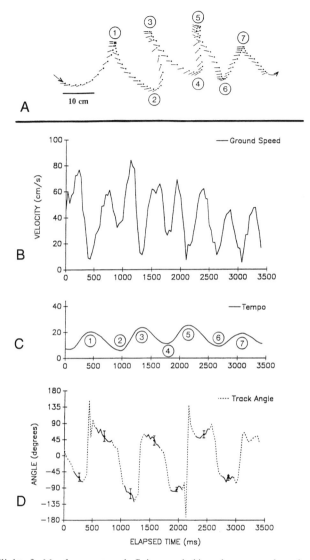

Figure 9. Flight of a *Manduca sexta* male flying upwind in a pheromone plume in a wind tunnel (61 cm wide). Wind direction from right to left. (A) Ground track and orientation of the body axis (dashes). Dots represent the moth's position each 33 ms. Circled numbers next to counterturns in the track correspond to labels of the apices of turns shown in (C). (B) Magnitudes of each 33 ms movement vector along the track shown in (A), plotted against time. (C) Horizontal excursions of the track (in cm), plotted against time, showing the "tempo" of the side to side movements. Circled numbers as in (A). (D) Track angles (see β in Fig. 6) measured each 33 ms (dashed line), plotted against time. Mean values (dots) and standard deviations (error bars) are depicted for the track angles distributed along each straight leg of the track. Modified from Willis and Arbas (1996), with authors' permission.

remains to be investigated. The nexus suggests that the signals of the odor receptors act on various stations of the central nervous system.

Computer simulations

The computer model of the nexus, consisting of mathematically defined elements, allows simulation of various types of odor plumes, including artificial ones, such as are used in wind-tunnel experiments. Within the odor plume, the "computer moth" receives brief odor pulses at random with an adjustable duration and average frequency. For simplicity, the frequency does not change with distance from the source. However, it is a crucial condition that each odor pulse is associated with a vector which points more or less accurately in the direction of the odor source. The actual angle (λ) of the movement of the odor packet with respect to the average wind direction must partially preserve the direction to the odor source (γ). The preservation p_γ ($= \lambda/\gamma$) used in the model decreases with increasing distance d from the odor source according to

$$p_\gamma = p_0\, h/(h + d) \qquad (12)$$

where h denotes the distance from the source at which a reference value p_0 (present at the odor source) is reduced by 50%. According to the computer model, the chance of the moth to arrive at the odor source is minimal if the odor packets are – artificially – aligned with the average wind direction.

The computer model allows the parameters to be varied so as to reproduce tracks of walking and flying moths (Fig. 8) that correspond to those observed in wind tunnels or in the field. The model produces a zig-zagging upwind walk (Fig. 8C) or flight (Fig. 8A, B, D) if the frequency of odor pulses is high enough (depending on the time constants of the low-pass filters). The moth responds with a surge upwind to every odor pulse. If the pulse frequency is too high, the anemotactic orientation is impaired. If the odor pulses fail to come, and if there is an internal turning tendency, the computer moth walks in circles (Fig. 8C), whereas, when flying, it exhibits (endogenous) casting (Fig. 8A, B). Exogenous casting occurs without an internal turning tendency and without flip-flopping (Fig. 8D). Interestingly, the simulation reveals properties of real flights which were not specifically implemented. For example, the ground speed remains fairly constant at different wind speeds, even when the moth changes from upwind flight to casting (see Figs 2 and 8B) (Kennedy, 1940; Willis and Baker, 1990; Zanen and Cardé, 1996) (see also Brady et al., 1989; Paynter and Brady, 1996). It remains to be investigated how sensitively the model reacts to variation of the various parameters. But the main performances, such as casting and upwind flight under odor stimulation, appear to be rather insensitive against variations.

Long-distance attraction

The mechanism of attraction to odor sources over long distances (several hundreds of meters or even kilometers) is not known. The distribution of odor released from punctuate odor sources has been investigated in plumes not beyond a few tens of meters (Murlis, 1996). The computer model of Kramer (1996) was designed for orientation a few meters downwind of the odor source. Whether it can also be applied for longer distances remains to be investigated. At a larger distance from the source, encounters with odor packets might occur with lower frequencies. Consequently, much larger time constants of the low-pass filters are needed to sustain positive anemotaxis.

For the dispersal of odor substances over greater distances, convection – wind and turbulence – is critical; however, it is not known for how far a distance inhomogeneities of the odor distribution are preserved. The region of space within which the odor concentration is suprathreshold for the receiver has been called the "threshold volume" (Bossert and Wilson, 1963) or the "active space" (Nakamura, 1976). Bossert and Wilson (1963) used an empirical formula to calculate the dimensions of the mean threshold volume of odor plumes. The extent of the threshold volume in the wind direction – that is, the maximum distance X_{max} over which the odor concentration is above threshold – is calcuated according to

$$X_{max} = \left(\frac{2Q}{K \, \pi \, C_y C_z w} \right)^{1/(2-n)}$$
(13)

(Sutton, 1953). This distance is certainly a function of the ratio Q/K, where Q (molecules/s) is the rate of release at the odor source and K (molecules/cm³) is the threshold concentration of the receiver. It is also thought to be a function of the wind velocity w (cm/s). The empirical values C_y (cm$^{1/8}$), C_z (cm$^{1/8}$) and n depend on the roughness of the ground and on the turbulence. These and also Q and K might depend on wind speed. For wind speeds of 1–5 m/s, values applicable to flat ground are (according to Sutton, 1953) n = 1/4, C_y = 0.4 cm$^{1/8}$ and C_z = 0.2 cm$^{1/8}$. Vegetation may not only reduce the wind speed, but can also influence distribution and average concentration of the odor (Aylor et al., 1976; Wall et al., 1981; Elkinton and Cardé, 1984; Perry and Wall, 1984; Baker and Haynes, 1989; Murlis et al., 1990).

The Sutton formula (Eq. 13) can be used, for example, to estimate the effective attraction distance X_{max} for the gypsy moth *Lymantria dispar*. Determinations of the release rate of single female gypsy moths gave peak values of Q = 10^{12} pheromone molecules/s, which is equivalent to about 5% of the contents of the female gland per s (Jacobson et al., 1960). Even lower release values (10- to 100-fold) were found when averaged over several hours or a day (Richerson and Cameron, 1974; Charlton and Cardé,

1982; Ma and Schnee, 1983). We take the threshold concentration K as 10^3 molecules/cm^3, as has been measured for bombykol in the silk moth *Bombyx mori* (Kaissling and Priesner, 1970), and a wind speed of 400 cm/s. With these values, according to Equation 13, the distance X_{max} would be 150 m.

Earlier studies report distances of about 4 km for the gypsy moth (Collins and Potts, 1932). In later studies, the distances found for the attraction of male moths to a single female were smaller, though not as small as expected from the Sutton formula and the measured release rates. The distances were still up to some hundreds of meters, possibly because pheromone packets of higher concentrations travelled beyond the borders of the mean active space (reviewed in Wall and Perry, 1987).

A value larger than 150 m can be expected, if one considers that the amount of odor molecules collected by the antennae not only depends on the odor concentration, but also on the airspeed of the flying moth (Kaissling, 1990, 1997). The threshold of *Bombyx mori* for bombykol was determined for an airstream velocity of 60 cm/s. The airspeed of the gypsy moth in the above experiments was much higher. Higher air speed might result in a larger adsorptive uptake of molecules by the antenna and, consequently, lead to a lower threshold in terms of pheromone concentration, and a larger X_{max}, respectively. Discrepancies between calculated and measured values can occur if important information about wind speed, temperature etc. is missing; such factors might influence the pheromone release Q, or the behavioral threshold K (Cardé and Roelofs, 1973; Charlton et al., 1993). Furthermore, the dispersal of odors is strongly influenced by the vegetation (Willis et al., 1994).

Interestingly, the active space can differ for the pheromone components of a given species. In some species, the major pheromone component is not only released at a higher rate compared with the minor components, but is, in addition, perceived most sensitively by the antennal receptor cells (Meng et al., 1989). Consequently, this compound could have a larger active space than the other components and elicit behavioral responses over a larger distance from the source.

Final remarks

This chapter has described some attempts towards a quantitative understanding of anemotactic behavior. The computer model of the nexus underlying anemotaxis proposed by E. Kramer has been constructed to assist the formulation of questions and the design of new experiments.

Certainly, the parameters of the model will have to be adjusted for every insect species. Ideally, the parameter values should be determined experimentally. This requires careful analysis of anemotactic responses to controlled odor stimuli, such as those suddenly terminated (Fig. 2), to single odor pulses (Fig. 4), or to repetitive odor stimuli (Fig. 1). For such analysis it seems useful also to record the direction of the body axis with respect to the average wind direction with sufficient time resolution (Fig. 9). Furthermore, it would be useful to know the actual wind direction at every moment. This is easier to control in locomotion compensators (Preiss and Kramer, 1986c) or in tethered flight (Preiss, 1991; Preiss and Kramer, 1986c). For studies on tethered flight in locusts see Preiss (1987, 1992, 1993), Preiss and Spork (1993), Baader et al. (1992), and Wortmann and Zarnack (1989, 1993). For analysis of optomotor responses and odor-controlled anemotaxis in flight simulators, see also Heisenberg and Wolf (1988) and Wolf and Heisenberg (1991).

Three-dimensional flight

The nexus described by Kramer (1996) is restricted to two-dimensional orientation. Only a few investigations have considered anemotactic flight in three dimensions. In fact, a flight track along the surface of an imaginary horizontal cylinder could be confused with zig-zagging if one monitors the flight track only from above. For three-dimensional flight the anemotactic model of Kramer (1996) needs to be modified. An extension of the model would also be necessary if it is to include the control of flight altitude, which influences the flow of visual ground patterns over the eye (David, 1982; Vickers and Baker, 1994b, 1996). For example, it has been shown that moths fly with lower ground speed at increasing pheromone concentrations (Kuenen and Cardé, 1994), and possibly nearer to the ground. In the presence of pheromone, male gypsy moths show optomotor responses in the vertical direction if visual patterns are moved in front of them. This reflex leads to stabilization of the flight altitude (Preiss and Kramer, 1983; Preiss and Futschek, 1985). In three-dimensional analyses of flight of freely flying honeybees, M.V. Srinivasan and S.W. Zhang (this volume) studied the control of ground speed and flight altitude.

Negative anemotaxis

Although, in moths, only positive (upwind) anemotaxis has been investigated so far, odor-controlled anemotaxis can also be negative. One example for negative (downwind) anemotaxis has been reported for trained honeybees walking on a locomotion compensator (Kramer, 1976). During the training phase, the bees were rewarded with sugar water when one of four pure odor substances was presented at a certain concentration. In the test situation, the odor concentration was adjusted continuously according to the position of the animal in the simulated odor field. The trained bees walked (in darkness) upwind for long as the odor concentration was between two critical values, below and above the concentration to which they had been trained. If the actual concentration was too low or too high – i.e., outside the critical values – the bees switched to walking downwind ("home"). These results led to the idea, that bees compare the momentary pattern of olfactory excitation with a pattern stored in the learning situation. They go upwind for as long as the two patterns are sufficiently similar. If the degree of similarity is too small, they go downwind.

Interestingly, the critical range of odor concentration narrowed after intake of sugar water and with increasing experience of the bees with the experimental situation. From further observations Kramer (1976) concluded that bees can identify the absolute odor concentration to within 10% or less of the learned concentration.

Outlook

It seems challenging to combine precise behavioral experiments with neurophysiological studies in order to understand how the nervous system perceives external stimuli and generates anemotactic behavior (see Egelhaaf and Borst, 1993; Arbas, 1996; Willis and Arbas, 1996). It would be fascinating to extend such studies to other insect groups, or other arthropods (see, for example, Weissburg, this volume), or even vertebrates that orientate within odor plumes released from small odor sources.

Acknowledgements
This paper would not have been possible without many intensive discussions with Ernst Kramer. I thank Horst Mittelstaedt for his interest and valuable suggestions, Arthur Vermeulen, Juergen Ziesmann, and Rudolf A. Steinbrecht for discussion, Ann Biederman-Thorson for help in translation of most parts of the manuscript and for many suggestions. I thank P. Goerner, K.G. Goetz, and P. Witzgall for critical comments. For technical help, I thank Ute Lauterfeld, Anka Guenzel and Carola Schmid. Last but not least I gratefully acknowledge the patience and help of Miriam Lehrer.

References

Arbas, E.A. (1996) Neuroethological study of pheromone-modulated responses. *In*: R.T. Cardé and A.K. Minks (eds): *Pheromone research: New directions*. Chapman & Hall, New York, pp 320–329.

Arn, H., Tóth, M. and Priesner, E. (1992) *List of sex pheromones of Lepidoptera and related attractants*, 2nd edition. OILB Pub Montfavet, France.

Aylor, D.E., Parlange, J.-Y. and Granett, J. (1976) Turbulent dispersion of disparlure in the forest and male gypsy moth response. *Environm. Entomol.* 5:1026–1032.

Baader, A., Schäfer, M. and Rowell, C.H.F. (1992) The perception of the visual flow field by flying locusts: A behavioural and neuronal analysis. *J. Exp. Biol.* 165:137–160.

Baker, T.C. (1989a) Sex pheromone communication in the Lepidoptera: New research progress. *Experientia* 45:248–262.

Baker, T.C. (1989b) Pheromones and flight behaviour. *In*: G.J. Goldsworthy and C. Wheeler (eds): *Insect Flight*, CRC Press. Boca Raton, Fla., pp 231–255.

Baker, T.C. and Haynes, K.F. (1989) Field and laboratory electroantennographic measurements of pheromone plume structure correlated with oriental fruit moth behaviour. *Physiol. Entomol.* 14:1–12.

Baker, T.C. and Vickers, N.J. (1996) Pheromone-mediated flight in moths. *In*: R.T. Cardé and A. Minks (eds): *Pheromone Research: New Directions*, Chapman & Hall, New York, pp 248–264.

Baker, T.C. and Vogt, R.G. (1988) Measured behavioral latency in response to sexpheromone loss in the large silk moth *Antheraea polyphemus. J. Exp. Biol.* 137:29–38.

Baker, T.C., Willis, M.A. and Phelan, P.L. (1984) Optomotor anemotaxis polarizes self-steered zigzagging in flying moths. *Physiol. Entomol.* 9:365–376.

Baker, T.C., Willis, M.A., Haynes, K.F. and Phelan, P.L. (1985) A pulsed cloud of sex pheromone elicits upwind flight in male moths. *Physiol. Entomol.* 10:257–265.

Baker, T.C., Hansson, B.S., Löfstedt, C. and Löfqvist, J. (1988) Adaptation of antennal neurons in moths is associated with cessation of pheromone-mediated upwind flight. *Proc. Natl. Acad. Sci. USA* 85:9826–9830.

Boeckh, J. and Ernst, K.-D. (1987) Contribution of single unit analysis in insects to an understanding of olfactory function. *J. Comp. Physiol.* 161:549–565.

Bossert, W.H., Wilson, E.O. (1963) The analysis of olfactory communication among animals. *J. Theoret. Biol.* 5:443–469.

Brady, J., Gibson, T. and Packer, M.J. (1989) Odour movement, wind direction, and the problem of host finding by tsetse flies. *Physiol. Entomol.* 14:369–380.

Cardé, R.T. (1979) Behavioral responses of moths of female-produced pheromones and the utilization of attractant-baited traps for population monitoring. *In*: R.L. Rabb and G.G. Kennedy (eds): *Movement of highly mobile insects: concepts and methodology in research.* North Carolina State Univ., pp 286–315.

Cardé, R.T. and Charlton, R.E. (1984) Olfactory sexual communication in Lepidoptera: strategy, sensitivity and selectivity. *In*: T. Lewis (ed.): *Insect Communication.* 12th Symp. Royal Entomol. Soc. Lond. Academic Press, pp 241:265.

Cardé, R.T. and Mafra-Neto, A. (1996) Mechanisms of flight of male moths to pheromone. *In.* R.T. Cardé and A. Minks (eds): *Pheromone research: New directions.* Chapman & Hall, New York, pp 275–290.

Cardé, R.T. and Minks, A. (eds) (1996) *Pheromone research: New directions.* Chapman & Hall, New York.

Cardé, R.T. and Roelofs, W.L. (1973) Temperature modification of male sex pheromone response and factors affecting female calling in *Holomelina immaculata* (Lepidoptera: Arctiidae). *Can. Ent.* 105:1505–1512.

Charlton, R.E. and Cardé, R.T. (1982) Rate and diel periodicity of pheromone emission from female gypsy moths, (*Lymantria dispar*) determined with a glass-adsorption collection system. *J. Insect Physiol.* 28:423–430.

Charlton, R.E. Kanno, H., Collins, R.D. and Cardé, R.T. (1993) Influence of pheromone concentration and ambient temperature on flight of the gypsy moth, *Lymantria dispar*, in a sustained-flight wind tunnel. *Physiol. Entomol.* 18:349–362.

Christensen, T.A. and Hildebrand, J.G. (1988) Frequency coding by central olfactory neurons in the sphinx moth *Manduca sexta. Chem. Senses* 13:123–133.

Christensen, T.A., Heinbockel, T. and Hildebrand, J.G. (1996) Olfactory information processing in the brain: encoding chemical and temporal features of odors. *J. Neurobiol.* 30:82–91.

Collins, C.W. and Potts, S.F. (1932) Attractants for the flying gypsy moth as an aid in locating new infestations. *USDA Tech. Bull.* 336:43.

Connor, W.E., Eisner, T., Van der Meer, R.K., Guerrero, A., Ghiringelli, D. and Meinwald, J. (1980) Sex attractant of an Arctiid moth (*Utetheisa ornatrix*): a pulsed chemical signal. *Behav. Ecol. Sociobiol.* 7:55–63.

David, C.T. (1982) Compensation for height in the control of ground-speed by *Drosophila* in a new, "barber's pole" wind tunnel. *J. Comp. Physiol.* 147:485–493.

David, C.T. and Kennedy, J.S. (1987) The steering of zigzagging flight by male gypsy moths. *Naturwiss.* 74:194–196.

David, C.T., Kennedy, J.S. and Ludlow, A.R. (1983) Finding of a sex pheromone source by gypsy moths released in the field. *Nature* 303:804–806.

Egelhaaf, M. and Borst, A. (1993) A look into the cockpit of the fly: visual orientation, algorithms, and identified neurons. *J. Neurosci.* 13:4563–4574.

Elkinton, J.S., Cardé, R.T. (1984) Odor Dispersion. *In*: W.J. Bell and R.T. Cardé (eds): *Chemical Ecology of Insects*. Chapman & Hall, London, pp 73–88.

Esch, H.E. and Burns, J.E. (1996) Distance estimation by foraging honeybees. *J. Exp. Biol.* 199: 155–162.

Fabre, J.-H. (1919) *Souveniers entomologiques*. Paris, 7e série Libr. Ch. DeLagrave.

Gewecke, M. (1975) The influence of the air-current sense organs on the flight behavior of *Locusta migratoria*. *J. Comp. Physiol.* 103: 79–95.

Hansson, B.S. (1995) Olfaction in Lepidoptera. *Experientia* 51: 1003–1027.

Heisenberg, M. and Wolf, R. (1988) Reafferent control of optomotor yaw torque in *Drosophila melanogaster*. *J. Comp. Physiol.* A 163: 373–388.

Hildebrand, J.G. (1996) Olfactory control of behavior in moths: central processing of odor information and the functional significance of olfactory glomeruli. *J. Comp. Physiol.* A 178: 5–19.

Holst, E. von and Mittelstaedt, H. (1950) Das Reafferenzprinzip. (Wechselwirkungen zwischen Zentralnervensystem und Peripherie). *Naturwissens.* 37: 464–476.

Jacobson, M., Beroza, M. and Jones, W.A. (1960) Isolation, identificaiton, and synthesis of the sex attractant of the gypsy moth (*Lymantria dispar*). *Science* 132: 1011–1012.

Kaissling, K.E. (1986a) Chemo-electrical transduction in insect olfactory receptors. *Ann. Rev. Neurosci.* 9: 121–145.

Kaissling, K.E. (1986b) Temporal characteristics of pheromone receptor cell responses in relation to orientation behaviour of moths. *In*: T.L. Payne, M.C. Birch and C.E.J. Kennedy (eds): *Mechanisms in Insect Olfaction*. Univ. Press, Oxford, pp 193–200.

Kaissling, K.E. (1987) *R.H. Wright Lectures on Insect Olfaction*. K. Colbow (ed.): Simon Fraser University, Burnaby, B.C., Canada.

Kaissling, K.E. (1990) Antennae and noses, their sensitivities as molecule detectors. *In*: V. Torre, L. Cervetto and A. Borsellino (eds): *Sensory Transduction*. Plenum Press, New York and London, pp 81–97.

Kaissling, K.E. (1997) Olfactory transduction in moths: II. Extracellular transport, deactivation and degradation of stimulus molecules. *In*. C. Taddei-Ferretti and C. Musio (eds): *From structure to information in sensory systems*. World Scientific, Singapore, New Jersey, London, Hongkong; *in press*.

Kaissling, K.E. and Kramer, E. (1990) Sensory basis of pheromone-mediated orientation in moths. Sensorische Grundlagen der pheromongesteuerten Orientierung bei Nachtfaltern. *Verh. Dtsch. Zool. Ges.* 83: 109–131.

Kaissling, K.E. and Priesner, E. (1970) Die Riechschwelle des Seidenspinners. *Naturwissens.* 57: 23–28.

Kaissling, K.E., Zack Strausfeld, C. and Rumbo, E. (1987) Adaptation processes in insect olfactory receptors: mechanisms and behavioral significance. *Int Symp. Olf. Taste IX. Ann NY Acad. Sci.* 510: 104–112.

Kaissling, K.E., Meng, L.Z. and Bestmann, H.-J. (1989) Responses of bombykol receptor cells to (*Z*,*E*)-4,6-hexadecadiene and linalool. *J. Comp. Physiol.* A 165: 147–154.

Kanzaki, R. (1996) Pheromone processing in the lateral accessory lobes of the moth brain: flipflopping signals related to zigzagging upwind walking. *In*: R.T. Cardé and A. Minks (eds): *Pheromone research: New directions*. Chapman & Hall, London, pp 291–303.

Kanzaki, R., Arbas, E.A., Strausfeld, N.J. and Hildebrand, J.G. (1989) Physiology and morphology of projection neurons in the antennal lobe of the male moth *Manduca sexta*. *J. Comp. Physiol.* A 165: 427–453.

Kanzaki, R., Ilkeda, A. and Shibuya, T. (1994) Morphological and physiological properties of pheromone-triggered flipflopping descending interneurons of the male silkworm moth, *Bombyx mori*. *J. Comp. Physiol.* A 175: 1–14.

Kennedy, J.S. (1940) The visual responses of flying mosquitoes. *Proc. Zool. Soc. London* A 109: 221–242.

Kennedy, J.S. (1983) Zigzagging and casting as a programmed response to wind-borne odour: a review. *Physiol. Entomol.* 8: 109–120.

Kennedy, J.S. (1986) Some current issues in orientation to odour sources. *In*: T.L. Payne, M.C. Birch and C.E.J. Kennedy (eds): *Mechanisms in Insect Olfaction*, Clarendon Press, Oxford, pp 11–25

Kennedy, J.S. and Marsh, D. (1974) Pheromone-regulated anemotaxis in flying moths. *Science* 184: 999–1001.

Kennedy, J.S., Ludlow, A.R. and Sanders, C.J. (1980) Guidance system used in moth sex attraction. *Nature* 288:475–477.

Kennedy, J.S., Ludlow, A.R. and Sanders, C.J. (1981) Guidance of flying male moths by windborne sex pheromone. *Physiol. Entomol.* 6:395–412.

Kodadová, B. (1996) Resolution of pheromone pulses in receptor cells of *Antheraea polyphemus* at different temperatures. *J. Comp. Physiol.* A 179:301–310.

Kodadová, B. and Kaissling, K.-E. (1996) Effects of temperature on silkmoth olfactory responses to pheromone can be simulated by modulation of resting cell membrane resistances. *J. Comp. Physiol.* A 179:15–27.

Kramer, E. (1975) Orientation of the male silkmoth to the sex attractant bombykol. *In*. D.A. Denton and J.P. Coghlan (eds): *Int. Symp. Olfaction and Taste V*, Acad. Press, New York, pp 329–335.

Kramer, E. (1976) The orientation of walking honeybees in odour fields with small concentration gradients. *Physiol. Entomol.* 1:27–37.

Kramer, E. (1986) Turbulent diffusion and pheromone-triggered anemotaxis. *In*: T.L. Payne, M.C. Birch and C.E.J. Kennedy (eds): *Mechanisms in Insect Olfaction*. Clarendon Press, Oxford, pp 59–67.

Kramer, E. (1992) Attractivity of pheromone surpassed by time-patterned application of two nonpheromone compounds. *J. Insect Behav.* 5:83–97.

Kramer, E. (1996) A tentative intercausal nexus and its computer model on insect orientation in windborne pheromone plumes. *In*: R.T. Cardé and A. Minks (eds): *Pheromone research: New directions*. Chapman & Hall, Ney York, pp 232–247.

Kuenen, L.P.S. and Cardé, R.T. (1994) Strategies for recontacting a lost pheromone plume: casting and upwind flight in the male gypsy-moth. *Physiol. Entomol.* 19:15–29.

Linn, C.E., Jr., Campbell, M.G. and Roelofs, W.L. (1986) Male moth sensitivity to multicomponent pheromones: critical role of female-released blend in determining the functional role of components and active space of the pheromone. *J. Chem. Ecol.* 12:659–668.

Linn, C.E., Jr., Campbell, M.G. and Roelofs, W.L. (1987) Pheromone components and active spaces: What do moths smell and where do they smell it? *Science* 237:650–652.

Linn, C.E., Jr. and Roelofs, W.L. (1989) Response specificity of male moths to multicomponent pheromones. *Chem. Senses* 14:421–437.

Ma, M. and Schnee, M.E. (1983) Analysis of individual gypsy moth sex pheromone production by sample concentrating gas chromatography. *Can. Ent.* 115:251–255.

Mafra-Neto, A. and Cardé, R.T. (1994) Fine-scale structure of pheromone plumes modulates upwind orientation of flying moths. *Nature* 369:142–144.

Mafra-Neto, A. and Cardé, R.T. (1995) Influence of plume structure and pheromone concentration on upwind flight of *cadra-cautella* males. *Physiol. Entomol.* 20:117–133.

Mafra-Neto, A. and Cardé, R.T. (1996) Dissection of the pheromone-modulated flight of moths using single-pulse response as a template. *Experientia* 52:373–379.

Mankin, R.W. and Hagstrum, D.W. (1995) Three-dimensional orientation of male *Cadra cautella* (Lepidoptera: Pyralidae) flying to calling females in a windless environment. *Environ. Entomol.* 24:1616–1626.

Marion-Poll, F. and Tobin, T.R. (1992) Temporal coding of pheromone pulses and trains in *Manduca sexta. J. Comp. Physiol.* A 171:505–512.

Masson, C. and Mustaparta, H. (1990) Chemical information processing in the olfactory system of insects. *Physiol. Rev.* 70:199–245.

Mayer, M.S. and Mankin, R.W. (1985) Neurobiology of pheromone perception. *In*: G.A. Kerkut and L.I. Gilbert (eds): *Comprehensive Insect Physiology, Biochemistry and Pharmacology, Vol. 9, Behaviour*. Pergamon Press, Oxford, New York, pp 95–144.

Meng, L.Z., Wu, C.H., Wicklein, M., Kaissling, K.E. and Bestmann, H.J. (1989) Number and sensitivity of three types of pheromone receptor cells in *Antheraea perny* and *A. Polyphemus. J. Comp. Physiol.* A 165:139–146.

Murlis, J. (1996) Odor plumes and the signal they provide. *In*: R.T. Cardé and A. Minks (eds): *Pheromone research: New directions*. Chapman & Hall, New York, pp 221–231.

Murlis, J. and Jones, C.D. (1981) Fine-scale structure of odour plumes in relation to insect orientation to distant pheromone and other attractant sources. *Physiol. Entomol.* 6:71–86.

Murlis, J., Willis, M.A. and Cardé, R.T. (1990) Odour signals: patterns in time and space. *Proc. 10th Int. Symp. Olfaction and Taste*, Univ. Oslo, pp 6–17.

Murlis, J., Elkinton, J.S. and Cardé, R.T. (1992) Odor plumes and how insects use them. *Ann. Rev. Entomol.* 37:505–532.

Nakamura, K. (1976) The active space of the pheromone of *Spodoptera litura* and the attraction of adult males to the pheromone source. *Proc. Symp. Insect Pheromones and their Applications*, Nagaoka and Tokyo, pp 145–155.

Olberg, R.M. (1983) Pheromone-triggered flip-flopping interneurons in the ventral nerve cord of the silkworm moth *Bombyx mori. J. Comp. Physiol.* 152:297–307.

Olberg, R.M. and Willis, M.A. (1990) Pheromone-modulated optomotor response in male gypsy moths, *Lymantria dispar L.*: Directionally selective visual interneurons in the ventral nerve cord. *J. Comp. Physiol.* A 167:707–714.

Palaniswamy, P., Underhill, E.W., Steck, W.F. and Chisholm, M.D. (1983) Responses of male redbacked cutworm, *Euxoa ochrogaster* (Lepidoptera: Noctuidae), to sex pheromone components in a flight tunnel. *Environ. Entomol.* 12:748–752.

Payne, T.L., Birch, M.C. and Kennedy, C.E.J. (eds) (1986) *Mechanisms in Insect Olfaction.* Clarendon Press, Oxford.

Paynter, Q. and Brady, J. (1996) The effect of wind speed on the flight responses of tsetse flies to CO_2: a wind-tunnel study. *Physiol. Entomol.* 21:309–312.

Perry, J.N. and Wall, C. (1984) A mathematical model for the flight of pea moth to pheromone traps through a crop. *Phil. Trans. R. Soc. Lond.* B 306:19–48.

Preiss, R. (1987) Motion parallax and figural porperties of depth control flight speed in an insect. *Biol. Cybern.* 57:1–9.

Preiss, R. (1991) Separation of translation and rotation by means of eye-region specialization in flying gypsy moths (Lepidoptera: Lymantriidae). *J. Insect Behav.* 4:209–219.

Preiss, R. (1992) Set point of retinal velocity of ground images in the control of swarming flight of desert locusts. *J. Comp. Physiol.* A 171:251–256.

Preiss, R. (1993) Visual control of orientation during swarming flight of desert locusts. *In*: K. Wiese, F.G. Gribakin, A.V. Popov and G. Renninger (eds): *Sensory Systems of Arthropods.* Birkhäuser Verlag, Basel, pp 273–287.

Preiss, R. and Futschek, L. (1985) Flight stabilization by pheromone-enhanced optomotor responses. *Naturwissens.* 72:435–436.

Preiss, R. and Kramer, E. (1983) Stabilization of altitude and speed in tethered flying gypsy moth males: influence of (+) and (–)-disparlure. *Physiol. Entomol.* 8:55–68.

Preiss, R. and Kramer, E. (1984) Control of flight speed by minimization of the apparent ground pattern movement: *In*: D. Varjú and H.-U. Schnitzler (eds): *Localization and Orientation in Biology and Engineering.* Springer-Verlag, Berlin, Heidelberg, pp 140–142.

Preiss, R. and Kramer, E. (1986a) Mechanism of pheromone orientation in flying moths. *Naturwissens.* 73:555–557.

Preiss, R. and Kramer, E. (1986b) Anemotactic orientation of gypsy moth males and its modification by the attractant pheromone (+)-disparlure during walking. *Physiol. Entomol.* 11:185–198.

Preiss, R. and Kramer, E. (1986c) Pheromone-induced anemotaxis in simulated free flight. *In*: T.L. Payne, M.C. Birch and C.E.J. Kennedy (eds): *Mechanisms in Insect Olfaction.* Clarendon Press, Oxford, pp 69–79.

Preiss, R. and Spork, P. (1993) Flight-phase and visual-field related optomotor jaw responses in gregarious desert locusts during tethered flight. *J. Comp. Physiol.* A 172:733–740.

Preiss, R. and Spork, P. (1994). Significance of reafferent information on yaw rotation in the visual control of translatory flight maneuvers in locusts. *Naturwissens.* 81:38–40.

Priesner, E. (1973) Artspezifität und Funktion einiger Insektenpheromone. *Fortschr. Zool.* 22:49–135.

Priesner, E. (1979) Progress in the analysis of pheromone receptor systems. *Ann. Zool Ecol. Anim.* 11:533–546.

Priesner, E. (1980) Sensory encoding of pheromone signals and related stimuli in male moths. *Neurotox 79, Soc. Chem. Ind., Lond.,* 359–366.

Priesner, E. (1984) Pheromone als Sinnesreize. *Verh. Ges. Dtsch. Naturf. und Ärzte* 113:207–226.

Priesner, E. and Witzgall, P. (1984) Modification of pheromonal behaviour in wild *Coleophora laricella* male moths by (Z)-5-decenyl acetate, an attraction-inhibitor. *Z. angew. Ent.* 98:118–135.

Rau, R. and Rau, N.L. (1929) The sex attraction and rhythmic periodicity in giant saturniid moths. *Trans. Acad. Sci.* 26:83–221.

Richerson, J.V. Cameron, E.A. (1974) Differences in pheromone release and sexual behavior between laboratory-reared and wild gypsy moth adults. *Environm. Ent.* 4:168–169.

Roelofs, W.L. and Cardé, R.T. (1977) Responses of Lepidoptera to synthetic sex pheromone chemicals and their analogues. *Ann. Rev. Entomol.* 22:377–405.

Rumbo, E.R. and K.E. Kaissling (1989) Temporal resolution of odour pulses by three types of pheromone receptor cells in *Antheraea polyphemus. J. Comp. Physiol.* A 165:281–291.

Schwinck, I. (1955) Weitere Untersuchungen zur Frage der Geruchsorientierung der Nacht-schmetterlinge: Partielle Fühleramputation bei Spinnermännchen, insbesondere am Seiden-spinner *Bombyx mori* L. *Z. Vergl. Physiol.* 37:439–458.

Sutton, O.G. (1953) *Micrometeorology, a study of physical process in the lowest layers of the earth's atmosphere.* McGraw-Hill, New York.

Vickers, N.J. and Baker, T.C. (1994a) Reiterative responses to single strands of odor promote sustained upwind flight and odor source location by moths. *Proc. Natl. Acad. Sci.* USA 91:5756–5760.

Vickers, N.J. and Baker, T.C. (1994b) Visual feedback in the control of pheromone-mediated flight of *Heliothis-virescens* males (Lepidoptera, Noctuidae). *J. Insect Behav.* 7:605–632.

Vickers, N.J. and Baker, T.C. (1996) Latencies of behavioral response to interception of fila-ments of sex pheromone and clean air influence flight track shape in *Heliothis virescens (F.)* males. *J. Comp. Physiol.* A 178:831–847.

Wall, C. and Perry, J.N. (1987) Range of action of moth sex-attractant sources. *Ent. Exp. Appl.* 44:5–14.

Wall, C., Sturgeon, B.M., Greenway, A.R. and Perry, J.N. (1981) Contamination of vegetation with synthetic sex attractant released from traps for the pea moth, *Gydia nigricana. Ent. Exp. Appl.* 30:111–125.

Wendler, G. (1975) Physiology and Systems Analysis of Gravity orientation in two insect species (*Carausius morosus, Calandra granaria). Fortschr. Zool.* 23:33–48.

Wendler, G. and Scharstein, H. (1986) The orientation of grain weevils (*Sitophilus granarius*): influence of spontaneous turning tendencies and of gravitational stimuli. *J. Comp. Physiol.* A 159:377–389.

Willis, M.A. and Arbas, E.A. (1996) Active behavior and reflexive responses: another perspec-tive on odor-modulated locomotion. *In*: R.T. Cardé and A. Minks (eds): *Pheromone research: New directions.* Chapman & Hall, New York, pp 304–319.

Willis, M.A. and Baker, T.C. (1984) Effects of intermittent and continuous pheromone stimu-lation on the flight behaviour of the oriental fruit moth, *Grapholita molesta. Physiol. Ento-mol.* 9:341–358.

Willis, M.A. and Baker, T.C. (1987) Comparison of manoeuvres used by walking versus *flying Grapholita molesta* males during pheromone-mediated upwind movement. *J. Insect Physiol.* 33:875–883.

Willis, M.A. and Baker, T.C. (1988) Effects of varying sex pheromone component ratios on the zigzagging flight movements of the oriental fruit moth, *Grapholita molesta. J. Insect Behav.* 1:357–371.

Willis, M.A. and Cardé, R.T. (1990) Pheromone-modulated optomotor response in male gypsy moths. *Lymantria dispar L:* upwind flight in a pheromone plume in different wind velocities. *J. Comp. Physiol.* A 167:699–706.

Willis, M.A., David, C.T., Murlis, J. and Cardé, R.T. (1994) Effects of pheromone plume structure and visual stimuli on the pheromone-modulated upwind flight of male gypsy moths (*Lymantria dispar*) (Lepidoptera: Lymantriidae) in a forest. *J. Insect Behav.* 7:385–409.

Witzgall, P. (1990) Attraction *of Cacoecimorpha pronubana* male moths to synthetic sex phero-mone blends in the wind tunnel. *J. Chem. Ecol.* 16:1507–1515.

Witzgall, P. (1996) Modulation of pheromone-mediated flight in male moths. *In*: R.T. Cardé and A. Minks (eds): *Pheromone research: New directions.* Chapman & Hall, New York, pp 265–274.

Wolf, R. and Heisenberg, M. (1991) Basic organization of operant behavior as revealed in *Drosophila* flight orientation. *J. Comp. Physiol.* A 169:699–705.

Wortmann, M. and Zarnack, W. (1989) On the so-called constant-lift reaction of migratiory locusts. *J. Exp. Biol.* 147:111–124.

Wortmann, M. and Zarnack, W. (1993) Wing movements and lift regulation in the flight of desert locusts. *J. Exp. Biol.* 182:57–69.

Wright, R.H. (1958) The olfactory guidance of flying insects. *Can. Ent.* 90:81–89.

Zack, C. (1979) *Sensory adaptation in the sex pheromone receptor cells of Saturniid moths.* Doctoral Thesis, Universität München.

Zanen, P.O. and Cardé, R.T. (1996) Effects of host-odour plume altitude and changing wind velocity on upwind flight manoeuvres of a specialist braconid parasitoid. *Physiol. Entomol.* 21:329–338.

Orientation and Communication in Arthropods
ed. by M. Lehrer
© 1997 Birkhäuser Verlag Basel/Switzerland

The evolution of communication and the communication of evolution: The case of the honey bee queen pheromone

R. Gadagkar

Centre for Ecological Sciences, Indian Institute of Science, Bangalore 560012, India, and Animal Behaviour Unit, Jawaharlal Nehru Centre for Advanced Scientific Research, Jakkur, Bangalore 560064, India.

Summary. Intraspecific chemical communication by means of pheromones is widespread in arthropods and is believed to have played a particularly important role in the evolution and the efficiency of social forms of life that have developed in several insect species. Using the honey bee queen pheromone as an example, this chapter discusses several ways in which the study of the evolution of chemical communication can potentially contribute to the resolution of a number of questions of vital importance for a better understanding of the evolution of sociality.

Introduction

Chemical communication is widespread in insects and has often reached impressive levels of sophistication even in solitary insects (*Drosophila*: Mayer and Doolittle, 1995; Moths: Ferveur et al., 1996; Kaissling, 1977; Schneider, 1984; Svensson 1996). In social insects, where communication between the members of a colony is by far more frequent and more critical than in solitary insects (Bell and Cardé, 1984; Free, 1987; Hölldobler and Wilson, 1990; Agosta, 1992; Winston, 1992), chemical communication plays a particularly significant role. For queens of many social insect colonies that need to rapidly and efficiently influence the behaviour of thousands of workers, chemical communication has perhaps no substitute. The honey bee *Apis mellifera* has been the subject of intense investigation in this regard (Bell and Cardé, 1984; Free, 1987; Winston, 1987; Winston and Slessor, 1992). We shall therefore use the honey bee queen pheromone as an example to highlight several general issues concerning the evolution of chemical communication. We will discuss several testable hypotheses that might contribute to our understanding of the evolution of pheromone-based communication. The aim of this chapter is to try to make the point that the evolution of chemical communication in social insects may communicate to use new insights concerning the evolution of sociality itself.

The honey bee queen pheromones

The honey bee queen produces a host of chemical substances that influence the behaviour and physiology of the workers in her colony. Because each colony consists of a single queen and many thousand workers, communication between the queen and her workers is, as expected, primarily mediated by chemicals. The well known effects of queen pheromones on workers include rapid detection of the presence or absence of the queen. A retinue of some eight to 10 workers, the composition of which changes every few minutes, feed and lick the queen and thereby acquire the queen pheromones and pass them on to other workers. The pheromones also inhibit the development of worker ovaries and stimulate building and foraging activities. Workers of a queen-right colony almost never lay eggs. Instead, they engage in building combs, feeding the larvae, grooming and feeding the queen, protecting the hive from intruders, foraging, and storing honey and pollen. The nearly complete sterility of the workers and their devotion to non reproductive activities, once considered paradoxical, are now interpreted as a strategy to maximize their "indirect fitness". By enhancing the queen's reproductive success (i.e., her "direct fitness"), workers enhance the transmission of their own genes, because their mother's offspring share with them many of these genes.

Which of the workers' responses are mediated by which subset of the chemical repertoire of the queen is not entirely clear. Indeed, the queen's chemical repertoire itself remains only partially known. However considerable progress has been made in recent years. Winston and Slessor (1992) have succeeded in identifying five of the most essential components of the queen pheromone which together elicit most of the important behavioural responses observed in the workers. One queen equivalent of this so-called queen mandibular pheromone (QMP) consists of about 200 µg of 9-keto-(E)2-decenoic acid (9ODA), about 80 µg of 9-hydroxy-(E)2-decenoic acid (9-HDA), of which about 56 µg is the (–) optical isomer and about 24 µg the (+) optical isomer, about 20 µg of methyl p-hydroxybenzoate (HOB) and about 2 µg of 4-hydroxy-3-methoxyphenylethanol (HVA) (Fig. 1). The latter two aromatic compounds are minor and, indeed, somewhat unexpected components. The aliphatic 9-ODA and 9-HDA are the major components whose involvement in the effects of the queen pheromones on worker bees has been known for a long time.

Queen control or queen signal?

The role that the queen's pheromones play in the workers' physiology and behaviour is evident from the finding that the various responses of the workers described above disappear upon removal or death of the queen. Most importantly from an evolutionary point of view, in the absence of the

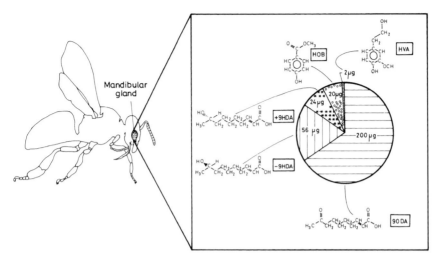

Figure 1. The queen mandibular pheromone preparation, containing well-defined amounts of five of the most important components of the mandibular gland secretion, elicits most of the responses expected from workers. See text for expansion of the names of the components.

queen, worker ovaries begin to develop and workers begin to lay small numbers of haploid eggs. Not surprisingly, this observation has been interpreted to mean that the pheromones are used by the queen to *control* the workers, *prevent* them from laying eggs and *force* them to build combs, forage and do all of those things that they do in the presence of the pheromone. Intrinsic to this concept of *queen control* of workers is the hidden assumption that the pheromone not only makes the workers do what they would not have done in the absence of the pheromone, but also that the pheromone makes the workers do what is *not good* for them in an evolutionary sense. In other words, the queen pheromone is thought to make the workers behave in a manner that is contrary to their inclusive fitness (= direct fitness + indirect fitness).

Keller and Nonacs (1993) have questioned this assumption. They argue that pheromonal queen control which can make the workers act against their own best interests has never really been demonstrated, and cannot possibly evolve. They argue that queen control has never really been demonstrated, because all phenomena hitherto interpreted as queen control bear alternative interpretations, the most logical of which is that workers are actually acting in a manner that maximizes their inclusive fitness. But can refraining from laying eggs be a way of maximizing their inclusive fitness? Yes, and for several reasons. First, as already mentioned above, the queen is the workers' mother and her offspring are therefore their siblings. Second, a healthy queen is by far a superior egg layer compared to any worker. Third, queens can lay both haploid and diploid eggs, whereas the workers (which are incapable of mating), can only lay male destined haploid eggs. Fourth, workers should prefer the queen eggs, even when it comes to haploid eggs, because when the queen is multiply mated, as is most often the case in nature, each worker is more closely related to the queen's sons (who would be related to her by 0.25) than she is to another worker's sons (who would only be related to her by 0.125 if the worker in question is her half sister) (Ratnieks, 1988, 1990; Ratnieks and Visscher, 1989). Keller and Nonacs (1993) argue that pheromonal queen control cannot possibly evolve, because a situation where the queen accomplishes what is good for her at the expense of the workers who are forced to accept what is not so good for them, is evolutionarily unstable. In such a situation, workers would be expected to evolve defenses against queen control and revolt against her hegemony. A mutation in the workers that can ignore the queen pheromone would be favoured by natural selection as it would then be able to accomplish what is good for the worker, from the point of view of its own inclusive fitness. If it is in the best interests of the workers to leave egg laying to the queen when she is present and healthy, but efficiently detect the absence of the queen or any deterioration in her health and start laying their own eggs, we would still see the same phenomena in response to the queen pheromone as we do now. However, the queen pheromone should then be thought of as a *signal* that the workers use to decide when they should let the queen reproduce and when they should take over egg laying on to themselves. Whether the queen is controlling the workers, making them do what is good for her but not so good for them, or whether the workers are using the queen pheromone as a signal to do what is best for them, which also happens to be approximately what the queen wants them to do, are not equivalent and it is not a matter of semantics either.

At the heart of this genuine dichotomy in the nature of chemical communication is a question of fundamental importance to the study of social evolution: are both queens and workers simultaneously maximizing their respective inclusive fitnesses, leading to an evolutionary stable equilibrium, or are queens successfully manipulating workers into acting against their inclusive fitness so that social insect colonies are inherently evolutionarily unstable? Keller and Nonacs favour the former interpretation which seems reasonable and might perhaps represent a new level of maturity in our approach to the study of social insect biology. The significance of this shift in interpretation of the function of the queen pheromone for our understanding of both the evolution of chemical communication and the evolution of sociality, is not trivial. A few additional comments about the view championed by Keller and Nonacs are perhaps necessary before we can fully appreciate this shift. Keller and Nonacs do not entirely rule out the possibility of pheromonal queen control at the expense of worker inclusive fitness. Their contention is that the available evidence can equally well be interpreted (and perhaps better interpreted, in conjunction with the argument of evolutionary stability) as a case of the workers using the queen pheromone as a signal to achieve their own interests. Indeed, they proceed to make several testable predictions that might help decide between the theory of pheromonal queen control and the signal hypothesis.

(i) The inhibitory effects of the queen on worker reproduction should be independent of her egg laying ability, as it would then prove that workers are being suppressed even when the queen is not such a good egg layer. The signal hypothesis, on the other hand, predicts that the level of suppression of worker reproduction be proportional to the queen's egg laying ability.

(ii) Queens laying only male-destined eggs should exhibit the same level of inhibition on worker reproduction as do queens producing both sons and daughters, as this would prove that workers are being inhibited both when it is good and when it is bad for them to reproduce on their own. The signal hypothesis, however, predicts that worker reproduction should be inhibited more strongly when the queen is producing daughters than when she is producing only sons.

(iii) In multi-queen colonies (as is often the case in ants), inhibition of worker reproduction should increase as queen number increases, because more pheromone from more queens should cause more inhibition. The signal hypothesis suggests, instead, that workers should be more likely to reproduce on their own when there are many queens, as the relatedness of the queens' offspring to themselves decreases with increase in queen number.

(iv) When there is a genuine queen-worker conflict such as is apparent in the optimal sex investment ratio (Trivers and Hare, 1976), the queen's preferred optimum should prevail over the workers' preferred optimum. A 1:1 female to male ratio (optimum for the queen), rather than a 3:1 ratio (that is optimum for the workers in a monogynous colony with a singly mated queen), would suggest that the queen successfully forces workers to act in ways that maximize her own interests at the expense of the workers' interest.

Because these predictions remain to be tested, Keller and Nonacs do not rule out the possibility of genuine queen control. They suggest, however, that whereas the term "pheromonal queen control" be reserved for situations where there is evidence that "workers or subordinate queens are chemically manipulated by queen(s) into pursuing actions that are contrary to their inclusive fitness", the term "pheromonal queen signal" be used for situations where "workers or subordinate queens react to queen pheromones in ways that increase their (and possibly the queens') inclusive fitness", and the term "pheromonal queen effect" be used "where the changes in the workers' or subordinate queens' behaviour have an unknown consequence on their inclusive fitness". It might be worth noting that Keller and Nonacs keep referring to subordinate queens because they are also concerned with many ant species where, unlike in case of the honey bee colony, there are several queens per colony. There is often a clear dominance hierarchy among the queens in polygynous ant colonies so that dominant queens must also influence subordinate queens in much the same way as they do workers. The nomenclature suggested by Keller and Nonacs implies that all effects of the queen pheromone on workers and subordinate queens should begin their life (in the literature) as pheromonal queen effects until they attain the status of either pheromonal queen control or pheromonal queen signal, after the appropriate experiments have been carried out and one or the other criterion satisfied. It may well happen that what starts out as pheromonal queen effect may sometimes end up as pheromonal queen control and at other times as pheromonal queen signal, depending on the particular effect being considered and on the particular species of social insect being investigated.

This dichotomy between *control* and *signal* in chemical communication implies a similar dichotomy between *exploitation* and *mutualism*, between evolutionary *instability* and *stability* in all relations between queens and workers or dominant and subordinate queens in social insect colonies (see Markl, 1985, for an illuminating discussion of communication dyads with different costs and benefits to senders and receivers of information). Understanding the evolution of chemical communication between queens and workers or subordinate queens will therefore have much to communicate to us about the evolution of sociality itself. When considerable further progress has been made in our understanding of chemical communication between queens and workers and between dominant and subordinate queens, and we will have discovered a large number of queen pheromonal effects and transferred most of them into either the control category or the signal category, we may be able to prepare a tally of cases of queen control versus queen signal in different species and with reference to different kinds of pheromonal effects, a tally that would promise to reveal a great deal about the nature of evolution of chemical communication and of evolution of sociality itself. In particular, it should help resolve the issue of exploitation versus mutualism, a dichotomy that has long plagued discussions of the evolution of eusociality (see Lin and Michener, 1972; Alexander, 1974; Trivers and Hare, 1976; Alexander and Sherman, 1977; Gadagkar, 1985). It would also be most relevant to a possible resolution of the virtually neglected question of whether or not selfishness and perhaps even solitary life can reemerge from the highly eusocial state (Gadagkar, 1997b, c).

Queen pheromone as poisonous prestige and handicap?

Zahavi and Zahavi (1997) have suggested that the queen pheromone is a poison, the possessing of which is a handicap, but that it is precisely the "prestige" associated with the ability to withstand this handicap that makes it possible for the queen to influence the workers, and at the same time to make it advantageous for the workers to help the queen. Because this suggestion may seem extraordinary to most of us, it is necessary to digress momentarily from our discussion of the honey bee queen pheromone and see it in the backdrop of Zahavi's handicap principle and his ideas about the evolution of honest signals (Zahavi, 1975, 1977, 1990, 1993). Zahavi is the most ardent individual selectionist around today who insists that an honest attempt should be made to explain all known biological phenomena in the framework of individual selection. He sees in the theory of kin selection the same problem of instability that most of us recognize in group selection. The cause of instability is, of course, the possibility that cheaters can garner the advantage of altruism exhibited by some members of the group without any investment on their own part. This would give an advantage to the cheaters who would increase in frequency and finally drive the altruists to extinction. True, the problem of such instability is less serious in the case of kin selection as compared to group selection. But there is no denying the fact that the problem of cheaters does not exist in the case of individual selection, simply because nobody is supposed to be altruistic; everybody is doing the best he or she can do under the circumstances. If a viable explanation can be found within the framework of individual selection for why it is advantageous to the honey bee worker to help the queen, rather than try to reproduce by herself, then it would certainly be more satisfactory than even a kin selectionist explanation that relies on the indirect advantage of helping. It is the extreme rarity of people who are willing to pursue an individual selectionist explanation to its logical conclusion, especially with respect to worker altruism in highly eusocial insects, that makes the Zahavis' point of view so valuable.

For over two decades, Zahavi (1975, 1977) has championed (amidst much skepticism) the so-called handicap principle. Initially, Zahavi's intention was to provide a satisfactory explanation for the elaborate and exaggerated secondary sexual characters and displays usually shown by males (such as the train of the peacock, the antlers of deer, or the songs of some birds). These traits are supposed to have evolved by sexual selection, a mechanism that even Darwin thought best to keep distinct from natural selection (Darwin, 1871). A widely accepted mechanism for the action of sexual selection is Fisher's run-away selection (Fisher, 1930) which postulates that, initially, the secondary sexual characters correlated well with male quality and hence females that had a preference for males with the elaborate traits had an advantage over other females. Fisher argued that, later in evolution, the very fact that females prefer elaborately ornamented males gives an added advantage to the males possessing the ornaments so that selection for the male secondary sexual characters goes beyond the level predicted by their correlation with fitness. In other words, male ornaments become a handicap, making it, for example, harder for their bearers to escape from predators. However, males with such characters are not easily eliminated by natural selection because females show a preference for such males, a preference that has persisted from the period when male ornamentation was correlated with male fitness. As opposed to Fisher (1930), Zahavi (1975) has argued that females will not prefer traits that are not correlated with fitness just because these traits used to be correlated with fitness at some time in the past. The handicap, Zahavi argues, is a true indicator of male quality, because if a male has survived despite the disadvantages involved, then he must possess very good genes. In other words, the handicap is an honest, reliable signal of male genetic quality. Although early models appeared to show that Zahavi's idea cannot work (Davis and O'Donald, 1976; Maynard Smith, 1976; Kirkpatrick, 1986), subsequent, more realistic models show that Zahavi's handicap principle is indeed plausible (Kodric-Brown and Brown, 1984; Nur and Hassan, 1984; Grafen, 1990a, b).

Apart from providing a potential explanation for exaggerated male secondary sexual characters, the handicap theory leads to another important conclusion, namely, that communication signals must be costly in order to be honest. If a peacock's quality is assessed by the length of his tail, then there is no way an inferior peacock can bluff and indicate a higher than true quality, because inferior peacocks can neither grow nor carry long tails. Thus, in principle, any signal can be trusted to be an honest signal if it is costly, because no male can provide such a signal unless he is capable of carrying it despite the handicap. If male quality were to be inferred by the females through some inexpensive signal that anybody could give, cheaters would get away by sending signals indicating a quality higher than that they actually possess, and the signaling system would soon break down (Zahavi, 1987, 1993). In addition to its compelling logic, and many convincing examples discussed by Zahavi, the prediction that a signal can only be honest if it is costly has already been verified through formal evolutionarily stable strategy (ESS) models (Grafen, 1990a, b; Godfrey, 1991; Maynard Smith, 1991; Johnstone and Grafen, 1992a, b).

The idea that only costly signals are honest signals, together with the handicap hypothesis from which this idea has been derived, lead Zahavi to the concept of prestige as a reward for altruism. This concept, based on his investigations of cooperative breeding in the Arabian Babbler (Zahavi, 1990, 1995), is radically different from all of the theories proposed previously. Because group selection, kin selection, as well reciprocal altruism (Trivers, 1971; Wilkinson, 1988), are all susceptible to cheating, Zahavi rejects them as possible explanations for why babbler helpers actually help. Instead, he explains the evolution of helping in babblers through "old fashioned" individual selection. He argues that all apparent cases of altruism are, as a matter of fact, acts of selfishness. By investing in the welfare of the group, taking risks in defending the group and behaving in apparently altruistic ways, individuals increase their prestige in the group. This prestige serves as an honest signal indicating the quality of this individual as a collaborator and as a rival in intra-group conflicts. The social prestige thus acquired helps the individual to increase its chances to reproduce, when the opportunity to do so arises. Zahavi's main evidence for this idea comes from his observation that his babblers are highly motivated to help, and do not ever try to get the benefits of group living without investing in its welfare. Indeed, babblers compete with each other in allo-feeding, in feeding the nestling, and in performing sentinel duties (Zahavi, 1990). Moreover, dominants often prevent subordinates from helping the group, thus keeping this privilege to themselves. If social prestige indeed helps the individual to increase its direct fitness, then this individual's altruism must be considered as one that has been motivated by a selfish design. Individual selection would thus be sufficient to explain the evolution of this type of altruism. None of the other theories, namely, group selection, kin selection and reciprocal altruism, can explain why there should be competition for being altruistic.

Looking at honey bees from their unabashed, individual selection bias, Zahavi and Zahavi (1997) argue that, given the inevitable superiority of the honey bee queen over the workers, the best that a worker can do is to bide her time until an opportunity arises for her to lay haploid eggs and produce some sons. But very few workers will actually get that chance. Zahavi and Zahavi (1997) suggest that the probability of getting that chance can be increased not by sulking and refraining from working, but, instead, by actually working for the colony and, by doing so, enhancing their prestige. In small social insect colonies, as in the babblers, the achievements of different individuals and their consequent prestige in the group can be known to all members of the colony. In a large colony such as the honey bee, however, individuals remain largely anonymous and a different mechanism to indicate prestige is required. It is here that Zahavi and Zahavi (1997) turn their attention to the honey bee queen pheromone. They suggest that the queen pheromone is a poison made by the queen to which workers are far more sensitive than is the queen herself and thus there is not much that workers can do about this. Hence, as long as the queen is alive and healthy, she will suppress them. This is equivalent to their using the queen pheromone as an honest signal to do what is best for them under the given circumstances. It is only when the queen becomes weak or she dies that the workers have a real opportunity to try to reproduce, and it is here that their prestige may come in handy.

If the queen pheromone is a poison, then workers are expected to exhibit some natural variability in their resistance to this poison. This ability can in turn be an honest signal of their quality. Like the babblers which compete with each other to perform altruistic acts, workers must compete with each other to acquire the queen pheromone, and they are indeed known to do so. Zahavi and Zahavi (1997) have also postulated that working hard for the welfare of the colony might in fact help the workers metabolize the queen pheromone faster and hence enhance their ability to deal with the poison. To quote Zahavi and Zahavi (1997): "...the very act of working for the hive may increase the worker's ability to carry pheromone in the same way that a college student may 'hold his liquor' better after a year spent as a construction worker than he did in his fraternity party days." The pheromone can thus be thought of as a handicap that queens and some workers (more than others) can tolerate on account of their superior physical condition. Just as the pheromone carried by the queen can be an honest signal indicating that she is better at egg laying than any worker, the quantity of queen pheromone carried by each worker can be an honest signal of how good and strong this worker is (both as a companion and as a rival) compared to other workers, a comparison that especially comes to the fore upon death of the queen. As Zahavi and Zahavi (1997) point out, the suggestion that the quantity of queen pheromone carried by an individual is a handicap which indicates prestige in a large colony and is therefore an honest signal of the individual's quality, is yet only a hypothesis, but it is an eminently testable one. Here I am not concerned with the possibility of its ultimate correctness (for idle speculation in this matter, without experiment, may be futile). Instead, I wish to examine its theoretical implications for the evolution of chemical communication and for the evolution of sociality.

Kin- and group selection versus individual selection

Let us first compare and contrast the point of view of Zahavi and Zahavi (1997) with that of Keller and Nonacs (1993) which we discussed in the previous sections. Both points of view consider the queen pheromone to be an honest signal that the workers find in their best interests to obey. But there is a subtle difference between the two points of view. For Keller and Nonacs, both queens and workers are maximizing their inclusive fitness and thus achieving a stable evolutionary equilibrium. Mutations making the workers lay some eggs in the presence of the queen will not be favoured, because the workers' inclusive fitness is maximized by suppressing their own reproduction and permitting the queen to lay eggs. For Zahavi and Zahavi (1997) it means that there is an inherent asymmetry between queen and worker (or between different workers). With the help of the queen pheromone as an honest signal, the inferior individual can correctly assess the quality of the superior individual and accept its own subordinate status, because there is nothing else that it can do. Workers fitness (direct fitness, which is all that Zahavi cares about!) may in fact be enhanced if they suppressed the queen and took over egg laying (in spite of their poor egg laying capacity), but they cannot compete with the queen in dealing with the poison that is in the pheromone. The main reason for this difference, in the arguments of Keller and Nonacs on the one hand and the Zahavi and Zahavi (1997) on the other, is that whereas the former use the framework of kin selection (maximization of inclusive fitness), the latter argue with the power of individual selection (maximization of classical individual fitness). According to Zahavi and Zahavi (1997), "the chance the worker has to reproduce within the colony is the cement that permitted the creation of large, stable partnerships, encompassing thousands of individuals, in which the reproductive success of the individual workers depends on the success of the queen".

Keller and Nonacs (1993), on the other hand, do not consider the possible role of individual selection. According to their theory, only kin- and perhaps group selection can explain the role that queen pheromone plays in communication. In the context of polygynous ant colonies, Keller and Nonacs worry that a queen attacking other queens chemically is also likely to affect herself, and that this is one of the reasons to take recourse to thinking of the queen pheromone not as a weapon of attack but rather as a signal. A fundamental tenet of Zahavi's theory of signal selection is, however, that no signal can be reliable and will therefore not be taken seriously by the receiver unless it is costly for the sender. Thus, the possibility that a queen using a pheromone to influence other queens may be attacking herself is not likely. Only a queen capable of withstanding the harmful effects of the pheromone can use it as a reliable signal of her superior status. Keller and Nonac's problem arises because they are implicitly assuming that all individuals (workers as well as queens) are similar in their ability to deal with the queen pheromone, which is in stark contrast to the ideas of Zahavi.

It should be noted that Keller and Nonacs propose that true control which makes the workers act against their best interests can also evolve in the case of physical control of workers by the queens (via aggressive behaviour that is particularly common in small colonies, see e.g., Premnath et al., 1996; Gadagkar, 1997b). According to Zahavi's theoretical framework there should be no real difference between physical and chemical control. Even in cases in which queens use physical methods to influence workers, there should evolve a system of costly, honest signals that would permit the interpretation that workers are responding to some queen's characters (whatever these may be; in the case of physical control, for example, enlarged body size) as signals to do what is best for them, i.e., letting the queen lay eggs when she is strong and healthy, but taking it upon themselves to do so when the queen is not in a good shape or when she has died.

Keller and Nonacs (1993) argued, in addition, that chemical control of workers by queens cannot have evolved, because it would become prohibitively expensive for queens to stay ahead of the workers in the chemical arms race. However, they did not consider the possibility that it might also become prohibitively expensive for workers to stay ahead, and therefore it might be profitable for them to obey the queen. The implication is that if the queen were to use the pheromone to *control* the workers and force them to do what is not in their best interests, then she may not necessarily succeed, because there is no guarantee that she wins the chemical arms race. Therefore, the pheromone must be considered to be effective in influencing the workers' behaviour solely on the basis that workers use it as a *signal* to do what is best for them. If, however, the queen pheromone is to be thought of as being not just a signal, but rather a *poison* the carrying of which is a handicap that queens can bear better than workers, as proposed by Zahavi and Zahavi (1997), then it would follow that queens inherently have an advantage over workers and should therefore be more likely to win the arms race. The assumption of inherent and inevitable differences between queens and workers is, indeed, an essential component of the individual selection argument of Zahavi and Zahavi (1997). They write: "The fact that the queen is able to raise daughters smaller and weaker than herself makes it possible for her to exploit them: it is precisely the inequality between queen and workers that limits the workers' options and makes the asymmetrical partnership so stable."

Should the Zahavis' hypothesis, proposing that the queen pheromone is a poison and thus a handicap serving to build prestige, be verified, then our picture of the evolution of chemical communication between queens and workers will be rather radically altered, and consequently our picture of the evolution of sociality will be an even more radically different one. In particular we might then be able to explain at least the maintenance of sociality in the honey bee (and perhaps other highly social insects) by individual selection without recourse to kin selection or group selection. Hence my assertion again, that our understanding of the evolution of chemical communication between queens and workers will have a great deal to communicate to us about the evolution of sociality itself.

Queen-worker dichotomy: A chicken and egg problem

Perhaps the most fascinating aspect of honey bee colonies is the differentiation of the bees into a sterile worker caste and a fertile queen caste. The question that stems from this fact relates to the possible differences between queens and workers in their pheromone blends and the mechanism of the origin of these differences. These are the questions that Plettner et al. (1996) address in a recent path-breaking paper. Workers, too, produce mandibular gland secretions that are added to the brood food and may serve as preservatives and nutrients. Instead of the two major components of the queen's secretions, namely 9-keto(E)2-deconoic acid (9-ODA) and 9-hydroxy-(E)2-decenoic acid (9-HDA), workers secrete acids hydroxylated at the 10th or ω-carbon atom, rather than the 9th or ω-1 carbon atom as in the case of the queen's acids. Instead of the queen's 9-HDA, workers secrete 10-hydroxy-(E)2-decenoic acid (10-HDA), and instead of the queen's 9-ODA, workers secrete the diacid acid derived from their 10-HDA. In other words, queens and workers differ essentially only in the position of the carbon atom that is hydroxylated. But how does this difference arise? Based on a series of experiments analyzing the fate of deuterated test compounds applied to excised queen and worker mandibular glands, using gas chromatography-mass spectrometry (GC-MS), Plettner et al. (1996), have proposed the following caste-specific, bifurcated three step biosynthetic pathway for the production of these compounds (Fig. 2).

The starting point is stearic acid, an 18-carbon, straight chain, saturated hydrocarbon, which is a very common intermediate step in the oxidation of lipids (see Mahler and Cordes, 1966, for a detailed account of lipid metabolism). In the first step of the proposed pheromone biosynthetic pathway, functionalization is achieved by the addition of a hydroxyl group on either the 18th (ω) or the 17th (ω-1) carbon atom. This functionalization which foreshadows the queen-worker differences depending on whether it happens at the ω or the ω-1 carbon atom is, however, itself not caste-specific; both ω and ω-1 functionalizations occur in both castes to about the same extent. In the second step, the 18-carbon hydroxy acids are shortened to give 10-HDA and 9-HDA by the standard chain-shortening cycles of β oxidation that normally occur during fatty acid metabolism. It is the β oxidation step that is caste-specific – queens preferentially channel the ω-1 compounds and workers the ω compounds into the oxidation pathway. In the final step, oxidation of the ω or ω-1 hydroxy group that was added

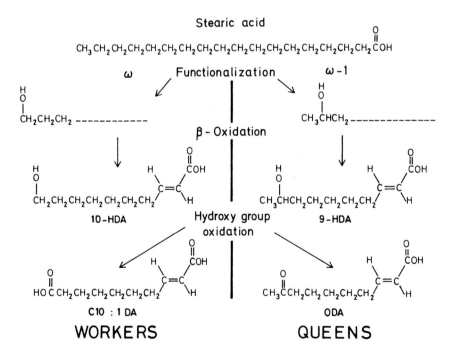

Figure 2. The caste-specific, three-step, bifurcated pathway for the biosynthesis of queen and worker pheromones, as proposed by Plettner et al. (1996).

in the first step, results in the formation of the diacid in the case of workers and the keto acid in the case of queens.

The caste-specific pheromone biosynthetic pathways elucidated by Plettner et al. (1996) permit us to explore yet another and rather different context in which understanding the evolution of chemical communication can tell us much about the evolution of sociality. Of social insect queens and workers, which one is ancestral and which derived? On the one hand, queens in social insect species can be thought of as being comparable to the undifferentiated (into queen or worker) adult insects in their solitary ancestors (or, equivalently, in other extant solitary taxa) and the workers can be thought of as being a new invention of sociality. After all, adults in solitary species are all potentially capable of reproducing, and it is the character of being sterile and merely working to rear another individual's brood that is a novel feature of social insects. On the other hand, workers in social species may be thought of as being comparable to their solitary ancestors or extant solitary counterparts, and the queens can be thought of as an invention of sociality. After all, adults in solitary species are all capable of nest building, foraging and brood rearing and it is the character of inhibiting reproduction of conspecifics and attempting to become the sole reproductive in a group, at the cost of losing foraging and brood rearing abilities altogether if necessary, that is a novel feature of social insects. A reasonable solution to this conundrum is to compromise and think of the solitary insects as queen and worker combined, because each individual is capable of reproduction as well as nest building, foraging and brood care. And this is a largely correct solution because both queens and workers, at least in the advanced social species, are considerably modified and exaggerated in their respective roles compared to solitary insects. Nevertheless, I believe that, if and when possible, we should try to make an objective assessment of whether queens are ancestral and workers are derived or whether it is vice versa. I will argue that the pheromone biosynthetic pathway elucidated by Plettner et al. (1996) provides one such opportunity.

I have recently hypothesized (Gadagkar, 1997a) that, because the pheromone biosynthetic pathway employed by the workers deviates relatively little from the typical lipid metabolism pathway, it might perhaps simply be adopted from there. The diacid they make can be relatively easily channeled into an energy generating role and its degradation products can be profitably fed into the Krebs cycle. On the other hand, I speculate that the pheromone biosynthetic pathway of the queens is quite a deviation from the standard lipid metabolism pathway. In particular, the keto acid is not something one would expect if energy generation is the immediate goal. The expense involved in further breaking down the keto acid makes it a poor candidate to be fed into the Krebs cycle. I therefore speculate that, in the course of making their pheromones, the workers are doing more or less what any solitary insect would do anyway for generating energy from lipids, and that their pheromone biosynthetic pathway is therefore the more ancestral one. Conversely, queens have considerably modified the ancestral lipid metabolism pathway in order to make a pheromone that has only lately (relatively speaking) become necessary. In order to do so they are prepared to make an end product such as the keto acid which is energetically unwise, but I argue that energy generation is not their motivation here. Surely they have other mechanisms of generating energy even from lipids. And even if their overall efficiency of generating energy from lipids is lower than that of workers, it does not matter that much because it is the foragers, not the queens, that have to fly great distances in search of food. The pheromone biosynthetic pathway of the queens appears therefore to be relatively more derived. One might also argue that the function of the worker pheromone, namely, to act as a preservative and nutrient, is also a more ancestral function, more likely to have been useful in the solitary condition. Conversely, the function of the queen pheromone appears to be more derived as it fulfills a relatively more recent requirement and hence is unlikely to have been of much use in the ancestral solitary condition. Workers thus seem to use an ancestral biochemical pathway to make a product that may also have been required in the ancestral condition. And queens seem to be using a rather derived form of the biochemical pathway to make a product that has a rather derived function. At least in this limited context, workers seem to be ancestral and queens seem to be derived. This one context, important as it is, cannot be thought to have solved our general problem of who came first, the queen or the worker. It would be prudent, even necessary, to be on the look out for more opportunities to classify queens and workers as ancestral or derived. Indeed, a new and highly derived function of the worker pheromone may yet be discovered which may alter our conclusion. Thus, we may well come up with different conclusions each time and only the relative scores for "ancestral" and "derived" that queens and workers accumulate in the long run can help us solve this conundrum in any general sense. But I believe this is a good beginning. We see once again that an understanding of the evolution of chemical communication can lead to important insights into the evolution of sociality itself.

The evolution of caste polymorphism

Yet another striking feature of the social insects, the highly social insects in particular, is the morphologial differentiation of queens and workers which may sometimes reach such proportions that, if encountered separately, queens and workers may get classified as different species (Wheeler, 1913). While there is considerable differentiation between honey bee queens and workers, the greatest intra-specific size variations has been recorded in the Asian ant *Pheidologeton diversus* where some workers weigh 500 times and have a head width 10 times larger compared to other workers (Moffett, 1987). Here the differentiation is not between queens and workers, but rather between the so-called major workers and minor workers. Whether it is between queens and workers or between major and minor workers, these extreme degrees of intra-species, intra-sexual dimorphism require an explanation. The fact that no solitary species seem to match these levels of differentiation suggest that the explanation is linked to the social habit of these insects.

I have recently offered a speculation (Gadagkar, 1994, 1996b, 1997d) which was inspired by the idea of evolution by gene duplication first suggested by Haldane (1932) and Muller (1935) and elaborated and championed by Ohno (1970). The idea is that redundant, duplicate copies of genes can accumulate potentially lethal mutations without killing the organism and can eventually give rise to novel genes coding for novel structures via pathways that would be inaccessible to an individual with a single copy of the gene. I have argued that a very similar consequence will accrue to social insects, although for a somewhat different reason. The evolution of altruistic sterile worker castes in the social insects was considered paradoxical until Hamilton proposed the theory of inclusive fitness (Hamilton, 1964a, b). Today it is common practice to recognize inclusive fitness as having two components, a direct component gained through production of offspring and an indirect component, gained through aiding close genetic relatives. Sterile worker castes are expected to gain fitness exclusively through the indirect component (Wilson, 1971, 1975; Hölldobler and Wilson, 1990), and in no other group is there a comparable level of dependence on the indirect component of inclusive fitness.

I argue that, when some individuals in a species begin to rely on the indirect component of inclusive fitness while others continue to rely on the direct component, as workers and queens in social insects do, different sets of genes in queens and workers will be liberated from previous epistatic constraints. These genes then become free to evolve in new directions. There is no gene duplication here in the conventional sense, but the consequence, namely, liberation from previously existing constraints (due to the action of stabilizing selection) and the opportunity to diversify in different directions (through the action of directional selection), is similar. To put it simply, an individual can evolve into a "super" egg layer if it does not have to simultaneously be a very good forager, or it can evolve into a "super" forager if it does not have to simultaneously be a very good egg layer.

I have speculated that, compared to solitary species, social insects are also in a better position to exploit the evolutionary advantages of conventional gene duplication (Gadagkar, 1997d). I argued in the previous section that the function of the worker pheromone and the biochemical pathway involved in its production are relatively more ancestral, and that the function of the queen pheromone and the biochemical pathway involved in its production are relatively more derived. If this is true, then it is not difficult to see the tremendous advantage of conventional gene duplication in developing the derived condition from the ancestral one. It seems likely that the enzymes involved in the β oxidation step (see Fig. 2) give rise to specificity for substrates hydroxylated at the ω or ω-1 positions. Imagine that the ancestor of the social insect species had a gene that coded for an enzyme which could deal only with the substrate that was hydroxylated at the ω position. The workers in the descendant social species can continue to use this gene and this enzyme to make worker pheromones which may perhaps have even been made by the ancestor. A duplication of the gene involved can permit the evolution of an alternate enzyme which can handle the substrate hydroxylated at the ω-1 position. We know that such a substrate must already have been available, because both kinds of hydroxylations occur to an equal extent in both queens and workers. The duplicated gene would now be free to evolve in new directions without reduced fitness due to the reduction in the efficiency of energy production through lipid metabolism. Thus, new directional evolution can sometimes give rise to substances with remarkable properties such as the queen pheromone. A similar chance occurrence of such a mutation could hardly have been utilized effectively by a solitary species. Because social insects set aside some individuals for the sole purpose of monopolizing reproduction and inhibiting and controlling all others, they are in a special position to exploit such a consequence of conventional gene duplication and evolve in directions that are not open to solitary species. Once again we see an intimate link between possible mechanisms of evolution of chemical communication and those of the evolution of sociality.

Conclusion

We have discussed six hypotheses concerning the evolution of chemical communication between the honey bee queen and her workers and considered their implications for the evolution of sociality (Tab. 1). Today we are not quite in a position to unambiguously ascertain the correctness or otherwise of any of these hypotheses. However, these hypotheses are testable and are therefore expected to lead eventually to a better understanding of the evolution of chemical communication and hence to a better understanding of the evolution of sociality. But if this is the situation today with respect to honey bee whose queen pheromone is the best studied one, imagine the situation with respect to other species of social insects. If details concerning the evolution of chemical communication such as exemplified by the hypotheses considered here vary between honey bees and other social insects, our conclusions regarding the evolution of insect sociality will necessarily have to be revised as new information becomes available from different species. The evolution of social life in insects and

Table 1. Hypotheses concerning the evolution of chemical communication between honey bee queens and workers and their implication for our understanding of the evolution of sociality. Hypotheses 1 and 2 based on Keller and Nonacs (1993), hypothesis 3 based on Zahavi and Zahavi (1997) and hypotheses 4–6 based on Gadagkar (1996, 1997d,e).

Hypothesis	*Implication*
(concerning the evolution of chemical communication)	(for the evolution of sociality)
1. Queen pheromone is a weapon used by the queens to control the workers and force them to act against their best interests	Social insect colonies are evolutionary unstable with scope for mutations that make the workers revolt against the hegemony of the queen.
2. Queen pheromone is a signal used by the workers to do what maximizes their (and perhaps the queen's) inclusive fitness.	Social insect colonies are evolutionary stable associations of queens and workers that maximize their own and each others inclusive fitness.
3. Queen pheromone is a poisonous handicap carrying of which gives an individual (queen or worker) prestige which translates into opportunities for reproduction.	Social insect colonies (including the apparent altruism of workers) are moulded by individual selection where each individual, queen or worker, is doing the best it can under the circumstances.
4. Queen pheromone biosynthetic pathway is derived by a modification of the worker pheromone biosynthetic pathway that already existed in the solitary ancestral species.	In social evolution, workers are ancestral and queens are derived, suggesting that groups of worker-like individuals came together and queens evolved later as a consequence of social life.
5. Worker pheromone biosynthetic pathway is derived by a modification of the queen pheromone biosynthetic pathway that already existed in the ancestral solitary species.	In social evolution, queens are ancestral and workers are derived suggesting that groups of queen-like individuals came together and workers evolved later as a consequence of social life.
6. The derived pheromone biosynthetic pathway (be it the queen's or the workers') is made possible by a gene duplication event at the locus coding for one of the enzymes of the ancestral pathway.	Social insects are in a unique position to utilize the consequences of gene duplication because in the same species queens and workers follow different developmental pathways and require different biochemical pathways to be active.

especially the evolution of altruism on the part of workers remains a major unsolved problem although it has received much theoretical and empirical attention. Perhaps we have reached something of a dead end because of a possible narrow approach to the problem. The ideas discussed in this chapter suggest that a new spurt of progress may be achieved by temporarily turning our attention away from the explicit consideration of the evolution of altruism and focusing, instead, on other aspects of sociality – the evolution of chemical communication being just one example.

Acknowledgements
I thank Miriam Lehrer for encouraging me to think about chemical communication in the Hymenoptera, Ram Rajasekharan, Photon Rao and Amitabh Joshi for many helpful discussions and Mark Winston, Keith Slessor, Christian Peeters and Wolfgang Kirchner for many helpful comments on portions of this manuscript. My work is supported by grants from the Department of Science and Technology, Ministry of Environment and Forests, Government of India and the Council of Scientific and Industrial Research. I thank *Current Science* for permitting to use, with modification, some passages from Gadagkar (1997a).

References

Agosta, W.C. (1992) *Chemical communication – the language of pheromones*. Scientific American Library, New York.

Alexander, R.D. (1974) The evolution of social behavior. *Annu. Rev. Ecol. Syst.* 5:325–383.

Alexander, R.D. and Sherman, P.W. (1977) Local mate competition and parental investment in social insects. *Science* 196:494–500.

Bell, W.J. and Cardé, R.T. (eds) (1984) *Chemical ecology of insects*. Chapman and Hall, London.

Darwin, C.R. (1871) *The descent of man, and selection in relation to sex*. John Murray, London.

Davies, G.W.F. and O'Donald, P. (1976) Sexual selection for a handicap. A critical analysis of Zahavi's model. *J. Theor. Biol.* 57:345–354.

Fisher, R.A. (1930) *The Genetical Theory of Natural Selection*. Oxford University Press, Oxford.

Ferveur, J., Cobb, M., Boukella, H. and Jallon, J. (1996) World-wide variation in *Drosophila melanogaster* sex pheromone: behavioural effects, genetic bases and potential evolutionary consequences. *Genetica* 97:73–80.

Free, J.B. (1987) *Pheromones of Social Bees*. Chapman and Hall, London.

Gadagkar, R. (1985) Evolution of insect sociality – A review of some attempts to test modern theories. *Proc. Indian Acad. Sci. (Anim. Sci.)* 94:309–324.

Gadagkar, R. (1994) The Evolution of Eusociality. *In*: A. Lenoir, G. Arnold and M. Lepage (eds): *Les Insectes Sociaux,* Proceedings of the 12th Congress of the International Union for the Study of Social Insects IUSSI, Paris, pp 10–12.

Gadagkar, R. (1996) The Evolution of Eusociality, including a review of the social status of *Ropalidia marginata. In*: S. Turillazzi and M.J. West-Eberhard (eds): *Natural History and Evolution of Paper-Wasps*. Oxford University Press, Oxford, pp 248–271.

Gadagkar, R. (1997a) What's the essence of Royalty – one keto group? *Curr. Sci.* 71:975–980.

Gadagkar, R. (1997b) *Cooperation and Conflict in Animal Societies*. Harward University Press, Cambridge, Massachusetts; *in press*.

Gadagkar, R. (1997c) A unified model for the evolution of eusociality in a primitively eusocial wasp. *Curr. Sci.; submitted*.

Gadagkar, R. (1997d) Social evolution – has nature ever rewound the tape? *Curr. Sci.* 72:950–956.

Gadagkar, R. (1997e) Caste polymorphism in social insects: a theory based on evolution by gene duplication. *J. Genetics; submitted*.

Grafen, A. (1990a) Biological signals as handicaps. *J. Theor. Biol.* 144:517–546.

Grafen, A. (1990b) Sexual selection unhandicapped by the Fisher process. *J. Theor. Biol.* 144:473–516.

Godfray, H.C.J. (1991) Signalling of need by offspring to their parents. *Nature* 352:328–330.

Haldane, J.B.S. (1932) *The Causes of Evolution*. Longmans Green, London.

Hamilton, W.D. (1964a) The genetical evolution of social behaviour. I. *J. Theor. Biol.* 7:1–16.

Hamilton, W.D. (1964b) The genetical evolution of social behaviour. II. *J. Theor. Biol.* 7:17–52.

Hölldobler, B. and Wilson, E.O. (1990) *The Ants*. Harvard University Press, Cambridge, Massachusetts.

Johnstone, R.A. and Grafen, A. (1992a) Error-prone signalling. *Proc. R. Soc. Lond.* B 248:229–233.

Johnstone, R.A. and Grafen, A. (1992b) The continuous Sir Philip Sidney Game: a simple model of biological signalling. *J. Theor. Biol.* 156:215–234.

Kaissling, K.-E. (1977) Control of insect behaviour via chemoreceptor organs. *In*: H.H. Shorey and J.J. McKelvey, Jr. (eds): *Chemical Control of insect behaviour: Theory and Application* Wiley Sons, Inc., New York, pp 45–65.

Keller, L. and Nonacs, P. (1993) The role of queen pheromones in social insects: queen control or queen signal? *Anim. Behav.* 45:787–794.

Kirkpatrick, M. (1986) The handicap mechanism of sexual selection does not work. *Am. Nat.* 127:222–240.

Kodric-Brown, A. and Brown, J.H. (1984) Truth in advertising: the kinds of traits favoured by sexual selection. *Am. Nat.* 124:309–323.

Lin, N. and Michener, C.D. (1972) Evolution of sociality in insects. *Quart. Rev. Biol.* 47:131–159.

Mahler, H.R. and Cordes, E.H. (1966) *Biological Chemistry*, Second Edition, Harper & Row Publishers, New York.

Markl, H. (1985) Manipulation, modulation, information, cognition: some of the riddles of communication. *In*: B. Hölldobler and M. Lindauer (eds): *Experimental Behavioral Ecology and Sociobiology*. Gustav Fischer Verlag, Stuttgart, New York, pp 163–194.

Mayer, M.S. and Doolittle, R.E. (1995) Synergism of an insect sex pheromone specialist neuron: Implications for component identification and receptor interactions. *J. Chem. Ecol.* 21:1875–1891.

Maynard Smith, J. (1976) Sexual selection and the handicap principle. *J. Theor. Biol.* 57:239–242.

Maynard Smith, J. (1991) Honest signalling – the Philip Sidney game. *Anim. Behav.* 42:1034–1035.

Moffett, M. (1987) Division of labor and diet in the extremely polymorphic and *Pheidologeton diversus*. *Nat. Georgr. Res.* 3(3):282–304.

Muller, H.J. (1935) The origination of chromatin deficiencies as minute deletions subject to insertion elsewhere. *Genetics* 17:237–252.

Nur, N. and Hasson, O. (1984) Phenotypic plasticity and the handicap principle. *J. Theor. Biol.* 110:275–297.

Ohno, S. (1970) *Evolution by Gene Duplication*. Springer-Verlag, Berlin.

Plettner, E., Slessor, K.N., Winston, M.L. and Oliver, J.E. (1996) Caste-selective pheromone biosynthesis in honeybees. *Science* 271:1851–1853.

Premnath, S., Sinha, A. and Gadagkar, R. (1996) Domiance relationship in the establishment of reproductive division of labour in a primitively eusocial wasp (*Ropalidia marginata)*. *Behav. Ecol. Sociobiol.* 39:125–132.

Ratnieks, F.L.W. (1988) Reproductive Harmony via Mutual Policing by Workers in Eusocial Hymenoptera. *Am. Nat.* 132:217–236.

Ratnieks, F.L.W. (1990) Assessment of Queen Mating Frequency of Worker in Social Hymenoptera. *J. Theor. Biol.* 142:87–93.

Ratnieks, F.L.W. and Visscher, P.K. (1989) Worker Policing in the Honeybee. *Nature* 342:796–797.

Schneider, D. (1984) Insect olfaction: deciphering system for chemical messages. *Science* 163:1031–1037.

Svensson, M. (1996) Sexual selection in moths: The role of chemical communication. *Biol. Rev.* 71:113–135.

Trivers, R.L. (1971) The evolution of reciprocal altruism. *Q. Rev. Biol.* 46:35–57.

Trivers, R.L. and Hare, H. (1976) Haplodiploidy and the evolution of social insects. *Science* 191:249–263.

Wheeler, W.M. (1913) *The Ants – their structure, devlopment and behaviour*. Columbia University Press, New York.

Wilkinson, G.S. (1988) Reciprocal Altruism in Bats and Other Mammals. *Ethol. Sociobol.* 9:85–100.

Wilson, E.O. (1971) *The Insect Societies*. The Belknap Press of Harvard University Press, Cambridge, Massachusetts.

Wilson, E.O. (1975) *Sociobiology: the New Synthesis*. Harward University Press, Cambridge, Massachusetts.

Winston, M.L. (1987) *The Biology of the Honey Bee*. Harvard University Press, Cambridge, Massachusetts.

Winston, M.L. (1992) Semiochemicals and Insect Sociality. *In*: M. Isman and B. Roitberg (eds): *Evolutionary Perspectives on Insect Chemical Ecology*. Chapman and Hall, New York, pp 315–333.

Winston, M.L. and Slessor, K.N. (1992) The Essence of Royalty: Honey Bee Queen Pheromone. *Am. Scientist* 80:374–385.

Zahavi, A. (1975) Mate Selection – A Selection for Handicap. *J. Theor. Biol.* 53:205–214.

Zahavi, A. (1977) The Cost of Honesty (Further remarks on the Handicap Principle). *J. Theor. Biol.* 67:603–605.

Zahavi, A. (1987) The theory of signal selection and some of its implications. *In*: V.P. Delfino, (ed.): *Proceedings of the International Symposium on Biological Evolution*. Adriatica Edetricia, Bari, pp 305–325.

Zahavi, A. (1990) Arabian Babblers: the quest for social status in a cooperative breeder. *In*: P.B. Stacey and W.D. Koenig (eds): *Cooperative breeding in birds*. Cambridge University Press.

Zahavi, A. (1993) The fallacy of conventional signalling. *Phil. Trans. R. Soc. Lond.* B 340: 227–230.

Zahavi, A. (1995) Altruism as a handicap – the limitations of kin selection and reciprocity. *J. Avian Biology* 26:1–3.

Zahavi, A. and Zahavi, A. (1997) *The Handicap Principle*. Oxford University Press, Oxford.

Subject Index

DATE DUE

DEC 2 0 1999

DEMCO, INC. 38-2971